Chaos, Order,
and Patterns

NATO ASI Series

Advanced Science Institutes Series

A series presenting the results of activities sponsored by the NATO Science Committee, which aims at the dissemination of advanced scientific and technological knowledge, with a view to strengthening links between scientific communities.

The series is published by an international board of publishers in conjunction with the NATO Scientific Affairs Division

A	**Life Sciences**	Plenum Publishing Corporation
B	**Physics**	New York and London
C	**Mathematical and Physical Sciences**	Kluwer Academic Publishers
D	**Behavioral and Social Sciences**	Dordrecht, Boston, and London
E	**Applied Sciences**	
F	**Computer and Systems Sciences**	Springer-Verlag
G	**Ecological Sciences**	Berlin, Heidelberg, New York, London,
H	**Cell Biology**	Paris, Tokyo, Hong Kong, and Barcelona
I	**Global Environmental Change**	

Recent Volumes in this Series

Series B: Physics

Chaos, Order, and Patterns

Edited by
Roberto Artuso

University of Milan
Milan, Italy

Predrag Cvitanović

Niels Bohr Institute
Copenhagen, Denmark

and
Giulio Casati

University of Milan
Milan, Italy

Plenum Press
New York and London
Published in cooperation with NATO Scientific Affairs Division

Proceedings of a NATO Advanced Study Institute
on Chaos, Order, and Patterns,
held June 25–July 6, 1990,
in Lake Como, Italy

Library of Congress Cataloging in Publication Data

NATO Advanced Study Institute on Chaos, Order, and Patterns (1990: Centro di
cultura scientifica "A. Volta")
    Chaos, order, and patterns / edited by Roberto Artuso, Predrag Cvitanović,
and Giulio Casati.
       p.   cm.—(NATO ASI series. Series B: Physics; vol. 280)
    "Proceedings of a NATO Advanced Study Institute on Chaos, Order, and Pat-
terns, held June 25–July 6, 1990, in Lake Como, Italy"—T.p. verso.
    Includes bibliographical references and index.

    1. Chaotic behavior in systems—Congresses. 2. Pattern percep-
tion—Congresses. I. Artuso, Roberto. II. Cvitanović, Predrag. III. Casati, Giulio,
1942-   . IV. North Atlantic Treaty Organization. Scientific Affairs Division. V.
Title. VI. Series: NATO ASI Series. Series B, Physics; v. 280.
Q172.5.C45N377   1990   91-42273
003'.7—dc20   CIP

ISBN 978-1-4757-0174-6   ISBN 978-1-4757-0172-2 (eBook)
DOI 10.1007/978-1-4757-0172-2

© 1991 Plenum Press, New York
Softcover reprint of the hardcover 1st edition 1991
A Division of Plenum Publishing Corporation
233 Spring Street, New York, N.Y. 10013

SPECIAL PROGRAM ON CHAOS, ORDER, AND PATTERNS

This book contains the proceedings of a NATO Advanced Study Institute
held within the program of activities of the NATO Special Program on
Chaos, Order, and Patterns.

SPECIAL PROGRAM ON CHAOS, ORDER, AND PATTERNS

PREFACE

This volume contains the proceedings of the NATO Advanced Study Institute held at Centro di Cultura Scientifica "A. Volta", Villa Olmo, Como, 25 June - 6 July 1990. R. Artuso, University of Milano, was the scientific secretary of the Institute, the director was P. Cvitanović, Niels Bohr Institute, while G. Casati, University of Milano, acted as the host and the co-director. Other members of the scientific organizing committee were R.E. Ecke, Los Alamos National Laboratory, M.J. Feigenbaum, Rockefeller University, and I. Procaccia, Weizmann Institute.

The attendence at the school consisted of 20 lecturers and 89 students. The term "student" covers here a broad range from a graduate student to a well-established professional, and indeed the Best Student prize was won by Eddie G.D. Cohen, a student well advanced. The organizers of the school would like to thank H.H. Rugh and R. Mainieri for running the very lively "student" seminar series, to our hosts at Villa Olmo for making our stay so pleasant, to the lecturers and seminar speakers for their valiant efforts to enlighten us, to Dipartimento di Fisica, Università di Milano for additional funding, and to R. Artuso for making this school a success.

The Feigenbaum lectures were written up by Z. Kovács, while A. Oliveira and S. Rugh assisted respectively R.S. MacKay and B.V. Chirikov in preparation of their lecture notes. Of the invited lecturers two (Coullet and Steinberg) never supplied manuscripts. In spite of the lack of sufficiently grand video projection screens, we have enjoyed their lectures, and hope that their lecture notes will meet some other deadline for some other proceedings.

Though this was the first school of the NATO Scientific Affairs Division new Non-linear program, in spirit it was a follow up on the 1987 Noto NATO "Nonlinear Evolution and Chaotic Phenomena" Advanced Study Institute. The organizers hope that this is a start of a series of schools within the field of nonlinear dynamics, that will come to play a role in training young nonlinear scientists similar to the role other series of schools (such as the Les Houches schools) have played in furthering other fields.

The physics of nonlinear systems is a confluence of lines of research emerging from the entire spectrum of traditional physics and mathematics specialities: nuclear physics, quantum field theory, statistical mechanics, condensed matter physics, astrophysics, ergodic theory, dynamical systems theory, ... The shared thread is the nonlinear nature of most realistic physical systems. Numerical experimentation with modern computers has made us appreciate the fact that for such systems the long time evolution is so irregular, and the asymptotic states so complicated, that the precise outcome might be unpredictable even though the laws of motion are deterministic. This complexity is already present in the simplest conceivable dynamical systems, and it teaches us that writing down the correct Lagrangian no longer suffices in itself to stake a claim on the theory of nature. However, the same computer experimentation has also given us an insight of fundamental import; large classes of dynamical systems, arising from wildly different physical situations or equations of motion, can give raise to the same universal phase–space structures. Central themes of the current physics of nonlinear phenomena research are "experimental" numerical searches for such universalities, and their theoretical encoding.

The Institute concentrated on some of the currently most active fronts of nonlinear physics, such as the spatio–temporal chaos, fluid turbulence, description of strange attractors in terms of unstable periodic orbits, measures of complexity, and so on.

Some of the problems that have marked this research during the past decade have entered maturity: much of the physics of low-dimensional transitions from regular to chaotic motion is now believed understood, and on this front the Institute had concentrated on the latest developments in the mathematical fundations of the renormalization theory.

Methods developed in the study of nonlinear systems are applicable to a wide range of problems; one particularly interesting recent application discussed by the lecturers was prediction in chaotic systems and nonlinear data smoothening methods.

There is still no coherent theory of turbulence, and a very important new effort, and the main focus of the Institute, was a head-on, hands-on numerical experimentation with spatio–temporal chaos: coupled-map lattices, cellular automata, neural networks and a variety of hydrodynamical turbulent phenomena. This field is still in its infancy, but a serious theoretical advance here is of potentially immense importance for its possible implications for classical hydrodynamics, statistical mechanics, gravity and other strongly nonlinear field theories.

Predrag Cvitanović
Niels Bohr Institute
Copenhagen, Denmark

Copenhagen, November 1990

CONTENTS

SCALING FUNCTION DYNAMICS

Mitchell J. Feigenbaum
lecture notes prepared by Zoltan Kovács

The Rockefeller University, New York, NY 10021, USA

1   Introduction

In September 1979, P. Hohenberg gave me a picture which showed the first preliminary results of A. Libchaber's experiment on liquid helium [1], the power spectrum of a measured signal (Fig. 1). It was immediately clear that the picture had something to do with period doubling, but how it was that one was supposed to understand a one-dimensional theory for a discrete dynamics in order to learn what a fluid was doing was in no way very clear. Over a period of a few months, I tried to understand the picture, and, in the end, was lead to an idea that I have called the scaling function [2]. In these lectures, I shall try to explain what the idea is that came out of this observation and while doing so, discuss the idea that goes under the name of "presentation functions" [3]. I will indicate what these notions mean and explain how from that picture you can determine what is actually the most interesting part of the dynamics.

2   The power spectrum

2.1   The organization of the spectrum

Let us look more carefully at what is in Fig. 1. The frequency scale is linear, so you see a frequency f together with $f/2$, $f/4$, $3f/4$, $f/8$, $3f/8$, $5f/8$, $7f/8$, etc. Although it is not obvious, the right answer in understanding the organization of this picture is that you are presented an object that comes in a number of parts which are sequences of spectral lines at the odd multiples of reciprocal powers of 2. That means the first part is the line $f/2$, the next part are the lines $f/4$ and $3f/4$, there is another part of four lines $f/8$, $3f/8$, $5f/8$, $7f/8$ in between, etc.

If you saw more lines in Fig. 1, it would be still harder to tell that they came in groups of lines because there is no very simple relation from one group of lines to the next. In some sense, it is true that there is a manner in which, for example, the set of four lines is just decreased in size from the set of two lines. In fact, there is a whole range of reductions of amplitude when you look through the set of such spectral lines,

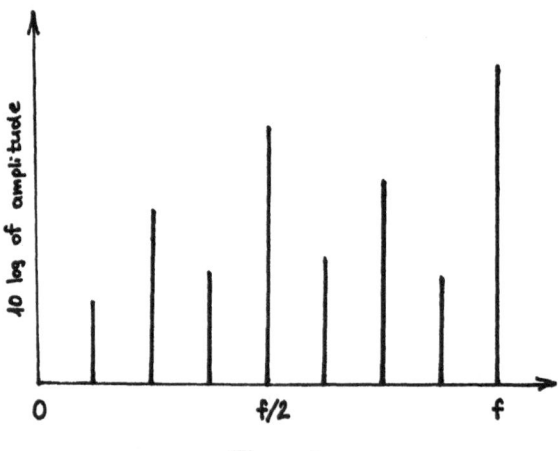

Figure 1

and when there are enough of them, it is no longer very clear exactly how to pick them apart.

One can now start to figure out how one relates the set of lines with a new higher power of 2 in the denominator to the previous ones. The idea is very simple. We have a continuous time signal $x^{(n)}(t)$ from some measurement on the fluid system, and its Fourier transform produced the picture shown in Fig. 1. The index n refers to the periodicity of the signal, that is, it is periodic with a period that, at least asymptotically, goes like 2^n times some fundamental period T_0. $x^{(n)}(t)$ is a periodic signal, so its Fourier transform has discrete spectral lines

$$\hat{x}_p^{(n)} = \frac{1}{2^n T_0} \int_0^{2^n T_0} dt \, x^{(n)}(t) e^{2\pi i p t / 2^n T_0} \tag{1}$$

where p is an integer index.

If you want to compare spectral components of half frequencies to the previous ones, it seems obvious to break (1) into two pieces:

$$\hat{x}_p^{(n)} = \frac{1}{2^n T_0} \int_0^{2^{n-1} T_0} dt \, \left[x^{(n)}(t) + x^{(n)}(t + 2^{n-1} T_0)(-1)^p \right] e^{2\pi i p t / 2^n T_0} \tag{2}$$

Now you immediately learn what the spectrum looks like. If p is odd, these are the frequencies which irreducibly have a denominator 2^n. If p is even, then the denominator is 2^{n-1}, so those are the frequencies that would appear if the signal had not had periodicity of $2^n T_0$ but rather had half that period. In fact, the basic frequencies that are signaling to you that the periodicity is truly $2^n T_0$ are the ones corresponding to odd values of p. Consider what (2) says for odd indices:

$$\hat{x}_{2p+1}^{(n)} = \frac{1}{2^n T_0} \int_0^{2^{n-1} T_0} dt \, \Delta^{(n)}(t) e^{2\pi i (2p+1) t / 2^n T_0} \tag{3}$$

where

$$\Delta^{(n)}(t) = x^{(n)}(t) - x^{(n)}(t + 2^{n-1} T_0) \tag{4}$$

2

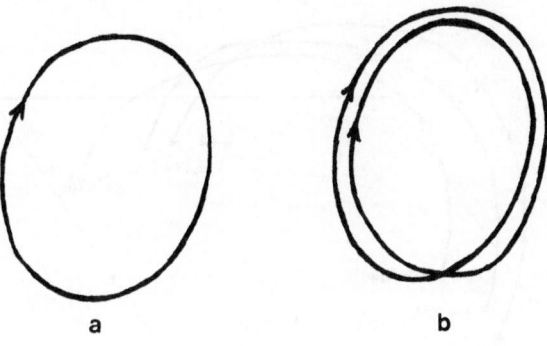

a   b

Figure 2

Equations (3) and (4) say that these spectral components would have identically vanished had the signal had periodicity $2^{n-1}T_0$, because then the two values of the signal in $\Delta^{(n)}(t)$ would be exactly the same. The new spectral lines that tells you that the period has actually doubled are not driven by the full signal; they are only driven by the difference of the signal $x^{(n)}(t)$ and its value half the period later: The strenghts of the differences determine the strengths of the new spectral components.

If one looks at $\Delta^{(n)}(t)$ more carefully, one can see that it is, in some sense, a small quantity. We have, in phase space, an orbit which, after a certain amount of time, closes on itself (Fig. 2a), and when the period doubles, it makes two almost identical copies of itself (Fig. 2b). If I look at it at time t and half a period later, I come around almost to the identical point again but there is a slight separation. This separation as I follow it around the orbit is precisely the quantity whose Fourier transform adds a new set of spectral lines.

If it should have turned out that these little distances all were of the same size and they all decreased from level to level by the same factor, then you would immediately discover that all of the new spectral lines would be one factor smaller than all the previous ones, and you would obtain a picture that is similar to Fig. 1 but a much simpler picture. In that case the spectral lines would come very regularly, every set of lines spaced by the same distance in strength below the previous level. However, it is not true that the difference is the same everywhere along the orbit, and this is the one special complication that we have to pay some attention to.

2.2   Scaling properties of close returns

What we have learned from the above simple considerations is that the Fourier transform of the signal whose periodicity is 2^n and its odd spectral components are given by the Fourier transform of the object $\Delta^{(n)}(t)$, the difference of the signal and its value half a period later. At this point, let us think more about these little differences, because there is an important principle to discern in them.

When one thinks about the orbits of a continuous system, one also thinks about making surfaces of sections and considering Poincaré maps. Originally the orbit gives one point on the Poincaré section and every period doubling doubles the number of points on it. There is obviously some self-similarity in the picture because every two steps it almost comes back to the same point. The closest return to the starting point occurs half a period later, and if the period is long enough, the point at a quarter of

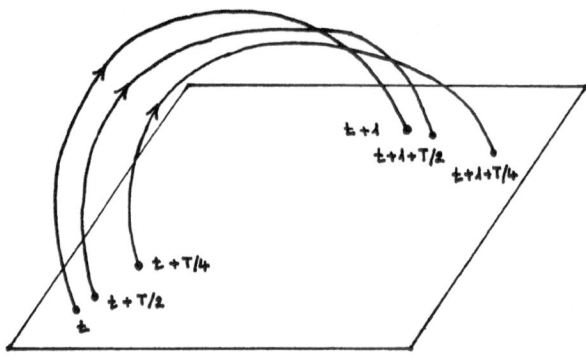

Figure 3

the period away also makes a reasonably close return. If the period doubles again, it does not make for completely different things, rather, each of these points splits apart again on a finer scale.

After many doublings, up to some period T, we have a point $x(t)$ on the Poincaré section, another point $x(t + T/2)$ nearby to it and a third one $x(t + T/4)$ less nearby (Fig. 3). One time step later (at the next return to the surface of section), somewhere far away there is a point $x(t + 1)$. Near to that point we find $x(t + 1 + T/2)$ and near to it again is $x(t + 1 + T/4)$. Let us assume that the ratio of the splittings around $x(t)$ is the same as that around $x(t + 1)$:

$$\frac{x(t) - x(t + T/2)}{x(t) - x(t + T/4)} = \frac{x(t + 1) - x(t + 1 + T/2)}{x(t + 1) - x(t + 1 + T/4)}$$

One can easily verify by linearizing the differential equations that if this statement is true in the surface of section, then the simple differential flow connecting the points of the Poincaré section in the phase space will preserve the fact that the ratios of these distances are the same along the orbit, provided there is no singularity in the flow. In other words, if we look at these differences, we expect them to have the preserved property that they *scale* from one level to the next. This property is rather interesting because it says if we understand the theory of these ratios in the Poincaré section, we already know what to do in the continuum problem, that is, there is no impediment in moving from discrete time to continuous time.

At this point we turn to the one-dimensional theory of period doubling, because it tells us the scaling properties of close returns. If we have a 1-d map $f(x)$ with a quadratic critical point at x_c, we can consider this map at the control parameter value where it has a superstable periodic orbit of length 2^n; the closest return to the critical point is its 2^{n-1}th iterate with a distance $\Delta^{(n)}$. Tuning the control parameter so that the map has a superstable orbit of length 2^{n+1}, we see the closest return to x_c at the 2^nth iterate with a distance $\Delta^{(n+1)}$. The theory tells us that the ratio $\Delta^{(n)}/\Delta^{(n+1)}$ is a fixed universal ratio given by the number α, a fundamental ingredient of the renormalization group treatment for period doubling [4]. We will see in Sec.3 that this theory is based on the idea that there is a certain operator R, the renormalization group operator, and it has a fixed point function g that defines a dynamics $x_{t+1} = g(x_t)$ with universal scaling properties.

2.3   Connection to the fluid system

The next question we must understand in producing a theory for the picture shown in Fig. 1 is how it is that a simple one-dimensional problem can tell us, in an embedding-free way, what the Fourier transform is of that unknown system of equations. A crucial idea for the connection between the behavior of the fluid and the 1-d problem was worked out by Collet and Eckmann in 1979 [5]. It says the fluid, a system with an infinite number of degrees of freedom, has a dynamics in which, if you look at the orbit and its stability, you see that the eigenvalues are situated in the complex plane in a nontrivial fashion. Namely, there is an innumerable number of eigenvalues all crowded around 0, they are transverse to the motion we are looking at and produce the collapse of the flow basically to a one-dimensional flow.

There is only one other eigenvalue inside the unit circle, and it is located on the real axis between -1 and $+1$. In the course of period doubling, this eigenvalue migrates from $+1$ to -1 along the real axis, and reaching -1, its double iterate returns to $+1$, where the period doubling cascade continues with this new eigenvalue. In summary, that means the eigenvalues ensure that locally the points on the Poincaré surface end up lining along one particular straight line along which we will see the identical kind of behavior that we have seen in the 1-d problem. In other words, if it turns out that period doubling is the kind of dynamical behavior you are witnessing, then any point in the phase space with its very close returns must end up lying on a little linear segment.

The straight line segments are embedded in the infinite dimensional phase phace of the fluid. Measuring one particular coordinate is a projection along one axis which makes angles with these segments. If our knowledge is a statement about how the one length on the segment scales to the other, then, whatever the angle of this little segment to the axis is, however it is embedded in the original high-dimensional phase space, that ratio is an invariant under the embedding. If I project along a generic axis, the linear scaling property will have exactly the same value in the projection as it does in the embedding.

In fact, if we consider these ratios, we abstract from the 1-d problem the set of considerations that are completely independent of what the original picture looks like. That means we can take any data we have and subject it to the same one-dimensional scaling treatment. The scaling properties that we can learn from the 1-d problem will tell us a set of internal relationships that we should be able to understand now in an arbitrary system.

We are now in a situation where, if it is true that there is a scaling preserved everywhere along the orbit, we can deduce, completely free of embedding, the scaling information from the 1-d problem. Knowing what the scalings are, we can go into an arbitrary time signal from an arbitrary system and determine from a finite time observation the total behavior of the system. Obviously, we must have some basic distance taken from the measurements because the theory does not tell us anyting about the particular embedding; we must approximately know how the system lives in its phase space. Once we know that, however, then we can take the basic distance and, multiplying it by its scalings, compute where the next close returns have to go.

2.4 Construction of the spectrum

The scalings in the 1-d case have the nice property that they will be invariant along the orbit. The reason for that is the following. If we take two little distances of close returns around one particular point and apply the map to them, they will be multiplied by almost the same number, that is, the eigenvalue of the linearized map; however, taking the ratio of the distances, this number cancels out. The only place that can fail is where the derivative vanishes.

The type of dynamics f or g defines comes about very crucially, because there is a critical point where the derivative is 0. So long as we keep away from the critical point, the scaling ratio α is preserved. However, when we make close passages to the critical point, this ratio can change very seriously. If we look at the distances one time step later, the corresponding intervals are mapped through the quadratic critical point, so their lengths are squared and the ratio becomes α^2. Therefore, it is certainly true that there are at least two different kinds of scalings that are present in this dynamics.

Now we can ask what other scaling values would come up, and the answer is that, actually, these two scales provide a good approximation to the scaling properties. Indeed, once we have gone through the critical point, so the scaling is α^2, we are on a nice linear part of the map and the ratio will not change. On the other hand, if we think about a piece with scaling α, we see that its preimages came through a linear part, so they had to have the same scaling α. One can loosely guess that this problem should be well described by an object which has basically two scales in it, the one is α and the other is α^2.

Taking the above observations into account, one now is in a position to do the calculation about the spectrum. The intervals $\Delta^{(n)}(t)$ can be written as the product of the nth level set of scalings σ_n, in principle different values for different values of t, and the previous intervals:

$$\Delta^{(n)}(t) \equiv \sigma_n(t/2^n T_0)\, \Delta^{(n-1)}(t) \tag{5}$$

It is clearly important what fraction of the orbit I am looking at, so I normalized the argument of σ by the period of the orbit $2^n T_0$. That says the scalings are defined as a function of how much through the orbit I am looking at. The Δ's change sign when I go halfway through the orbit (see (4)), thus the function σ has the property that it is periodic with period 1 and the second half of the period differs from the first half in its sign only:

$$\sigma\left(\tau + \frac{1}{2}\right) = -\sigma(\tau) \tag{6}$$

I will use this function to compute the power spectrum.

It is going to be a problem that I have odd numbers over 2^n in the exponent of (3), so I define a smooth interpolation for the odd spectral components by introducing a continuous frequency variable

$$\omega = 2\pi\, \frac{2p + 1}{2^n T_0} \tag{7}$$

The nth interpolation now is simply

$$\hat{x}^{(n)}(\omega) = \frac{1}{2^n T_0} \int_0^{2^{n-1} T_0} dt\, \Delta^{(n)}(t) e^{i\omega t} \tag{8}$$

It is virtuous to introduce the interpolation through half of the integral rather than through the original integral going up to $2^n T_0$, because smaller values of t yield smaller frequencies, so it will produce a smoother interpolation. Actually, (8) is the smoothest interpolation based on ignorance of $\Delta^{(n)}(t)$ that I can write down.

Using (5), I rewrite (8) as

$$\hat{x}^{(n)}(\omega) = \frac{1}{2^n T_0} \int_0^{2^{n-1} T_0} dt\, \sigma(t/2^n T_0) \Delta^{(n-1)}(t) e^{i\omega t} \tag{9}$$

Now, based on the ideas discussed at the end of Sec.2.2, I make the assumption that σ has the following set of values:

$$\sigma(\tau) = \begin{cases} \alpha^{-2} & 0 \leq \tau < 1/4 \\ -\alpha^{-1} & 1/4 \leq \tau < 1/2 \end{cases} \tag{10}$$

This gives me all the information I need, and I simply inject that into the calculation. I break the integral in (11) into two pieces:

$$\hat{x}^{(n)}(\omega) = \frac{1}{2^{n-1} T_0} \int_0^{2^{n-2} T_0} dt\, \frac{\sigma(0) - \sigma(1/4) e^{i\omega 2^{n-2} T_0}}{2} \Delta^{(n-1)}(t) e^{i\omega t} \tag{11}$$

Using (8) we can rewrite it as

$$\hat{x}^{(n)}(\omega) = \frac{\sigma(0) - \sigma(1/4) e^{i\omega 2^{n-2} T_0}}{2} \hat{x}^{(n-1)}(\omega) \tag{12}$$

That is the end of the calculation. What it says basically is that the spectral lines $\hat{x}^{(n)}(\omega)$ are Fourier transform of $\Delta^{(n)}(t)$ given by (5) as a product $\sigma \Delta^{(n-1)}$, so the Fourier transform of $\Delta^{(n)}$ is the convolution of the Fourier transforms. In this approximation, when $\sigma(\tau)$ has approximately two different values, what would have been a convolution simply turns out to be a multiplicative property in the Fourier transform as well.

Now we analyse (12) for different values of ω. One special value we care about is $\omega = 2\pi(2p+1)/2^n T_0$. where the new spectral lines are supposed to be. We insert this into (12) and discover that

$$\hat{x}^{(n)}_{2p+1} = \frac{\sigma(0) - i\sigma(1/4)(-1)^p}{2} \hat{x}^{(n-1)}(\omega) \tag{13}$$

The 1-d theory tells us that α is about -2.5; using this value, the modulus of the prefactor in (13) turns out to be about 0.22. Taking its logarithm based 10, it produces -6.6 db in amplitude, which gives -13.2 db for the power spectrum. This is the number that was first produced by Nauenberg in a numerical calculation simply by looking at the fixed point function [6].

Another computation we can make is $\hat{x}^{(n)}_{2(2p+1)}$. These are the frequencies that came in at the previous level, now looking one level later. We obtain

$$\hat{x}^{(n)}_{2(2p+1)} = \frac{\sigma(0) + \sigma(1/4)}{2} \hat{x}^{(n-1)}(\omega) \tag{14}$$

That gives a factor of $0.28 \approx -5.5$ db, thus these lines are decreasing less quickly. Finally, we can look at the lines that came in two levels ago:

$$\hat{x}^{(n)}_{2^2(2p+1)} = \frac{\sigma(0) - \sigma(1/4)}{2} \hat{x}^{(n-1)}(\omega) \tag{15}$$

7

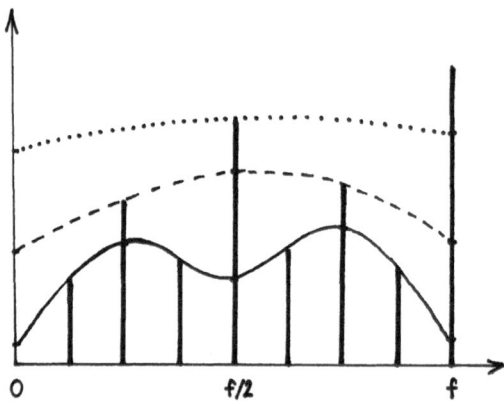

Figure 4. The first several interpolations to obtain the spectral lines at $f/2$ (dotted line), $f/4$ and $3f/4$ (dashed line), $f/8, \ldots, 7f/8$ (full line).

This number turns out to be -0.12. This is a serious decrease in strength, it is about -9 db.

In fact, this is not a particularly simple spectrum. To construct it, we can follow the iteration procedure defined by (12) and the values given above. We start off at the line at frequency f and draw a smooth interpolation to obtain the line at $f/2$ (Fig. 4). In the second step, we want to find the interpolation that gives the lines $f/4$ and $3f/4$. We can do that in the following way. Equation 14 tell us to drop the line at $f/2$ by a small amount of 5.5 db, while the component with frequency f must be decreased by a large amount of 9 db. Based on these points, we can draw a smooth interpolation that determines the new lines at $f/4$ and $3f/4$. Now I continue doing these interpolations again and again.

The large reduction factor in (15) keeps burning holes into the smooth interpolation curves at the places of the old spectral lines, and the new lines have just the medium height between the bottom of a hole on one side of it and the top of the hill on the other. The new sets of lines determined by the successive interpolations are to be added to the previous ones, thus producing a rather complicated spectrum. Indeed, the new spectral lines come as a unit but they are not, in any sense of the world, just dropped by a constant amount. There is growing variance, because the amount by which the lines are dropped depends on when they came into existence.

In the course of thinking about the picture of Fig. 1, what you should most importantly have learned is that there is a feature that allows you to go from the 1-d case to the very general circumstance. That feature is something that is invariant under all transformations, it is a very interesting object, it tells you the full set of scaling properties everywhere along this attractor, and one would like to understand generally what the theory for this so-called scaling function is.

3   The scaling function

In the previous section, I explained in qualitative ways how one understands scaling properties that appear in these infinitely renormalizable systems. Now let me proceed and say in more formal fashions the information that I discussed in Sec.2.

3.1 The invariance of scalings

In the previous discussion I was thinking about a system that was continuous in time. There was a certain basic period asymptotically that was called T_0. From this point onwards I will simply consider the one-dimensional discrete dynamics and take T_0 as one time step of the dynamics. The basic point is that if one produces the scaling information from the 1-d dynamics, then one can use it in a continuous sytem.

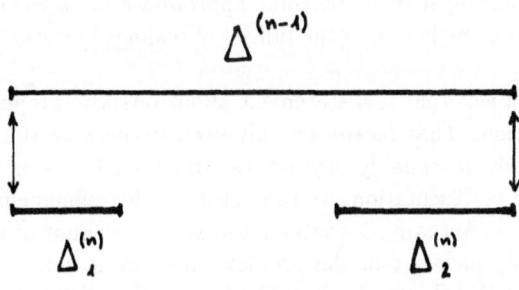

Figure 5

One possible way to carry out the analysis is to consider the system at a control parameter value at which it produces a periodic time signal $x_t^{(n)}$ with periodicity 2^{n+1}. (If the signal is not periodic, e.g. at the accumulation point of period doubling, we can make it periodic by hand by taking a finite piece from the orbit and repeat it infinitely many times, i.e. $x_t^{(n)} \equiv x_{t \bmod 2^{n+1}}$. In the limit $n \to \infty$ we obtain correct results.) There are 2^n little differences $\Delta_t^{(n)}$ of the orbit points and their close returns halfway through the orbit:

$$\Delta_t^{(n)} \equiv x_t^{(n)} - x_{t+2^n}^{(n)} \qquad\qquad t = 0, 1, \ldots, 2^n - 1 \qquad (16)$$

If we add 2^n to t, then the terms in the difference exchange their roles, so

$$\Delta_{t+2^n}^{(n)} = -\Delta_t^{(n)} \qquad\qquad (17)$$

The scalings defined at the nth level are simply

$$\sigma_t^{(n)} \equiv \sigma_n(t/2^{n+1}) = \frac{\Delta_t^{(n)}}{\Delta_t^{(n-1)}} = \frac{x_t^{(n)} - x_{t+2^n}^{(n)}}{x_t^{(n-1)} - x_{t+2^{n-1} \bmod 2^n}^{(n-1)}} \qquad (18)$$

An important feature of this definition is that the time indices in the first terms of the numerator and denominator are exactly the same, which means that these terms refer to the same point. The reason that is important is that if we subject x to an arbitrary coordinate transformation, i.e. replace x by $h(x)$ where h is a smooth invertible function, then the differences of very close points in the numerator and the denominator are multiplied by approximately the same number, namely, the derivative $h'(x)$. Taking the ratio in (18), the derivative cancels out, thus leaving the scaling unchanged, at least in the limit of large n when the differences are infinitesimally small, so this expression is invariant under all smooth coordinate transformations.

9

There is another important consequence of the fact that the time indices in the first terms agree. When refining the object, we are constructing a Cantor set: we have a parent interval that is fractured into two child intervals. Due to our definition in (16), the outer endpoints of the children will line up with the endpoints of the parent interval (Fig. 5). This means the new pieces are well coordinated, they have an intelligently well defined gap, which, for example, includes unstable periodic points within it. Once we know the scaling ratios, then we also know the ratio of the gap itself, therefore we can understand the whole geometry. (There is a subtlety here which is worth mentioning. The scalings can be regarded as a measure of the strengths of some periodic orbits, so we might imagine replacing scalings by some appropriate cycle eigenvalues. However, this information seems to be less than the full set of scalings because it does not include the gap information.)

Now we have an object that is a statement about intrinsic properties, not any one particular representation. That means we only need to measure the differences in our system that are already reasonably asymptotic and then by using σ, which is independent of coordinate representation, we can calculate the refinements in whatever the real coordinatization is. Actually, σ contains the largest amount of invariant information one could possibly pick out in the problem, because it essentially allows one to completely reconstruct the object. In general, this is only asymptotic information. We cannot say what the whole object is by this deduction, we can only say how to refine the object, so if we understand it for short periods of time, then we can figure what will happen for arbitrarily long periods of time.

It is worth noting that the flow induced by a system of differential equations is a smooth coordinate transformation, so you can also use this scaling information in a continuous system. However, it is no use in taking the period doubling fixed point g and Fourier transforming its dynamics, because that spectrum will not, in general, agree with the spectrum of the fluid system. It is these pieces of refinement information that are independent of embedding.

3.2 Return time expansions

t has a discrete set of values, nevertheless, we can consider σ_n as a function of a continuous variable $\tau \in [0, 1]$, and $\sigma_n(\tau)$ is simply defined to have piecewise constant values

$$\sigma_n(\tau) = \sigma_t^{(n)} \qquad \text{for} \quad \frac{t}{2^{n+1}} \leq \tau < \frac{t+1}{2^{n+1}} \tag{19}$$

Now the argument of σ_n is independent of n. If I can verify that the limit $n \to \infty$ for σ_n exists, then this will produce intervalwise the definition of a function $\sigma(\tau)$ on the interval $[0, 1]$. As a consequence of (17), the scalings have the symmetry

$$\sigma_n\left(\tau + \frac{1}{2}\right) = -\sigma_n(\tau) \tag{20}$$

from which it immediately follows that σ is periodic of period 1, but it is sufficient to look at it for $\tau \in [0, 1/2)$ (i.e. for $t = 0, 1, 2, \ldots, 2^n - 1$) to determine all the information we can compute in this fashion. One can then use that function, for example, in a Fourier transform integral to compute the power spectrum.

It will turn out to be useful to write τ in binary notation:

$$\tau = .0\varepsilon_1\varepsilon_2\ldots\varepsilon_r\ldots \tag{21}$$

To find the argument t, we multiply it by 2^{n+1}, which gives

$$t = 2^{n+1}\tau = 2^{n-1}\varepsilon_1 + 2^{n-2}\varepsilon_2 + \cdots + 2^{n-r}\varepsilon_r + \cdots \tag{22}$$

This point is of absolute importance in all of the considerations about scaling. When one considers where in time one is looking at something, one writes that time as an expansion in terms of *return times*. In our problem one makes close returns at every power of 2, so we write the amount of time in a binary expansion. There is no impediment basically to carrying this idea through now for any other problem. In particular, if you look at a quasiperiodic motion, e.g. golden mean rotation, one establishes that the relevant return times are Fibonacci numbers, so one does all these same considerations but one expand time in terms of combinations of Fibonacci numbers [7].

In the discrete dynamics, $x_t^{(n)}$ is defined as the map applied t times to a starting point. The underlying strategy in the return-time expansion is that a map iterated to a return time can be controlled and has nice scaling properties. It is also crucial the fact that the ε's in (22) have been reversed, because it turns out that the short-time (i.e. high-index) bits are going to be irrelevant for the discussion, it is only the high-time (low-index) bits that are important. The reason is very clear: a short-time bit is a small change in t, a small number of applications of this smooth dynamical system, but we know that the ratio of differences in (18) is invariant under a smooth coordinate transformation. The only time the short-time bits can have an important impact on the ratios is if adding 1 to a very faraway bit produces an avalanche and you see a high-order bit changing. (These comments seem obviuos, nevertheless, to really make them rigorous is difficult in the extreme.)

The high-order bits have to do with the scaling information, while the low-order bits simply deposit the scaling structure globally over the entire dynamics. The picture loosely is that we have a little piece on the Poincaré section with very nice scaling behavior in it, it is copied in one step to the next region and then copied again and again. One time step on the Poincaré surface is a smooth transformation that preserves the invariant scaling information, so it is the same in all these regions. As the dynamics keeps copying it around, slow distortions are building up, and by the time you make a close passage to the starting place you discover that the local geometric scaling factor has somewhat changed, there is an internal substructure in the original structure with a slightly different scaling from what one might expect. What one sees is an object that in space has a very complicated, interpenetrated character, it is deployed in many places all over the space. What is simple is that if you look at the scaling structure and you follow it according to the dynamics, then that structure will be more or less constant as you pursue it in time. To actually discuss what this structure is like in a given region of space is a rather complicated proposition because of these different interpenetrated scalings.

Let us return now to the scalings and express them in the binary notation. Remember, there are two things I have to compute in (18), $t + 2^n$ in the numerator and $t + 2^{n-1}$ in the denominator. The first bit of t is 0 because I only have to look at the first half of the period, so adding 2^n will just put 1 into this bit. When we add 2^{n-1} in the denominator, we have to add 1 to ε_1 and then take it mod 2^n. It is easy to see that the result is

$$(t + 2^{n-1}) \bmod 2^n = 2^{n-1}\overline{\varepsilon}_1 + 2^{n-2}\varepsilon_2 + \cdots + 2^{n-r}\varepsilon_r + \cdots \tag{23}$$

i.e we simply replace ε_1 by its binary complement $\overline{\varepsilon}_1$. The scaling is then

$$\sigma(.0\varepsilon_1 \ldots \varepsilon_r \ldots) = \frac{x_{0\varepsilon_1 \ldots \varepsilon_r \ldots} - x_{1\varepsilon_1 \ldots \varepsilon_r \ldots}}{x_{\varepsilon_1 \varepsilon_2 \ldots \varepsilon_r \ldots} - x_{\overline{\varepsilon}_1 \varepsilon_2 \ldots \varepsilon_r \ldots}} \tag{24}$$

11

We omitted here the supersripts of the x's since the number of bits in the index also gives the periodicity information.

3.3 The scalings and the fixed point function

In order to go further and see what this quantity looks like, I have to tell you something about the actual dynamics and the renormalization group theory. The basic qualitative statements are that the powers of 2 are return times and return times produce geometric scaling. That means if the dynamics is generated by a map f possessing a unique quadratic maximum at $x = 0$ and we compose the map with itself a return time number of time steps, then, starting at $x = 0$, it will bring back very close to 0, and each return time we come a fixed geometric factor α closer to the starting point. The point of the renormalization group theory is that if we blow up f^{2^n} n times by the factor α around $x = 0$, i.e take $\alpha^n f^{2^n}(x/\alpha^n)$, then, as $n \to \infty$, this procedure will converge to a definite nice function g which is independent of f. You can immediately verify that, if this statement is true, g satisfies the following functional equation:

$$g = \alpha g g \alpha^{-1} \tag{25}$$

The operation of taking the function, composing it with itself, magnifying it up by a certain number is called the renormalization group operator R, and (25) says the function g is the fixed point of this operator.

We are now in a position to start to investigate what comes out through the close return times. First, we have to identify the points x_t with the dynamics so that

$$x_t \equiv f^{t+1}(0) \tag{26}$$

The point $x = 0$, the critical point of the map, is of special importance because the dynamics is singular around it, and it is also the centre of the magnification process. However, the actual definition of x_t is not the tth iterate of the critical point $x = 0$, rather, it is the tth iterate of the critical value $f(0)$, which is chosen to be 1. It will be clear later why we use this definition. The algebraic singularity around 0 is, in a typical case, quadratic, so

$$1 - f(x) \sim x^2 \tag{27}$$

Now we can rewrite (24) with the definition of x_t:

$$\sigma(.0\varepsilon_1 \ldots \varepsilon_r \ldots) = \tag{28}$$
$$\frac{f\left(\ldots (f^{2^{n-r}})^{\varepsilon_r} \ldots (f^{2^{n-1}})^{\varepsilon_1}(0)\right) - f\left(\ldots (f^{2^{n-r}})^{\varepsilon_r} \ldots (f^{2^{n-1}})^{\varepsilon_1} f^{2^n}(0)\right)}{f\left(\ldots (f^{2^{n-r}})^{\varepsilon_r} \ldots (f^{2^{n-1}})^{\varepsilon_1}(0)\right) - f\left(\ldots (f^{2^{n-r}})^{\varepsilon_r} \ldots (f^{2^{n-1}})^{\overline{\varepsilon}_1}(0)\right)}$$

It is very important in this expression that we group the compositions so that first we take the largest number of iterations and save the smallest number for last (the very smallest is the extra $+1$ in (26)). The reason for that will also become clear. Before the next step, we assume that we have a large but finite number of digits and n is asymptotically large, much larger than the number of digits. If that is true, then we want to replace the asymptotically large number of iterates of f by a magnification, i.e we multiply $f^{2^{n-r}}$ by α^{n-r} and its argument by α^{-n+r}. In order not to change anything, we must put an extra α^{-1} between each term, and we obtain

$$\sigma(.0\varepsilon_1 \ldots \varepsilon_r \ldots) = \frac{f(\{0\varepsilon_1\}) - f(\{1\varepsilon_1\})}{f(\{\varepsilon_1\}) - f(\{\overline{\varepsilon}_1\})} \tag{29}$$

where the shorthands for the arguments read

$$\{0\varepsilon_1\} = \ldots (\alpha^{n-r}f^{2^{n-r}}\alpha^{-n+r})^{\varepsilon_r}\alpha^{-1}\ldots(\alpha^{n-1}f^{2^{n-1}}\alpha^{-n+1})^{\varepsilon_1}\alpha^{-1}(0)$$

$$\{1\varepsilon_1\} = \ldots (\alpha^{n-r}f^{2^{n-r}}\alpha^{-n+r})^{\varepsilon_r}\alpha^{-1}\ldots(\alpha^{n-1}f^{2^{n-1}}\alpha^{-n+1})^{\varepsilon_1}\alpha^{-1}(\alpha^n f^{2^n}\alpha^{-n})(0)$$

$$\{\bar{\varepsilon}_1\} = \ldots (\alpha^{n-r}f^{2^{n-r}}\alpha^{-n+r})^{\varepsilon_r}\alpha^{-1}\ldots(\alpha^{n-1}f^{2^{n-1}}\alpha^{-n+1})^{\bar{\varepsilon}_1}\alpha^{-1}(0)$$

$$\{\varepsilon_1\} = \{0\varepsilon_1\}$$

Each term in the arguments goes, asymptotically in n, to the function g, thus we can rewrite the right-hand side of (29) as

$$\frac{f\left(\alpha^{-p}\ldots g^{\varepsilon_r}\alpha^{-1}\ldots g^{\varepsilon_1}\alpha^{-1}(0)\right) - f\left(\alpha^{-p}\ldots g^{\varepsilon_r}\alpha^{-1}\ldots g^{\varepsilon_1}\alpha^{-1}g(0)\right)}{f\left(\alpha^{-p}\ldots g^{\varepsilon_r}\alpha^{-1}\ldots g^{\varepsilon_1}\alpha^{-1}(0)\right) - f\left(\alpha^{-p}\ldots g^{\varepsilon_r}\alpha^{-1}\ldots g^{\bar{\varepsilon}_1}\alpha^{-1}(0)\right)} \tag{30}$$

(There is obviously a subtlety here, we have to understand very well how quickly we can take this limit. Some of the correct thinking about that question was discussed by Sullivan in his lecture [8].)

The number of digits terminates at some number, so we had to put a large negative power of α into the arguments of the four final f's in (30) (and in (29) as well) to take this into account. That factor will make these arguments very small, so we can use (27) to see what effect f has in (30). Equation (27) says f is quadratic near 0 with a maximum height of 1. By taking differences the 1's go away and the second derivative cancels out in the ratio, thus we are left with the expression

$$\sigma(.0\varepsilon_1\ldots\varepsilon_r\ldots) = \tag{31}$$

$$\frac{\left(\ldots g^{\varepsilon_r}\alpha^{-1}\ldots g^{\varepsilon_1}\alpha^{-1}(0)\right)^2 - \left(\ldots g^{\varepsilon_r}\alpha^{-1}\ldots g^{\varepsilon_1}\alpha^{-1}(1)\right)^2}{\left(\ldots g^{\varepsilon_r}\alpha^{-1}\ldots g^{\varepsilon_1}\alpha^{-1}(0)\right)^2 - \left(\ldots g^{\varepsilon_r}\alpha^{-1}\ldots g^{\bar{\varepsilon}_1}\alpha^{-1}(0)\right)^2}$$

where we used the $g(0) = 1$ normalization for the fixed point function. This is simply a formal writing in terms of the fixed point function g of exactly what the scalings are. One can see that it was computed for the original problem, whatever the dynamics was, not for the dynamics of the fixed point. Nonetheless, the scaling information turns out to be universally computable simply in terms of the fixed point function g. Thus, at least at the level of this calculation, this is an affirmation that the scaling information is indeed independent of the particular choice of coordinatization.

In principle, it can cause problems that we do a lot of compositions in (31). Since we allow only a finite string of ε's and let $n \to \infty$, then we can control the computation and all the considerations are correct. But there is a question if I have correctly accounted for what happens for infinite strings. I simply assert that the correct answer is that there is an exponential fall-off on the high-index digits, so we can complete the construction by continuity.

3.4 Properties of the scaling function

Now we make some easy computations to see the nature of the behavior of σ. First of all, it is easy to compute $\sigma(0)$, because all the ε's vanish and we have no functions:

$$\sigma(0) = \frac{0^2 - (\alpha^{-1})^2}{0^2 - 1^2} = \alpha^{-2} \tag{32}$$

This is a rather obvious calculation because the basic statement of the renomalization group theory says that, when we look near 0, we see a scale reduction by α^{-n} for a close return at time 2^n. Now 0 is a quadratic critical point which means if distances reduce by α near 0, then quadratically imaged, they must be decreased by α^2.

We can do a slightly more complicated calculation for the case when we have a lot of 0's and a 1 as the rth digit. We just read off from (31) what that produces for large r:

$$\sigma(.0\ldots01) = \frac{(g(0))^2 - (g(\alpha^{-r}))^2}{(g(0))^2 - (g(\alpha^{-r+1}))^2} \xrightarrow{r \to \infty} \alpha^{-2} \tag{33}$$

where we used the fact that g is also a quadratic map. If you know more about g, you can also specify how, as this bit of 1 moves further and further off to the right, you converge to the value α^{-2}.

You can do similar considerations about any particular value and estimate how important these farover bits are. The general answer is that they fall off exponentially, so you discover that the function σ has the property that it has weaker and weaker dependence on the low-lying bits. This is not surprising because the ratio that we are looking at is unchanged under smooth coordinate transformations such as a small number of time steps corresponding to faraway bits in the time index.

The reason that I choose $t + 1$ iteration in the definition of x_t is that otherwise I would have had a slightly anomalous result. Had I chosen to have just t in (26) and not $t+1$, then, instead of having the original f in the terms of (30), I would have thrown out that, so (31) would have been the same expression except with terms unsquared. That means the calculation of $\sigma(0)$ would have ended up by producing α^{-1} while the result of (33) for $\sigma(0^+)$ would not have changed. Thus I would have been in a situation that changing one bit very far off to the right would have resulted in a big discontinuity. My definition cannot eliminate the discontinuity either, however, it moves that one step earlier, which means that I obtain α^{-1} at $\sigma(0^-)$, i.e for a infinite string of 1's. The point is that there is a discontinuity that is going to be encountered when we look at objects terminating at some number of binary digits, which means the scalings for the strings $\varepsilon_1 \ldots \varepsilon_k 0111 \ldots$ and $\varepsilon_1 \ldots \varepsilon_k 1000 \ldots$ are different. However, in this case adding 1 to a very faraway bit of the former string starts an avalanche resulting in a change of ϵ_{k+1}, so the change in the scaling is not surprising. The extra $+1$ in the definition of x_t simply sticks the discontinuity at prior values to t, not after it. This is just a technical detail which makes the computation come out more nicely and more precisely, and it makes the dependence on the lower order bits an exact statement.

The next observation is that we see in (31) a composition of g's and factors α^{-1}. Let $k - 1$ be the number of 0 bits preceding $\varepsilon_m = 1$, then the corresponding g in the composition is preceded (in the order we apply the functions) by α^{-k}; we will see that the worst case is g simply preceded by α^{-1} corresponding to a long tail of 1's in the index. The function $g\alpha^{-k}$ is a contraction, because the factor α^{-k} expands the scale on the x axis so the function is flattened out around the quadratic maximum. The larger k is the closer this function is to a constant. In (31), we apply a composition of functions that are at least as contracting as $g\alpha^{-1}$. We can see that it is only the beginning terms (those which are applied first) of the compositions that are varying, so we apply a contracting function to slightly different starting points, and that is why the scaling ratio can change very much.

It is easy to establish a few nice properties of the function $c(x) \equiv g\alpha^{-1}(x)$. Let $\bar{x}$ denote the place in $(0, 1)$ where g vanishes. If you substitute it into the functional equation (25), you can see that $\bar{x}$ is a fixed point of c:

$$0 = g(\bar{x}) = \alpha g(g(\bar{x}/\alpha)) \implies (g\alpha^{-1})(\bar{x}) = \bar{x} \tag{34}$$

14

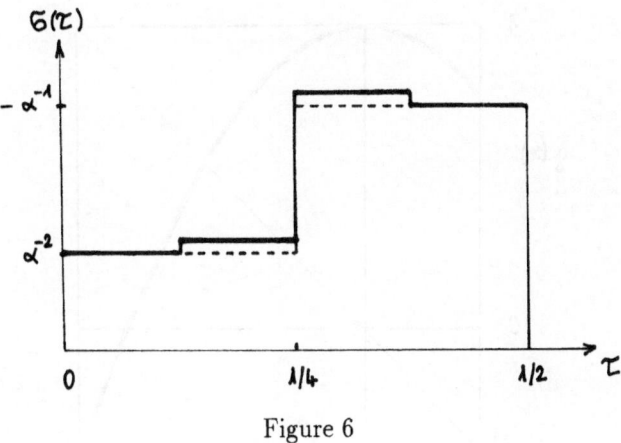

Figure 6

If you differentiate this, you discover that

$$g'(\overline{x}) = g'(g(\overline{x}/\alpha))g'(\overline{x}/\alpha) \implies g'(\overline{x}/\alpha) = 1 \tag{35}$$

which says the derivative of c at its fixed point is α^{-1}. Now we understand why the worst case is when the string has a long tail of 1's, it means one is watching a contraction into the fixed point going through a definite contractive value α^{-1}, and one has to look at the second derivative, the nonlinearity, to see exactly what changes are occuring in the scaling.

The last comment that I make at this technical level is that if I look at the value $\tau = .0\varepsilon_1\varepsilon_2\ldots\varepsilon_m\ldots$ terminating after some large number of bits, the scaling expression (see (31)) is given as the ratio of differences of squares:

$$\sigma(\tau) = \frac{[\ ]^2 - [\ ]^2}{[\ ]^2 - [\ ]^2}$$

Now I can ask what happens one time step earlier. That means I subtract 1 from $t = 2^{n+1}\tau$, so I do not have to do the final application of f in the composition, therefore I obtain, for the value infinitesimally before τ, the same expression except the terms are not squared:

$$\sigma(\tau - 0) = \frac{[\ ] - [\ ]}{[\ ] - [\ ]}$$

Equation 31 is a ratio of differences of squared terms, which is equal to the ratio of the differences times the ratio of the sums. Each of the terms appearing in the expression is essentially the same by the contractive argument, so the ratio of the sums is asymptotically 1. That says $\sigma(\tau)$ and $\sigma(\tau-0)$ become asymptotically identical except when τ is a diadic value, i.e. it has a finite expansion to the base 2. That means σ turns out to be a function which has derivative 0 except just prior to the diadic values.

The important point is you do not need very many ε's to get a very accurate value of σ. The computation that produced $\sigma(0) = \alpha^{-2}$ tells you, by the above argument, that $\sigma(0^-) = \alpha^{-1}$ instead of α^{-2}. Paying attention to the symmetry of σ, (20) says $\sigma(1 - 0) = \alpha^{-1}$, subtracting 1/2 you obtain $\sigma(1/2 - 0) = -\alpha^{-1}$, and that tells you

15

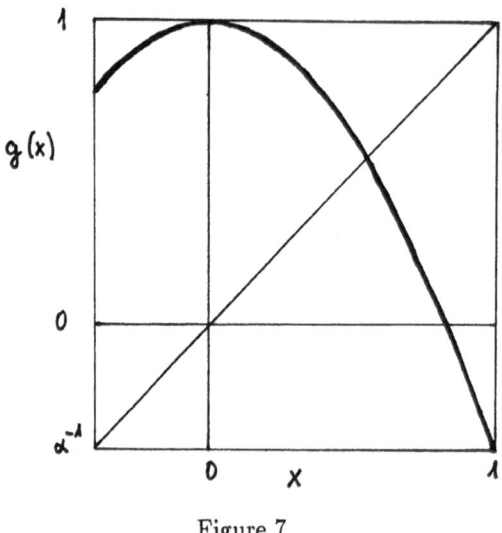

Figure 7

basically what the scaling function looks like (Fig. 6). In a first approximation when we only pay attention to the leading digit ε_1, it consists of two constant pieces. By putting in ε_2, we get some little decoration, and if we know the fixed point function g, we can compute σ at any precision with a finite number of ε's and estimate some overall geometric convergence for its dependence on the low-order bits. Once we know the scaling function, by the arguments of the previous section, we can figure out what other important quantities, e.g. the power spectrum of the system, looks like.

4 Presentation functions

In the previous two sections I presented a work that is about 10 years old. Now I will tell you more up-to-date thoughts about what one can do with the machinery I have presented. I will just continue to explain what we understand in terms of the scaling function matters. Let me proceed by trying to understand some of the topological properties that are implied by the considerations that I have been going through. It turns out that one can understand a rather tight organization as to how, for example, points on an orbit are deployed, to understand orderings on which scaling should be considered and then, in the end, to figure out directly what equations should be that will determine these scalings. At the end of this section I present an alternative method of computing the scaling function.

4.1 Generating the orbit from a backward dynamics

As I showed you in the previous section, you can directly compute the scaling function if you know the period doubling fixed point g. Nevertheless, it will also turn out that if one can directly compute these scalings, then the scaling information can be turned around backwards and reproduce the renormalization group fixed point. There is a set of considerations that lead to an alternative method of understanding renomalization group problems. How far that can go, that remains to be seen.

First of all, I want to pay more attention to the dynamics. One knows by numerical simulations and other ways what the period doubling fixed point g looks like (Fig. 7).

In this section, we define x_t as the $(t+1)$th iterate under the map g of the critical point $x = 0$: $x_t = g^{t+1}(0)$. The first point I want to pay some attention to is that when we look at these iterates, they come in two pieces, I_1 on the left between x_1 and x_3 including all the odd iterates and I_0 on the right between x_2 and x_0 including all the even iterates. The dynamics is such that one is alternating back and forth between these two intervals around the unstable fixed point.

As I pointed out in the previous sections, there is some rich scaling information, in particular, there is the scaling function I discussed. It is easy to pick apart some of the inherent information. If I consider the points x_{2t+1} and substitute in (25), I obtain

$$x_{2t+1} = g^{2t+2}(0) = \alpha^{-1} g^{t+1}(0) = \alpha^{-1} x_t \qquad (36)$$

which gives

$$x_{2t+1} = \alpha^{-1} x_t \qquad (37)$$

This is the first important statement. Directly in consequence of the equation of motion g obeying a scaling equation (25), the orbit itself inherits one very simple scaling property. (Had all the point obeyed (37), this would be a trivial problem.)

The situation would be nicer still if we could compute the points x_{2t}, because if we know the points x_t up to, say, $2^n - 1$ and know how to construct x_{2t} and x_{2t+1} from x_t, then we can immediately construct another 2^n points. That says just starting with the first point x_0 we have the entire orbit. The problem is that if I look at x_{2t}, I get g^{2t+1}, but we only know how to relate g^2 back to g. Because of that I consider $\alpha g(x_{2t})$, which says

$$\alpha g(x_{2t}) = \alpha g^{2t+2}(0) = x_t \qquad (38)$$

that is

$$x_{2t} = (\alpha g)^{-1}(x_t) \qquad (39)$$

That says finding the even-indexed points is, in some sense, equally easy, we just consider the inverse function applied to the point x_t.

At this point, I have to be careful in saying what I mean by the inverse of g, because there are, in general, two inverses. However, what we are interested in here is αg working on the even-indexed points, i.e. on I_0. If you look at Fig. 7, you see that that part is almost linear. It is its almost linear inverse that I mean by $(\alpha g)^{-1}$.

It is convenient to make up some names for these operations. I call the functions in (37) and (39) the presentation functions and denote them by $F_1(x_t)$ and $F_0(x_t)$, respectively. The indexes 0 and 1 simply stand for the fact that the lowest indexed point is x_1 for F_1 and x_0 for F_0. F_0 produces the even-indexed points, F_1 produces the odd-indexed points, so the index 0 or 1 is the time index mod 2. That suggests generally we should paste them together and we can simply write that

$$x_{2t+\varepsilon} = F_\varepsilon(x_t) \qquad\qquad \varepsilon = 0, 1 \qquad (40)$$

F_0 and F_1 are functions, I cannot paste them together as one function because it would be double-valued. Instead, I simply view them as the two *inverse branches* of a genuine function E (Fig. 8)

$$E(x) = \begin{cases} \alpha x & x_1 \leq x \leq x_3 \\ \alpha g(x) & x_2 \leq x \leq x_0 \end{cases} \qquad (41)$$

A reasonably straightforward calculation based on the nature of g and the value of α shows that the map E is expanding.

Now we have an alternative method of generating the orbit. x_0 is the fixed point of F_0, and if we apply the F's to x_0, i.e. construct its preimages under the map E, we

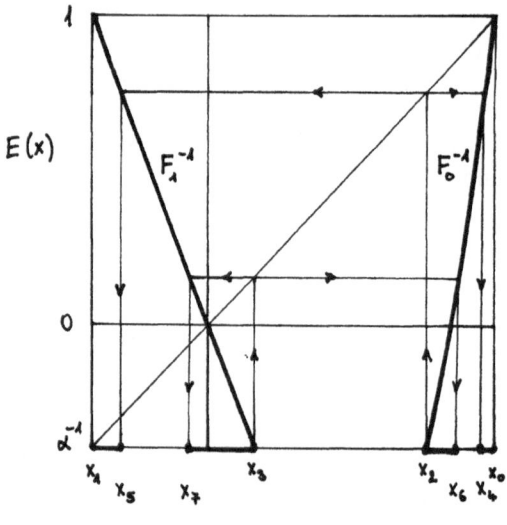

Figure 8

start producing the rest of the orbit (Fig. 8). It is easy to see what is happening: we obtain x_1 as the other preimage of x_0, its two preimages define x_3 and x_2 that confine a gap, and as we start pursueing this gap backwards, we start burning holes into places that previously were full intervals, and finally we produce the period doubling Cantor set.

There is something of very high interest in this picture that should truly be a surprise. This problem is superficially rather difficult, its difficulty lies in the fact that there is simultaneous expansion at the edges of the interval g acts upon, but there is a very strong contraction leading to a singularity at its quadratic maximum. The forward dynamics under g is exactly the same as the backward dynamics under the map E. E is completely expanding, it is hardly different from two straight lines, and that says all of the complicated dynamics is exactly available from this rather simple hyperbolic problem. Later I will try to replace the right branch of E by a straight line that will produce a highly interesting problem.

Having said that, one can ask immediately how this system knows, if I simply replace E by straight lines, that there was a quadratic maximum in g. The answer is that there is a coordination between the eigenvalues of the two fixed points of E which determines for you what the order of the extremum of g is. The order z of the extremum means that locally, for small x

$$1 - g(x) \sim |x|^z \qquad (42)$$

$z = 2$ is the normal quadratic case. (In principle, z can be any number, however, the problems related to period doubling terminate for $z = 1+0$, and there is a very different family of behaviours when $z < 1$.)

Remember that g obeys (25). If I differentiate it, it gives

$$g'\left(g(x/\alpha)\right) g'(x/\alpha) = g'(x) \qquad (43)$$

Now I want to know what $\alpha g'(x_0)$ is ($x_0 = g(0) = 1$):

$$\alpha g'(1) = \lim_{x \to 0} \frac{\alpha g'(x)}{g'(x/\alpha)} = |\alpha|^z \qquad (44)$$

18

where I used the derivative of (42). αg is just the right branch of E, so (44) gives $E'(1) = |\alpha|^z$, the eigenvalue of the right fixed point of E. On the left side of E the relevant formula is simply αx, so the eigenvalue of the left fixed point is just α. We can see now that there is a coordination that says if the eigenvalue of the right fixed point is the zth power of the other eigenvalue, then the corresponding dynamics has a zth order extremum. Thus, while this picture looks very simple with linear parts, it automatically encodes the algebraic singularity through the coordination of the two fixed point eigenvalues.

4.2 Intervals and scalings

The scaling function is related to looking at the ratio of small intervals. The first question is what the intervals look like in this picture. At the lowest level of resolution we just have the whole interval $[x_1, x_0]$. At the next level of resolution $n = 1$ we have two subintervals, the supports of the two parts of E. If I now take E^2, that will give me a map with four more or less linear branches. E^2 will take the four intervals $[x_1, x_5]$, $[x_7, x_3]$, $[x_2, x_6]$ and $[x_4, x_0]$ onto the whole interval, so I burn out two extra holes at $[x_5, x_7]$ and $[x_6, x_4]$. Consider now E^n, it takes 2^n intervals onto the full interval $[x_1, x_0]$. In fact, we even know how it does it, because we know F_0 is monotone increasing and F_1 is monotone decreasing. If, in constructing a particular piece of E^n, we use an odd number of F_1's, then it is a monotone decreasing part, if there is an even number of them, then it is a monotone increasing part.

The 2^n intervals E^n defines on $[x_1, x_0]$ can be encoded through a set of binary strings of length n:

$$\Delta^{(n)}(\varepsilon_n \ldots \varepsilon_1) \equiv F_{\varepsilon_1} \circ \ldots \circ F_{\varepsilon_n}[x_1, x_0] \tag{45}$$

i.e. we identify a particular interval according to which choice of F's we have taken. Observe that I have written the indices as $n, ..., 1$ in the left-hand side of (45) and I have written $1, ..., n$ in the right-hand side. The reason for it is that $x_{2t+\varepsilon} = F_\varepsilon(x_t)$, i.e. when I hit an index with F_ε, the index is multiplied by 2, the string is shifted one step to the left, and then the lowest significant bit, the last one, is added to it.

I can now answer immediately what are the points that lie in one of the intervals defined by (45). The point x_i belongs to the interval $\Delta^{(n)}(\varepsilon_n \ldots \varepsilon_1)$ if its lowest-order n bits are fixed and coincide with the indeces of the interval: $x_i = x_{\ldots \varepsilon_n \ldots \varepsilon_1}$. That means two points whose time indices agree mod 2^n belong to the same such interval. The reason is that the low-order bits are small time iterates, they move points over large distances all around the place, while high-order bits are close return times moving you in fine scale within a given piece of interval.

The next question is the size of the intervals. We can give the answer to that, because we know each one of the branches of E^n maps one of the intervals completely to $[x_1, x_0]$ in a monotone fashion. That means the two endpoints of an interval are obtained by applying the corresponding F's to x_0 and x_1, so the length of the interval is given by

$$\left| \Delta^{(n)}(\varepsilon_n \ldots \varepsilon_1) \right| = (x_{0\varepsilon_n \ldots \varepsilon_1} - x_{1\varepsilon_n \ldots \varepsilon_1})(-1)^{\sum_{i=1}^n \varepsilon_i} \tag{46}$$

The extra sign factor follows from the fact that the relevant branch of E^n is monotone decreasing, reversing order, if there is an odd number of F_1's in it, and the number of the F_1's used is simply the sum of the ε's. Now we know that (46) always yields a positive number. That will be important because we will quickly learn that the scaling function as it was defined in (18) and (24) must always be a positive number. That is an important piece of information which ends up selecting out some unique solutions, as we will see later.

Now I can immediately compute the scalings. The length of an $(n-1)$th level interval is

$$\left|\Delta^{(n-1)}(\varepsilon_{n-1}\ldots\varepsilon_1)\right| = (x_{0\varepsilon_{n-1}\ldots\varepsilon_1} - x_{1\varepsilon_{n-1}\ldots\varepsilon_1})(-1)^{\sum_{i=1}^{n-1}\varepsilon_i} \tag{47}$$

There is a more convenient way to write this: I replace the highest-order 0 in the first term by an ε_n, the 1 in the second term by an $\bar{\varepsilon}_n$ and I let the sum of the indices now go up to n:

$$\left|\Delta^{(n-1)}(\varepsilon_{n-1}\ldots\varepsilon_1)\right| = (x_{\varepsilon_n\ldots\varepsilon_1} - x_{\bar{\varepsilon}_n\ldots\varepsilon_1})(-1)^{\sum_{i=1}^{n}\varepsilon_i} \tag{48}$$

It is rather clear that if $\varepsilon_n = 0$, then this modification does not change the order of the terms in (47) and the sum only goes up to $n-1$, since ε_n is 0. On the other hand, had it been 1, I have cured it by adding in an extra -1 in the sign factor.

The point in doing this is that the sign factors in (46) and (48) are now exactly the same. If I take the quotient to compute the scaling, the sign factor drops out:

$$\sigma(\varepsilon_n\ldots\varepsilon_1) = \frac{x_{0\varepsilon_n\ldots\varepsilon_1} - x_{1\varepsilon_n\ldots\varepsilon_1}}{x_{\varepsilon_n\ldots\varepsilon_1} - x_{\bar{\varepsilon}_n\ldots\varepsilon_1}} \tag{49}$$

so I have an object that must be positive; that is a bit of imformation that we have not known before.

4.3 Linear restrictions on F_0

I now want to return to the question what happens if we replace F_0, the curved part of E^{-1}, by a straight line. The assertion is that if I replace F_0 by 2^n linear segments, then the scaling function, done to all orders, has precisely 2^{n+1} constant values. The idea is the following. Consider E^{n+1}, which breaks the even part I_0 into 2^n intervals. (I do not have to pay too much attention to the odd points because E is linear there.) The points of these intervals have indices with the last $n+1$ digits $\varepsilon_n\ldots\varepsilon_0$ fixed, and we also know that $\varepsilon_0 = 0$, because these intervals are created by F_0 acting on the 2^n intervals defined by E^n at the previous level.

Now I replace F_0 by a set of restrictions that it will map the intervals $\Delta^{(n)}(\varepsilon_n\ldots\varepsilon_1)$ linearly into the pieces of I_0. My assertion is that for such an F_0 the scaling function has precisely 2^{n+1} constant values. The reason for that is very simple. Let me now define

$$\sigma_n(\varepsilon_n\ldots\varepsilon_0) = \frac{x_{0\varepsilon_n\ldots\varepsilon_0} - x_{1\varepsilon_n\ldots\varepsilon_0}}{x_{\varepsilon_n\ldots\varepsilon_0} - x_{\bar{\varepsilon}_n\ldots\varepsilon_0}} \tag{50}$$

This nth level σ has 2^{n+1} values defined for it. Consider now σ_{n+1}

$$\sigma_{n+1}(\varepsilon_{n+1}\ldots\varepsilon_0) = \frac{x_{0\varepsilon_{n+1}\ldots\varepsilon_0} - x_{1\varepsilon_{n+1}\ldots\varepsilon_0}}{x_{\varepsilon_{n+1}\ldots\varepsilon_0} - x_{\bar{\varepsilon}_{n+1}\ldots\varepsilon_0}} \tag{51}$$

If I inspect this formula, I realize that

$$\sigma_{n+1}(\varepsilon_{n+1}\ldots\varepsilon_0) = \frac{F_{\varepsilon_0}(x_{0\varepsilon_{n+1}\ldots\varepsilon_1}) - F_{\varepsilon_0}(x_{1\varepsilon_{n+1}\ldots\varepsilon_1})}{F_{\varepsilon_0}(x_{\varepsilon_{n+1}\ldots\varepsilon_1}) - F_{\varepsilon_0}(x_{\bar{\varepsilon}_{n+1}\ldots\varepsilon_1})} \tag{52}$$

But now notice that the final n ε's are exactly the same in every argument of the F's. If the final n ε's are the same, that means the same linear restriction of F_0. Taking a difference eliminates the constant value, the slope cancels out in the fraction, so (52) can be rewritten as

$$\sigma_{n+1}(\varepsilon_{n+1}\ldots\varepsilon_0) = \sigma_n(\varepsilon_{n+1}\ldots\varepsilon_1) \tag{53}$$

This is the final result. The statement is that if F_0 is constructed of 2^n linear restrictions, then the scaling function to all orders is just the 2^{n+1} values determined by the nth level set of scalings.

In particular, notice that however many ε's I have, it is only the first $n+1$ of them— the refinement information, corresponding to higher powers of 2— that σ depends upon. At this level of 2^n linear restrictions, the fall-off is not exponential, it is absolutely strict, there is no memory after $n+1$ levels, so this is a replacement of the continuum problem by a problem with a finite amount of memory. This statement essentially finishes the computation of the scaling function. If you understand well what is in this statement you can immediately write down the equations that determine the scaling function, and you no longer need the renormalization group theory.

4.4   Reconstructing the dynamics

There is one missing piece of information concerning whether I had any right to mangle with the F's at all. The answer to that is very simple. If I draw a picture with the topology of Fig. 8— a map with a monotone decreasing and a monotone increasing part providing two inverses and a gap between them—then, whatever the actual details are, it will produce a period doubling fixed point. That is, this description turns out to be invertible: I can easily go from the scaling function σ to the presentation functions and the period doubling dynamics g. In fact, the 2^{n+1} values of σ completely parametrize the 2^n linear restrictions of F_0 and then I can go backwards and reconstruct an appropriate g.

The reconstruction is very easy. The dynamics is the so-called adding machine, it simply adds 1 to the index; therefore, if we find a transformation that takes an index i into $i+1$, we have found the dynamics that moves one point to the next. First, consider a point x_{2t} and try to find its image under the dynamics. It is easy to do that because I can apply F_0^{-1} on x_{2t} sending it to x_t and then apply F_1 yielding x_{2t+1}. This results in one restriction of g that I call g_0:

$$g_0 = F_1 F_0^{-1} \qquad \text{on} \quad [x_2, x_0] \tag{54}$$

and it takes the even points into the odd points. That says if I know F_0 and F_1, that determines the rightmost part of g that works on the even-indexed points. The other parts come about as follows.

Look at a point the index of which ends with a 0 and r 1's: $x_{...01...1}$. One time step later the index will contain a 1 and r 0's as its last digits. This is the general circumstance, and I can equally well write it down. First of all, I hit the point with F_1^{-r} that eliminates the 1's and yields an index ending with a 0. This is an even point, I can apply now g_0 to it to convert the last 0 to 1. Finally, I hit it with F_0^r, thus producing the last r 0's. The operation can be written as

$$g_r = F_0^r g_0 F_1^{-r} \qquad \text{on} \quad \alpha^{-r}[x_2, x_0] \tag{55}$$

The statement is that if I have an F_0 and an F_1, I can construct a complete dynamics adding 1 to the index t, I can, however, only give it to you in certain restrictions: g_0 was defined on the interval $[x_2, x_0]$, the even indexed points, while the functions g_r are defined on $F_1^r[x_2, x_0] \equiv \alpha^{-r}[x_2, x_0]$.

In this way, I can reconstruct the dynamics in pieces with gaps between them, so I will only obtain the function defined on a Cantor set. By the time I have refined the linear approximation of F_0 many times, I have pointlike supports and I just reconstruct

g on its orbit. The point of this is that the presentation functions F_0 and F_1 are exactly the same information in the linear polygonal approximation as the set of scalings, and from those I can reconstruct, at least on the orbit itself, what the actual dynamics is. Thus, the scaling information is the complete information that specifies the process.

At this point, one may ask what kind of g the particular F's will produce. In fact, it does not have to have any nice property. However, the relevant question to ask is what the nature of F_0 and F_1 should be so that the restrictions g_r piece themselves together to make a smooth map with, say, a quadratic extremum. One thing you know is that F_0 and F_1 must be chosen to have the correct coordination in their eigenvalues, but, of course, that is not enough to be met. The direct solution of this problem is decidedly tedious, however, it is very easy to figure out the answer, which is contained in the following statement. It turns out that you get some rather complicated equations that, when you play with them, end up producing very simple equations for the σ's. Thus, it is the parametrization of the F's in terms of the σ's that turns out to be most natural.

Consider what this statement says in details. Remember, a σ does not depend on anything below the leading $n + 1$ ε's. Let us look at a certain σ,

$$\sigma_n(\varepsilon_n \ldots \varepsilon_0) = \frac{x_{0\varepsilon_n \ldots \varepsilon_0 0 \ldots 0 \ldots} - x_{1\varepsilon_n \ldots \varepsilon_0 0 \ldots 0 \ldots}}{x_{\varepsilon_n \ldots \varepsilon_0 0 \ldots 0 \ldots} - x_{\bar{\varepsilon}_n \ldots \varepsilon_0 0 \ldots 0 \ldots}} \tag{56}$$

Now I have put a very large number of 0's into the indices, still this is exactly the nth level scaling of (50) since this scaling only depends on the leading $n+1$ ε's. Now imagine that I am really looking at such a fine resolution, such a large value of t, that I really have as many 0's at the end of the indices as I want. I now apply the dynamics which is simply adding 1's starting at the rightmost digit. I can continue the run around on the dynamics until all the bits on the right of ε_0 are 1's. That still leaves the value of the scaling unchanged:

$$\sigma_n(t) = \frac{x_{0\varepsilon_n \ldots \varepsilon_0 1 \ldots 1 \ldots} - x_{1\varepsilon_n \ldots \varepsilon_0 1 \ldots 1 \ldots}}{x_{\varepsilon_n \ldots \varepsilon_0 1 \ldots 1 \ldots} - x_{\bar{\varepsilon}_n \ldots \varepsilon_0 1 \ldots 1 \ldots}} \tag{57}$$

where t stand for the string $\varepsilon_n \ldots \varepsilon_0$.

Now I can ask what happens one time step later. To answer this question, I must apply g to the x's of (57). They have a long tail of 1's which means they resulted from a large number of applications of F_1, therefore they must be very close to 0, the critical point of g. If I have correctly synthesized g, then it is quadratic, and applying it to these x's yields the same expression as (57) except the terms are now squared. As a last step, I can cut off the tail of 1's by taking into account that $x_{\ldots \varepsilon 1} = F_1(x_{\ldots \varepsilon}) = \alpha^{-1} x_{\ldots \varepsilon}$. These powers of α drop out from the scaling, thus I finally have the statement that

$$\sigma_n(t + 1) = \frac{(x_{0\varepsilon_n \ldots \varepsilon_0})^2 - (x_{1\varepsilon_n \ldots \varepsilon_0})^2}{(x_{\varepsilon_n \ldots \varepsilon_0})^2 - (x_{\bar{\varepsilon}_n \ldots \varepsilon_0})^2} \tag{58}$$

with the exception $\sigma_n(2^n) = 1 - x_2^2$. That means if we have an F_0 with 2^n linear restrictions, the set of $\sigma_n(t)$'s can be expressed by the x's so that the one-index advanced σ depends on the x^2's exactly in the same way as the unadvanced σ depends on the x's. As a consequence, you can eliminate the x's between the two expressions and it gives you a complete set of equations for the 2^{n+1} values of σ_n. That says the problem is done.

As we have seen at the end of Sec.3, the scaling function is almost constant, i.e. $\sigma_n(t) \approx \sigma_n(t + 1)$. Thus, in some sense, the scaling is something which is not changing in time; in an exact sense, (57) and (58) form a full set of equations which determines the values of σ through the fact that the orbital scalings are as constants as possible. Numerically, for any finite n, it is very clear that there is a unique solution of positive σ's for these equations.

Finally, once you know σ, you can construct, through the F's, the period doubling fixed point g, and, as we have seen, you can obtain any other information that you want to know about your system. In some sense, the real solution of the problem is σ, because it provides the embedding-free information about how the dynamics is built up that a simple simulation based on the equations of motion cannot give you, and it is this property that makes the scaling function so important.

References

[1] A. Libchaber and J. Maurer, *J. Phys. (Paris) Coll.* **41**, C 3–51 (1980).

[2] M. J. Feigenbaum, *Phys. Lett.* **74A**, 375 (1979); *Commun. Math. Phys.* **77**, 65 (1980).

[3] M. J. Feigenbaum, *J. Stat. Phys.* **52**, 527 (1988).

[4] M. J. Feigenbaum, *J. Stat. Phys.* **19**, 25 (1978); **21**, 669 (1979).

[5] P. Collet, J.-P. Eckmann, and H. Koch, *J. Stat. Phys.* **25**, 1 (1981);
P. Collet and J.-P. Eckmann: *Iterated Maps on the Interval as Dynamical Systems* (Birkhäuser, Boston, 1980).

[6] M. Nauenberg and J. Rudnick, *Phys. Rev.* **B24**, 439 (1981).

[7] M. J. Feigenbaum, *Nonlinearity* **1**, 577 (1988).

[8] D. Sullivan, in this volume.

RENORMALIZATION, ZYGMUND SMOOTHNESS AND THE EPSTEIN CLASS

D. Sullivan

THES
Bures sur Yvette
France

To randomize a deck of n-cards one may turn over one of the split stacks before shuffling. The resulting permutation of order n if irreducible is called a *folding permutation* because it may be accomplished by a continuous mapping f of the real line to itself which folds the line once. The orbit of the turning point is finite and f restricted to this finite orbit is the folding permutation.

In fact there are two nontrivial theorems about this realization.

i) Any smooth unimodal shape for a graph can be ajusted vertically as a graph to realize any folding permutation (Sharkovski, 1959, Milnor-Thurston 1975 et al.)

ii) For the parabolic shape the vertical positioning is *unique* for each folding permutation. In particular, the geometry of the finite critical point orbit is determined uniquely up to affine equivalence by the combinatorical structure of the folding permutation. (Douady, Hubbard, Milnor, Sullivan, Thurston, ...).

The first theorem uses the order structure of the real line. To the best of my knowledge, all proofs of the second statement use holomorphic dynamics and quasi conformal mappings.

Physicists working numerically discovered surprising and far reaching generalizations of this geometric rigidity in the context of dynamical systems defined by smooth folding mappings. Certain infinite limiting versions of geometric rigidity suggested by example ii) occur in the general smooth families of i). Geometric structures that should depend on infinitely many parameters - like a Cantor set critical orbit up to smooth change of coordinates - only depend on a few, such as the power law at the critical point and the combinatorics (the vertical parameter).

These physicists, Feigenbaum in the U.S. and Coullet and Tresser in France introduced the language and scenarios of statistical physics to describe these phenomena because they were reminiscient of the universal exponents and renormalization group of critical points and phase changes in statistical mechanics. [F1], [F2], [CT].

Chaos, Order, and Patterns, Edited by R. Artuso *et al.*
Plenum Press, New York, 1991

Let us first say what "renormalization" means. Start with a folding mapping preserving some interval I and suppose some power of f preserves some interval RI containing the original turning point figure 1a). The graph of the inset box is that of a new folding mapping of the interval called a renormalization of f. If n and RI are taken to be minimal then we denote this renormalization Rf. Renormalization is a partially defined mapping from the space of all folding mappings of the interval, into itself.

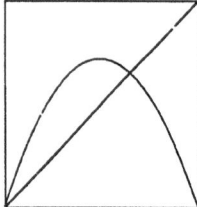

folding mappings of I

If the original f is varied by a vertical parameter say, RI and Rf will be defined in between the two extreme cases, Figure 1b

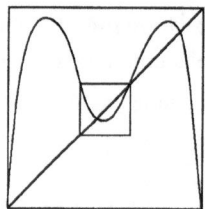

Figure 1a

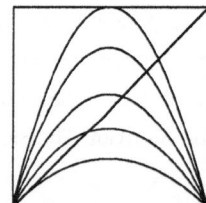

Figure 1b

As Rf varies between these two extremes all folding permutations occur . We may have that for $f_1 = Rf$ there are interior disjoint intervals in $I_1 = RI$ permuted by f_1 according to a second permutation σ_1. We may consider $f_2 = Rf_1$ etc. If the renormalization process is always possible we generate $f, f_1, f_2, \ldots, f_n$, and permutations $\sigma_0 \sigma_1 \ldots \sigma_n \ldots$ and we say f is *infinitely renormalizable* with combinatorial type $\sigma = (\sigma_0, \sigma_1, \ldots, \sigma_2, \ldots)$.

Let τ denote the unique permutation of order 2. One finds an infinitely renormalizable f of combinatorial type $(\tau, \tau, \tau, \ldots)$ at the limit of the cascade of period doubling bifurcations. The orbit closure of the critical point for such an f is a binary cantor set which was discovered numerically to have a universal geometric structure characterized by a countable set of self similarity ratios. Some aspects and consequence of this geometry for example the celebrated $\delta = 4.6692...$ ratio for the cascade itself were measured in fluid experiments of Maurice Libchaber. The presence of this numerical value in a system of many dimensions is explained by the n-dimensional perturbation of the Feigenbaum renormalization picture (Collet, Eckman, Koch [CEK]).

The physicists'scenario was as follows. For the $(\tau, \tau, ...)$ combinatorics, iteration of renormalization leads one to a special mapping $F(\tau, \tau, ...)$ which is fixed by renormalization. The geometry of any f of type $(\tau, \tau, ...)$ at deep levels becomes that of $F(\tau, \tau, ..., \tau ...)$ since $R_\tau ... R_\tau R_\tau f \to F(\tau, \tau, ...)$. The δ above measured the rate of expansion of R_τ in the space of folding mappings transverse to the codimension one manifold of maps of type $(\tau, \tau, ...)$. This picture works beautifully in the numerical experiments of Feigenbaum. This numerics was proved rigourously in a computer assisted discussion in a open neighborhood of some $F(\tau, \tau, ...)$ in Lanford [L1].

The existence of an $F(\tau, \tau, ...)$ was verified mathematically by Epstein [E] working in the space of folding mappings $h \cdot q = $ (diffeomorphism) $\cdot$ (quadratic polynomial) where h^{-1} has an injective complex analytic extension to $\mathbb{C}$-(x real but not in an open neighborhood of this dynamical interval}. We say these mappings have the Epstein form and write $f \in E(J)$ where J is the open neighborhood of the dynamical interval. Epstein also treated the other critical exponents in the $(\tau, \tau, ...)$ case.

In the rest of this paper we discuss why this Epstein class $E(J)$ is important for renormalization of dynamical systems. Namely the theorem of § 3 states any limit of renormalization starting from sufficient finite smoothness must lie in Epstein class $E(J)$ for some interval strictly larger than the dynamical interval. We also characterize in terms of the technique used (cross ratio distortion) exactly what smoothness class is required. It is : if $f = hQ = $ diffeomorphism . quadratic polynomial, then log h' satisfies the Zygmund condition § 1.

In the complete paper [S1] we prove generalizations of the scenario described above. In particular the Cantor set of any mapping f of type $(\sigma_0, \sigma_1, ...)$ where degree σ_i is uniformly bounded has universal asymptotic ratios. These ratios can be computed from canonical analytic functions $F(... \sigma_{-2}, \sigma_{-1}, \sigma_0, \sigma_1, \sigma_2, ...)$ which are precisely the limits of infinite renormalization starting from the Zygmund class of smoothness.

The theorem presented here about the Epstein class is the first step in this program. For the complete discussion see [S1].

1 Poincaré length distortion and smoothness class one plus Zygmund

We want to study the smoothness required for a diffeomorphism h to only distort cross ratios of small standard 4-tuples by an amount commensurable to the size of the 4-tuple.

One cross ratio $[a, b, c, d]$ can be computed by

$$-\log [a, b, c, d] = \iint\limits_{S} \frac{dx\,dy}{(x - y)^2} \, , \quad a < b < c < d ,$$

where S is the square $\{(x,y) \mid a \le x \le b , c \le y \le d\}$.

Thus the distortion by h , given by

$$\log \frac{[ha, hb, hc, hd]}{[a, b, c, d]} , \text{ equals } \int\limits_{S} \mu - (hxh)^* \mu$$

where μ is the measure, $\dfrac{dx\,dy}{(x - y)^2} = \mu$.

Calculating the integral we get $\left(\dfrac{1}{(x - y)^2} - \dfrac{h'x\,h'y}{(hx - hy)^2} \right)$, or

$$\frac{1}{(x - y)^2} \left[1 - \frac{h'x\,h'y}{[h']_{xy}^2} \right]$$

where $[h']_{xy}$ = average h' is the average of h' over the interval $[x,y]$.

Because we are assuming $b - a = c - b = d - c$ for every point (x,y) in the square S the factor $\dfrac{1}{(x - y)^2}$ is commensurable to $1/\text{area } S$. Thus a small bound ε on $\log \dfrac{h'x\,h'y}{[h']_{xy}^2}$

yields the bound ε on the distortion of cross ratio, $\log \dfrac{[ha, hb, hc, hd]}{[a, b, c, d]}$.

Calculating this $\log$ we get
$\log h'x + \log h'y - 2 \log [h']_{xy}$.

Let us replace the last term with the average taken after the $\log$ to obtain a) where

a) is $(\log h'x + \log h'y - 2 [\log h']_{xy})$ with an error of twice b), where

b) is $(\log \text{average } (h') - \text{average } (\log h'))$.
$\quad\quad\quad [x,y] \quad\quad\quad\quad [x,y]$

Let us say h satisfies the *local Koebe condition* if for $|x - y|$ sufficiently small one of the equivalent conditions hold

$$1) \left[1 - \frac{h'x\,h'y}{[h']_{xy}^2} \right] = 0(|x - y|) \quad\quad\quad \text{(here } 0(s) \text{ means a term at most } k.s \text{ as } s \to 0.)$$

$$2) \log \frac{h'x\,h'y}{[h']_{xy}^2} = 0(|x - y|)$$

Note. if both a) and b) are $0(|x - y|)$, then by the above 1) and 2) hold.

Proposition. *If h satisfies the local Koebe condition then the h distortion of cross ratios of small standard 4-tuples is commensurable to the size of the 4-tuple.*

Proof. The above calculation.

Expression a) suggests the Zygmund condition on continuous functions

$$Z: \quad \varphi(x) + \varphi(y) - 2\,\varphi(\tfrac{x+y}{2}) = O(|x - y|)\,.$$

Proposition. *If* φ *satisfies* Z *on an interval* J *then average* φ *over* J *is the value of* φ *at the midpoint with an error* $O(length\ J)$.

Proof. Think of the uniform measure on J as two dirac masses moving out uniformly from the center. Use the Z condition to replace the average of φ at the moving points by the value at the center. Q.E.D.

Corollary. *If* $\log h'$ *is Zygmund then expression* a) *is* $O(|x - y|)$.

Proof. Use the proposition, then the definition of Z again.

There is a converse to the corollary.

Say φ satisfies the *average property* if $\underset{[x,y]}{\text{average}}\ \varphi = \tfrac{1}{2}(\varphi x + \varphi(y)) + O(|x - y|)$.

Proposition. *If* φ *satisfies the average property for all intervals* $J \subset I$, *then* φ *satisfies the Zygmund property for all pairs* x,y *in* I.

Proof. Apply the average property to the intervals $\left[x, \dfrac{x+y}{2}\right], \left[\dfrac{x+y}{2}, y\right]$, and $[x,y]$ and combine averages of averages to get the Z-property for x,y.

Corollary. *The Zygmund property is equivalent to the average property.*

Proof. Propositions above.

Conclusion A. *expression a) is* $O(|x - y|)$ *iff* $\log h'$ *is Zygmund.*

Now we consider when expression b) is $O(|x - y|)$. We are concerned with small intervals J and we assume h' is continuous. Then h' varies only a little from one of its values $h'(x_0) = a$. The expression b) is unchanged if we multiply $h'x$ by $1/a$. Write $1/a\,h'$ on J as $1 + \varepsilon$ where ε is a small function. Expand the two terms of b)

$$\log \frac{1}{|J|} \int_J 1 + \varepsilon \ - \ \frac{1}{|J|} \int_J \log 1 + \varepsilon$$

$$= \left(\frac{1}{|J|} \int_J \varepsilon - \frac{1}{2}\left(\frac{1}{|J|}\int_J \varepsilon\right)^2 \cdots\right) - \left(\frac{1}{|J|}\int_J \varepsilon - \frac{\varepsilon^2}{2}\cdots\right)$$

$$= -\frac{1}{2}\left(\frac{1}{|J|}\int_J \varepsilon\right)^2 + \frac{1}{|J|}\int_J \varepsilon^2 \big/ 2 \cdots$$

Now the first term could be zero so there would be no cancellation. Thus we are forced to estimate each brutally with absolute values. Assume ε is Holder of order 1/2 on J,

$|\varepsilon(x) - \varepsilon(y)|^2 \le C_J |x - y|$. Since ε is zero at x_0 we get the estimate $C_J \cdot$ length J for the sum of the absolute values. Also if $C_J \cdot$ length J is sufficiently small the higher order terms can be ignored.

Conclusion B. expression b) is $O(|x - y|)$ if h' is Holder of order $1/2$. The coefficient for $|x - y| < \varepsilon$ is estimated by the *normalized* $\frac{1}{2}$ *Holder norm* : take the sup over all intervals J of length $\le \varepsilon$ of C_J above where $1 + \varepsilon = h'(x) / h'(x_0)$ for convenient x_0 in J and we assume C_J llength J l is sufficiently small.

Let us note that Zygmund functions are Holder α for all $\alpha < 1$. However, the α-Hölder constants are not determined by the Z-norm. Let us also note the normalized $\frac{1}{2}$ Holder norm of h' can be estimated by the usual $\frac{1}{2}$ Holder norm of log h' - the best C such that

$$\text{llog h'x - log h'yl}^2 \le C |x - y| .$$

Now we can summarize the above by the

Theorem. a) *If* log h' *is Zygmund then* h *satisfies the local Koebe distortion condition. The coefficient is controlled by the Zygmund norm of* log h' *and the* $\frac{1}{2}$ *Holder norm of* log h' . *Conversely,*

b) *if* log h' *is* $\frac{1}{2}$ *Holder, then the local Koebe condition for* h *implies* log h' *is Zygmund.*

Proof. The above discussion has been a proof of a). For part b) recall from above the local Koebe inequality implies expression a) plus expression b) is $O(|x - y|)$. The $\frac{1}{2}$ Holder implies expression b) is $O(|x - y|)$. Thus expression a) is $O(|x - y|)$. But this implies log h' is Zygmund by the third little proposition.   Q.E.D.

Problem. Derive necessary and sufficient conditions for the integral distortion to be commensurable to the linear scale. (In the above discussion we have estimated the integral by the integrand.) For the sketch of the solution see [S1].

2   The Koebe distortion argument of Denjoy, de Melo-Van Strien, Swiatek, Yoccoz, et al and Zygmund smoothness

Consider a composition g of many diffeomorphisms f_i between tiny intervals J_i all lying disjointly in some big interval I , $f_i : J_i \to J_{i+1}$.

The *classical Denjoy argument* estimates log lg'x / g'yl $x,y \in$ domain g in terms of the $\sum_1$ total variation $\text{llog } f_i'|$. This will be finite say if $f_i = f / I_i$ and log f' has bounded variation on I . The proof is the chain rule.

The new argument called the *Koebe principle* for one dimensional real dynamics treats the case when the factors can be divided into two groups so that relative to some coordinate system on I

i) for one group a Denjoy type argument can be used at least to study cross ratios.

ii) the factors in the other group decrease Poincaré length (a type of cross ratio) (because of a positive Schwarzian condition) even though $\log f_i'$ has unbounded variation.

Here if L, M, R is a partition of an interval T into 3 consecutive subintervals (the left, the middle, and the right) the *Poincaré length of* M *in* T is $\log(1 + \frac{MT}{LR})$. It is the length of M in the Riemannian metric on T = [a,b] corresponding to the form $|dx| / x - a + |dx| / b - x$.

The additive change of P-length along a composition is additive over the factors. *In a decomposition such as* i) ii) *above, the increase in* P-length is controlled by the factors of type i) *because there is a decrease for the factors of type* ii). *This is the first idea cf. Swiatek* [SW].

The *second idea is the four intervals argument*. Let J, L, M, R be contiguous equal length intervals and let h be a homeomorphism of the union into the real line so that one of hL and hM is *much smaller* than other. Discard from the original 4 intervals the outer interval next to the one of L, M called s made smaller. Let T denote the remaining three L, M, X and let $\ell \subset T$ be the one of L or M made larger. The P-length of $\ell \subset T$ is $\log 4$. The P-length of $h(\ell) \subset hT$ is very large because $h(\ell)$ is much larger than $h(s)$ and $h(T)$ is of course greater than hX. Thus one has the analogue of complex Koebe distortion :

Real Koebe distortion. *If a homeomorphism* $h : I \to$ reals *does not increase unit P-lengths too much the asymmetric distortion for interior symmetric triples is controlled.*

More precisely if $x, y \in I$ satisfy $|x - y|$ is as small as the distance to ∂I and $z = x + y / 2$, then $\frac{1}{M} \le (h(x) - h(z)) / (hy - hz) \le M$ where M can be calculated from the bound B on the additive increase of Poincaré length of unit Poincaré length subintervals $J \subset T$ where $T \subset I$, i.e. the B defined by

(P-length of $hJ \subset hT$ - P-length $J \subset T)_+ \le B$

for all $J \subset T$ so that P-length $J \subset T = 1$.

Remark. The point here as in Koebe distortion for schlicht mappings is we go from one analytic condition (in that case holomorphic ; in this case positive Schwarzian or controlled P-length increase) to interior control on the non-linearity.

We describe the dynamical Koebe distortion principle for a rather general class of dynamical systems. Let M be a compact one-manifold provided with a differentiable structure where overlap homeomorphisms $h_{\alpha\beta}$ are continuously differentiable and the $\log h_{\alpha\beta}'$ have bounded Zygmund norm (see §).

Suppose $f : M \to M$ is a smooth mapping with finitely many critical points where $f' = 0$. At a non singular point assume $\log f'$ is Zygmund. At a singular point c_i suppose there are coordinate systems in the (1 + Zygmund) structure so that f takes the form $x \to |x|^{r_i} + v_i$ or $x \to (\text{sign } x)(|x|^{r_i}) + v_i$ where $r_i > 1$.

Assume we have a long composition g of diffeomorphisms $f_i : J_i \to J_{i+1}$ where the J_i are disjoint in M and $f_i^{-1} = f$ restricted to J_{i+1}.

Theorem. *For the composition g the increase in Poincaré length and therefore the interior non linearity of g in domain g is controlled by constants of the coordinate systems and local models of f and are independent of the length of the composition g.*

Proof. We first need a lemma.

Lemma. If h is a diffeomorphism of the unit interval I and $\log h'$ is Zygmund, $T \subset I$ is a tiny interval $J \subset T$ has unit Poincaré length, then the Poincaré length of $hJ \subset hT$ is $1 + O(\text{length } T)$. The coefficient is controlled by the Zygmund norm of $\log h'$ and the Holder $\frac{1}{2}$ norm of $\log h'$ squared.

Proof of the lemma. We have proved this in § 1 when J sits in the middle of T. In general J may be tiny and near one end of T. We have to calculate the integral of §1 over the rectangle R of figure

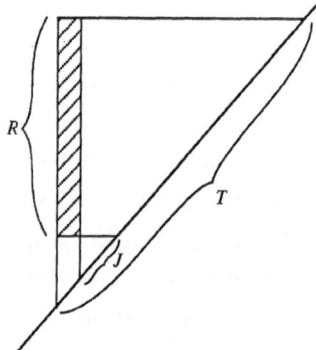

Figure 2

Using the local Koebe condition, and the fact that for a point in R the distance to the diagonal and the vertical distance to the diagonal are equivalent, the integral takes the form

$$a \cdot \int_a^b \frac{1}{t^2} O(t) \, dt$$

where $a \sim \text{length } J \sim \text{distance } (J, \partial T)$, $b \sim \text{length } T$. This yields a $\log b/a$ which has order b when a and b are commensurable. This is the case already discussed. Otherwise if $a << b$, $a \log b/a$ is much smaller than b. This proves the lemma.

Proof of Theorem. i) As we go along the composition a Poincaré length is decreased if we are entirely within one of the coordinate systems for the singular point models because f^{-1} has positive schwarzian there and maps of positive schwarzian decrease Poincaré length (de Melo and Van Strien) [MW]. ii) There are finitely many possible transitional cases for long intervals which don't fit inside one model or the other. We won't discuss these further. They are finite. iii) Finally we have the factors where the lemma applies. We view the lemma as saying intervals $J \subset T$ of any P-length ≥ 1 cannot increase by more than the multiplicative *factor* $1 + O(\text{length } T)$. By disjointness of the orbit of T this effect is controlled by the total length of M. Q.E.D.

3   Renormalization limits and schlicht mappings

Write unimodal mappings as Qh where Q is a quadratic polynomial and h is diffeomorphism with log h' bounded in $\frac{1}{2}$ Hölder, Zygmund sense of §1.

Theorem. *For the sequence of renormalizations* $R^n f = h_n Q$, *the Zygmund* $\frac{1}{2}$ *Hölder size of* log h'_n *is bounded. Thus* $R_n f$ *is precompact for the topology of uniform convergence on* I . *Any* C^0 *limit* g_∞ *of* $\{R_n f\}$ *belongs to the Epstein class* $E(J)$ *for some interval* J *containing* I *plus definite space on either side.*

For the proof see [S1]

REFERENCES

[CT]   P. Coullet and C. Tresser. "Iteration d'endomorphismes et groupe de renormalisation". J. de Physique Colloque C 539, C5-25 (1978). CRAS Paris 287 A, (1978).

[CEK]   P. Collet, J.-P. Eckmann and H. Koch. "Period-doubling bifurcations for families of maps on $\mathbf{R}^n$ ". J. Stat. Phys. 25, 1-14 (1980).

[E]   H. Epstein. "New proofs of the existence of the Feigenbaum functions". Commun. Math. Phys., 106, 395-426 (1986).

[F1]   M.J. Feigenbaum. "Quantitative universality for a class of non-linear transformation". J. Stat. Phys. 19. 25-52 (1978).

[F2]   M.J. Feigenbaum. "Universal metric properties of non-linear transformations". J. Stat. Phys. 21, 669-706 (1979).

[MV]   W de Melo and S. Van Strien. "Schwarzian derivative and beyond". Bull Amer Math Soc 18, 159-162 (1988).

[S1]   D. Sullivan D. "Bounds, quadratic, differentials, and renormalization conjectures". To appear in AMS volume (2) (1991) celebrating the Centennial of the American Mathematical Society.

[S2]   D. Sullivan. "Quasiconformal homeomorphisms in dynamics, topology, and geometry". ICM Berkeley 1986.

[S3]   D. Sullivan. "Differentiable structures on fractal like sets". In : Non-linear Evolution and Chaotic Phenomena, G. Gallavotti and P. Zweifel, eds., New-York, Plenum (1988).

See also Herman Weyl Centenary Volume.

[SW]   G. Swiatek "Critical Circle Maps". Commun. Math. Phys. 119. 109-128 (1988).

TORUS MAPS

R. S. MacKay
notes written with the assistance of A. Oliveira

Nonlinear Systems Laboratory
Mathematics Institute
University of Warwick
Coventry, CV4 7AL
ENGLAND

1. INTRODUCTION

Mathematically, the study of torus maps here developed is concerned with the dynamics of diffeomorphisms of the two-dimensional torus, isotopic to the identity.

Physically, this study is concerned with non-Hamiltonian coupled oscillators, dissipative systems having several modes of oscillation, and the Ruelle-Takens route to turbulence.

Consistent with the theme of this school, we will find chaos, order and patterns. We will find *order* in the persistence of quasiperiodic motions and the phenomenon of mode-locking (Section 2), *patterns* in the intricate global bifurcation diagrams (Section 3), and several forms of *chaos*, including a new one : toroidal chaos (Section 4).

The books of Guckenheimer & Holmes [GH] and Arnol'd [Arn] are highly recommended as background reading for this course.

1.1. Mathematical framework

Let us explain some basic terms.

Chaos, Order, and Patterns, Edited by R. Artuso *et al.*
Plenum Press, New York, 1991

By *diffeomorphism* we mean a smooth invertible map with smooth inverse. The smoothness required depends on the context, but we ask for a minimum of C^1 (differentiable with continuous derivative).

The *2-dimensional torus*, denoted $\mathbb{T}^2$, is the real plane modulo integer translations, i.e., $\mathbb{R}^2/\mathbb{Z}^2$. More generally, $\mathbb{T}^d = \mathbb{R}^d/\mathbb{Z}^d$ denotes the d-dimensional torus.

A diffeomorphism is *isotopic to the identity* if it can be continuously deformed to the identity via diffeomorphisms. For example, the diffeomorphism of $\mathbb{T}^2$ given by :

$$(*) \qquad \begin{cases} x' = x + \Omega_x - \dfrac{a}{2\pi}\sin 2\pi y \\[2mm] y' = y + \Omega_y - \dfrac{a}{2\pi}\sin 2\pi x \end{cases}$$

where $|a| < 1$, which we shall use as our example in numerics.

The *dynamics* of a diffeomorphism f is the qualitative behaviour of its orbits, $f^n(x)$, $n \in \mathbb{Z}$.

More generally, a *dynamical system* on a space M is a map $\varphi : (x,t) \mapsto \varphi_t(x) \in M$, where $x \in M$, $t \in \mathbb{R}$ or $\mathbb{Z}$, such that $\varphi_{t+s}(x) = \varphi_t(\varphi_s(x))$. The case $t \in \mathbb{R}$ is called a *flow*, and the case $t \in \mathbb{Z}$ a *map*. A vector field $\dot{x} = f(x)$ on a manifold M generates a flow.

1.2. From coupled oscillators to torus maps

Now we explain the relevance of torus maps to coupled oscillators.

In this context, an *oscillator* means a dynamical system $\dot{x} = f(x)$, $x \in M$, where M is a manifold, with an attracting periodic orbit γ (e.g. Figure 1). One can choose local coordinates $(\theta,z) \in \mathbb{R}/\mathbb{Z} \times \mathbb{R}^{m-1}$, where m is the dimension of M, so that $\gamma = \{(\theta,z) : z = 0\}$ and $\dot{\theta} = \omega \in \mathbb{R}\setminus\{0\}$.

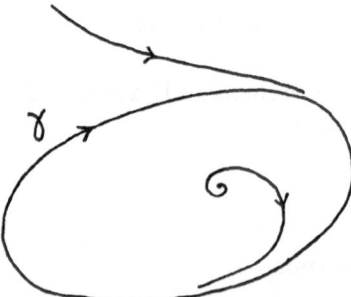

Figure 1. A two-dimensional example of an oscillator

By *coupled oscillators* we mean a system of the form :

$$\dot{x}_i = f_i(x_i) + \varepsilon g_i(x_0,...,x_d), \quad i = 0,...,d$$

where each of the systems $\dot{x}_i = f_i(x_i)$ is an oscillator, with an attracting periodic orbit γ_i, and $\dot{\theta}_i = \omega_i$.

For $\varepsilon = 0$, there exists an attracting invariant $(d+1)$-torus, $\gamma_0 \times ... \times \gamma_d$. The flow on it is $\dot{\theta} = \omega$, constant, where $\theta = (\theta_0,...,\theta_d)$ and $\omega = (\omega_0,...,\omega_d)$.

From the theory of normal hyperbolicity, e.g. [Shub], for ε small enough there exists an attracting invariant $(d+1)$-dimensional torus, $\mathbb{T}^{d+1}$, varying smoothly with ε, and the flow on it is close to $\dot{\theta} = \omega$.

Being close to $\dot{\theta} = \omega \neq 0$, the flow on the $(d+1)$-torus has a cross-section $\Sigma \cong \mathbb{T}^d$ (e.g. $\{\theta \in \mathbb{T}^{d+1} : \theta_0 = 0\}$) and the return map to Σ is a diffeomorphism of Σ isotopic to the identity (see Figure 2 for the case d= 1).

<u>Note</u> : A *cross-section* is a codimension-1 submanifold transverse to the flow such that the orbit of every point crosses it.

1.3. <u>Dissipative systems with several modes of oscillation</u>

There are other ways that flows on a torus can arise. They can be created by bifurcation from simpler situations. For example, an attracting equilibrium point, $\mathbb{T}^0$, can become an attracting periodic orbit, $\mathbb{T}^1$, through a Hopf (Andronov-Poincaré) bifurcation and then an attracting 2-dimensional torus, $\mathbb{T}^2$, through a Neimark–Sacker bifurcation (e.g. [GH, Arn]). This can continue to $\mathbb{T}^3$, ..., $\mathbb{T}^d$, under suitable conditions on tangential and normal spectra (e.g. [Sell]). Such behaviour can arise on forcing a system with several dampled modes of oscillation (e.g. [HJ,BH]). Also certain steps in the sequence can be jumped by undergoing codimension-2 bifurcations, e.g. [Lang, IL].

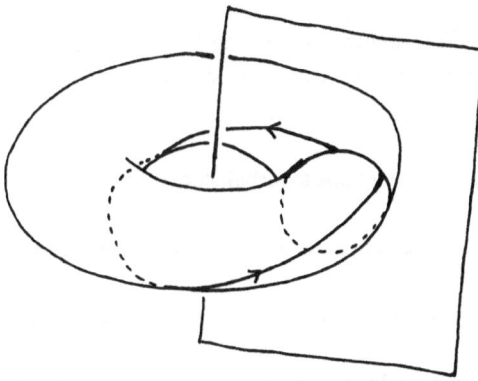

Figure 2. A flow on a 2-torus and a cross-section

1.4. Ruelle–Takens route to turbulence

Ruelle and Takens suggested that the onset of turbulence in fluids might be describable as the formation of a "strange attractor" in the appropriate space of velocity fields. In particular, they suggested that the equations of motion could develop an attracting invariant d-torus, $d \geq 3$, by the process of Section 1.3, and it could have a strange attractor on it. Let us start by giving a definition of a strange attractor, based on [Ru], though we emphasise that there is as yet no universally accepted definition.

Definition : A *strange attractor* for a dynamical system φ on a space M is an ergodic invariant probability measure μ on M for which (i) the set G of generic points has positive Lebesgue measure, $\lambda(G) > 0$, (ii) the Lyapunov exponent χ is positive.

Notes : 1) μ is *ergodic* if for every measurable invariant set A, $\mu(A) = 0$ or 1. Equivalently, for all $\psi \in L^1$,

$$\frac{1}{T} \int_0^T \psi(\varphi_t(x))\, dt \to \int \psi\, d\mu \quad \text{as } T \to \infty,$$

for μ-almost every $x \in M$, that is, the time average of ψ along the orbit of x converges to the ensemble average of ψ (e.g. [Wal]).

2) x is a *generic point* for (φ,μ), if the above is true for all continuous functions ψ. For μ ergodic, $\mu(G) = 1$ [Ox].

3) For an invariant measure μ, the limit $\chi(x) = \lim_{T \to \infty} \frac{1}{T} \log\|D\varphi_T x\|$ exists μ-almost every-

where. It measures the rate of divergence of infinitesimally close trajectories. For μ ergodic, $\chi(x)$ has the same value μ-almost everywhere, and the *Lyapunov exponent* of (φ,μ) is defined to be this almost everywhere value.

4) μ ergodic does not follow from $\lambda(G) > 0$, e.g. the *neutral necklace* of [BGKM].

5) This definition includes some things which are not particularly strange, e.g. a delta measure on a saddle point for a planar flow with an attracting homoclinic cycle, but on balance it seems to me to be a good one.

Ruelle and Takens, [RT], showed that arbitrarily close to any constant flow on $\mathbb{T}^d$, $d \geq 4$, there are flows with a strange attractor. This result was extended with Newhouse, [NRT], to $\mathbb{T}^3$ (in the C^2 topology).

They proposed the following route to turbulence :

$$\text{equilibrium} \to \text{periodic cycle} \to \mathbb{T}^2 \to \mathbb{T}^3 \text{ with a strange attractor.}$$

Some evidence in support of this scenario has been claimed [GB,ML] (see Figures 3 and 4).

Our aim is to see what really happens for typical families of flows on $\mathbb{T}^3$ near constant flows, equivalently, families of torus maps close to translations.

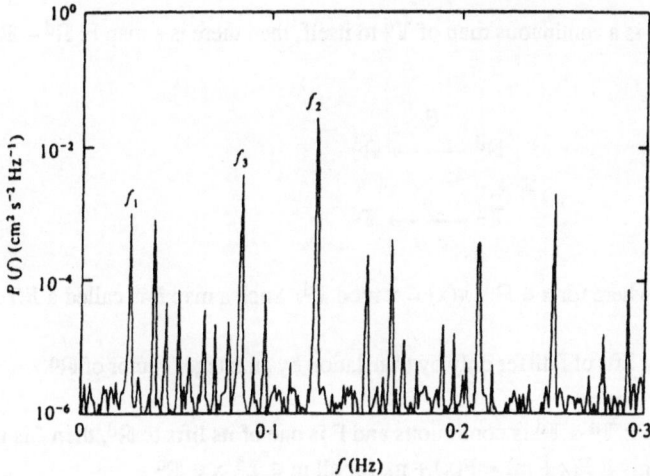

Figure 3. Spectrum showing the presence of 3 incommensurate frequencies in convection in water (from [GB]).

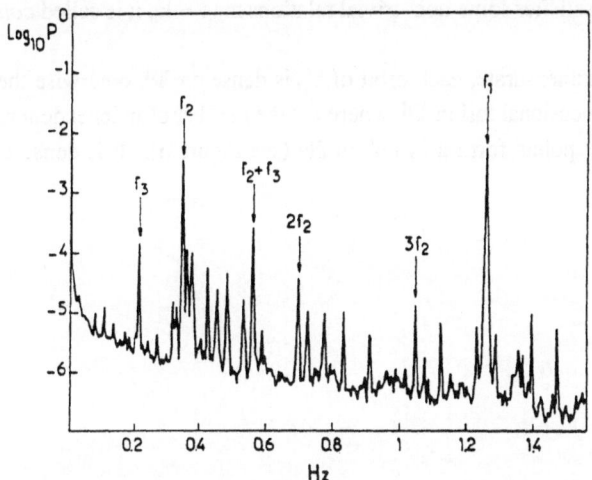

Figure 4. Fourier spectrum for the output of a liquid Helium convection cell of aspect ratio α = 2.8 and Prandtl number Pr = 0.62, near the onset of turbulence, with three frequencies present, and the locking state $f_1 - 3f_2 = f_3$ (from [ML]).

2. ORDER

2.1. KAM theory

It turns out that for most parameter values in families of torus maps, $f_\Omega : \mathbb{T}^d \to \mathbb{T}^d$, close to translations, the motion is quasiperiodic, conjugate to incommensurate translations. Since the theory works in any dimension we give the statements for arbitrary d. First we need some general results.

<u>Definition</u> : If f is a continuous map of $\mathbb{T}^d$ to itself, then there is a map F: $\mathbb{R}^d \to \mathbb{R}^d$, continuous, such that :

$$
\begin{array}{ccc}
 & F & \\
\mathbb{R}^d & \longrightarrow & \mathbb{R}^d \\
\pi \downarrow & & \downarrow \pi \\
\mathbb{T}^d & \longrightarrow & \mathbb{T}^d \\
 & f &
\end{array}
$$

i.e., $f \circ \pi = \pi \circ F$, where for $x \in \mathbb{R}^d$, $\pi(x) = x \mod \mathbb{Z}^d$. Such a map F is called a *lift* of f.

Different lifts of f differ only by translation by an integer vector of $\mathbb{R}^d$.

<u>Proposition</u> : If f: $\mathbb{T}^d \to \mathbb{T}^d$ is continuous and F is one of its lifts to $\mathbb{R}^d$, then f is isotopic to the identity if and only if $F(x + m) = F(x) + m$, for all $m \in \mathbb{Z}^d$, $x \in \mathbb{R}^d$.

We denote the family of translations by $T_\Omega(x) = x + \Omega$, where $x, \Omega \in \mathbb{R}^d$.

<u>Definition</u> : A point $\rho \in \mathbb{R}^d$ is said to be *incommensurate* if $m \in \mathbb{Z}^d$, $k \in \mathbb{Z}$, $m.\rho = k$ implies $m = 0, k = 0$. If ρ satisfies some non-trivial relations $m_i.\rho = k_i$, it is called *commensurate*.

For ρ incommensurate, each orbit of T_ρ is dense on $\mathbb{T}^d$, otherwise the closure of each orbit is a $(d-\ell)$-dimensional tori in $\mathbb{T}^d$, where ℓ is the number of independent relations satisfied. The commensurate points form a "web" in $\mathbb{R}^d$ (see Figure 5). It is dense but occupies zero measure.

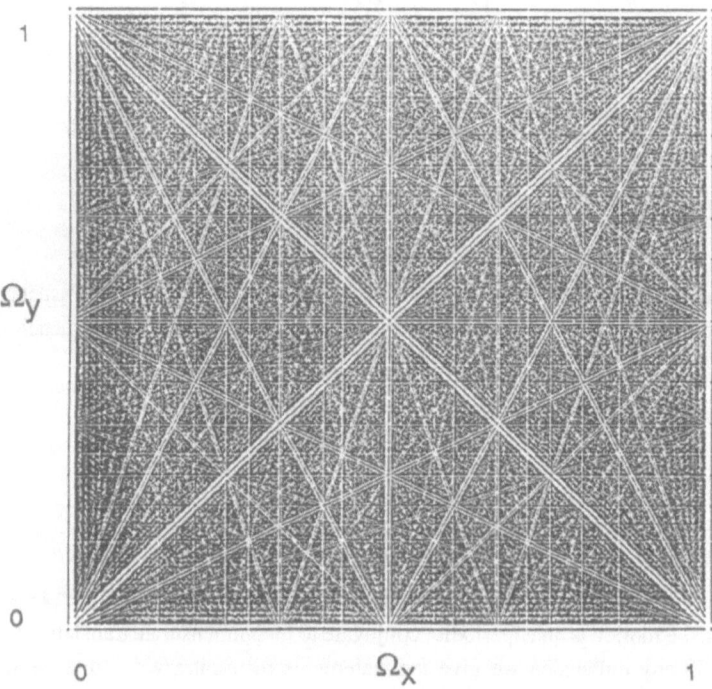

Figure 5. All rational points of the unit square with denominator less than 100.

KAM theory (named after Kolmogorov, Arnol'd and Moser) shows that for families F_Ω close to T_Ω, for most parameter values Ω, the dynamics remains equivalent to incommensurate translations (see [H,BMS] for clear presentations). To state the result we need the following definition.

<u>Definition</u> : A point $\rho \in \mathbb{R}^d$ satisfies a *Diophantine Condition*, denoted $\rho \in D_{C,\tau}$, if there are $C > 0$ and $\tau \geq d$ such that : $\forall\, m \in \mathbb{Z}^d \backslash 0$ and $k \in \mathbb{Z}$

$$| m.\rho - k | \geq \frac{C}{\| m \|^\tau} \quad , \text{ where } \| m \| = \sum_{i=1}^{d} | m_i | .$$

It can be shown that :

      (i) For $\tau \geq d$, C small enough, $D_{C,\tau} \neq \emptyset$

      (ii) For $\tau > d$ and A a bounded open set, $\lambda(A \backslash D_{C,\tau}) \to 0$ as $C \to 0$, where λ is the Lebesgue measure.

<u>Theorem</u> : Let $C > 0$, $\tau \geq d$, $r > 2\tau$ and A be a bounded open set of $\mathbb{R}^d$. If F_Ω is sufficiently close to T_Ω in C^r then there exists a map $\tilde{\Omega} : D_{C,\tau} \cap A \to \mathbb{R}^d$ such that $F_{\tilde{\Omega}(\rho)}$ is $C^{r-\tau}$-conjugate to T_ρ. And for $\tau > d$, $\lambda(A \backslash \tilde{\Omega}(D_{C,\tau} \cap A)) \to 0$ as $F_\Omega \to T_\Omega$.

<u>Note</u> : 1) The *C^r distance* between two families of diffeomorphisms f_μ and g_μ is given by :

$$\sup_{x,\mu} \| D^j f_\mu x - D^j g_\mu x \| \quad \text{for } j = 0,...,r.$$

2) f is *C^k-conjugate* to g if there exists a C^k diffeomorphism h such that f∘h = h∘g.

We illustrate the theorem by Figure 6, for families of the form $F_\Omega(x) = x + \Omega + ag(x)$, $d = 2$, where g is a multiply-periodic function with period 1 in each component of x, e.g. the family (*).

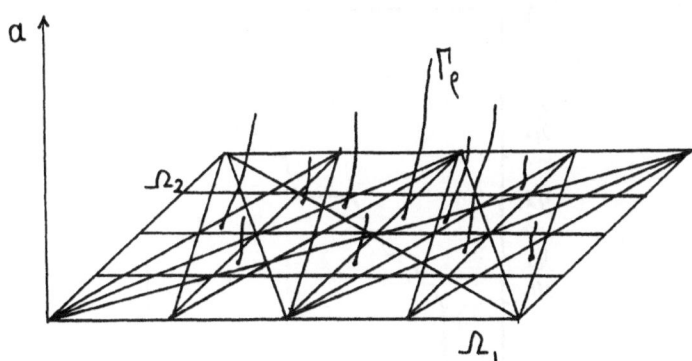

Figure 6. Commensurate lines in the plane a=0, and curves Γ_ρ on which $F_{\Omega,a}$ is smoothly conjugate to T_ρ.

We call the set of parameter values for which the map is conjugate to an incommensurate translation, the *KAM set*. So the question remaining is to understand what happens for parameter values in the complement of the KAM set.

2.2. Diffeomorphisms of the circle

The most basic concept for orbits of torus maps isotopic to the identity is that of rotation vector. Let us first recall the idea for the case d = 1, circle diffeomorphisms.

Definition : If $F : \mathbb{R} \to \mathbb{R}$ is a lift of a circle diffeomorphism, the *rotation number* of F is given by :

$$\rho(x,F) = \lim_{n \to \infty} \frac{F^n(x) - x}{n}$$

This is well defined, as Poincaré showed the limit above exists and is the same for all values of x (e.g. [GH,Arn]). It measures the average rate at which orbits of the lift advance along $\mathbb{R}$.

If ρ is irrational and F is C^2 then F is topologically conjugate to T_ρ (e.g. [GH,Arn]). Furthermore, the condition of small perturbation from the translations is not required for KAM theory when d = 1 [H].

If $\rho = p/q$ is a rational number then there exist periodic points of type (p,q), i.e., $F^q(x) = x + p$, and each orbit is backwards and forwards asymptotic to one such.

If F_Ω depends continuously on Ω, then $\rho(F_\Omega)$ is continuous. F_Ω is said to be *monotone* if for $\Omega' > \Omega$, $F_{\Omega'}(x) > F_\Omega(x)$; then the function $\rho(F_\Omega)$ is also non-decreasing.

For such a family, given $\omega \in \mathbb{R}$ irrational, the set $\{\Omega : \rho(F_\Omega) = \omega\}$ is a single point, e.g. [RhT], while $\{\Omega : \rho(F_\Omega) = p/q\}$ is typically an interval, because if F_Ω has a non-degenerate periodic orbit of type (p,q) then so will $F_{\Omega'}$ for all close enough Ω'.

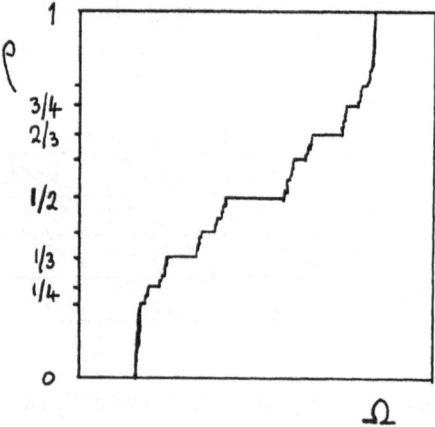

Figure 7. A devil's staircase

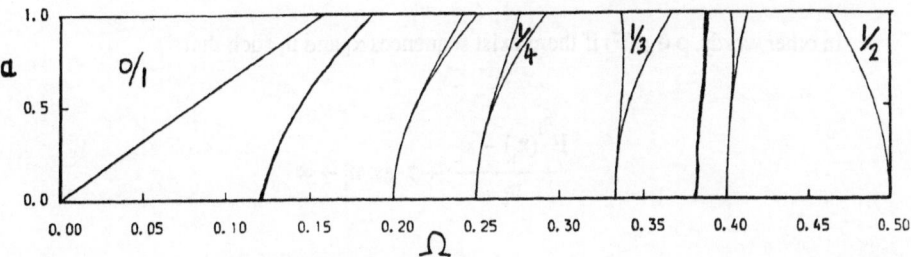

Figure 8. Typical mode-locking tongues and irrational curves in parameter space for a family of circle diffeomorphisms.

Note : A periodic point x of type (p,q) is said to be *non-degenerate* if $(F^q T_{-p})'(x) \neq 1$.

Hence $\rho(F_\Omega)$ is typically a *devil's staircase*, that is a continuous function which is constant for a whole interval at every rational value (Figure 7). But by KAM theory it is *incomplete*, that is the measure of the set of Ω such that ρ is irrational is positive. One obtains a diagram in parameter space like Figure 8, for families of the form $F(x) = x + \Omega + ag(x)$.

Note that the behaviour for non-invertible circle maps homotopic to the identity is considerably more complicated (e.g. [MT]), but this will not concern us here.

2.3. Rotation vector and rotation set

For higher dimensional torus maps, there are some new features when one tries to define the rotation vector.

Definition : If F is a lift of a map of $\mathbb{T}^d$ to $\mathbb{R}^d$, the *rotation vector* of $x \in \mathbb{R}^d$ is the vector in $\mathbb{R}^d$ given by :

$$\rho(x,F) = \lim_{n \to \infty} \frac{F^n(x) - x}{n}$$

The problems now are : (i) the limit need not exist.
(ii) it may be different for different values of x.

We want a notion of rotation set which captures all the possible rotation vectors. The one we use was introduced by Misiurewicz and Ziemian [MZ].

Definition : The *rotation set* of a lift F of a torus map is a subset of $\mathbb{R}^d$ given by :

$$\rho(F) = \{ \text{limit points of sequences } \frac{F^{n_i}(x_i) - x_i}{n_i} , x_i \in \mathbb{R}^d, n_i \in \mathbb{N} \text{ and } n_i \to \infty \text{ as } i \to \infty \}$$

43

In other words, $\rho \in \rho(F)$ if there exist sequences x_i and n_i such that

$$\frac{F^{n_i}(x_i) - x_i}{n_i} \to \rho \text{ as } n_i \to \infty$$

Misiurewicz and Ziemian [MZ] obtained the following results about $\rho(F)$.

Theorem : The rotation set, $\rho(F)$, is compact and connected. If d=2, it is also convex.

Theorem : For each extreme point of $\rho(F)$ there exists an orbit of F with that rotation vector.

Note : A set A in $\mathbb{R}^d$ is *compact* if it is closed and bounded. And it is *convex* if for any two points x, y $\in$ A, the convex combination $\lambda x + (1-\lambda)y \in A$, for $\lambda \in [0,1]$. A point x $\in$ A is an *extreme point* of A if it can not be written as $\lambda y + (1-\lambda)z$, for any y, z $\in$ A and $\lambda \in (0,1)$.

Although the definition of the rotation set sounds somewhat weak, it has many dynamical consequences, at least for d=2. For example, for p $\in \mathbb{Z}^d$, q $\in \mathbb{N}$, say x is a *periodic point of type (p,q)* if $F^q(x) = x + p$. Then

Theorem [Fra] : For d=2 and all rational points p/q in lowest terms in the interior of $\rho(F)$, there exists a periodic orbit of type (p,q).

We restrict attention from now on to the case d=2.

2.4. Mode-locking

Let $F : \mathbb{R}^2 \to \mathbb{R}^2$ be a lift of a diffeomorphism of $\mathbb{T}^2$, isotopic to the identity.

Say F is *mode-locked* if there exists m $\in \mathbb{Z}^2 \backslash 0$ and k $\in \mathbb{Z}$ such that m.ρ = k, for all ρ $\in \rho(F)$. F is *fully mode-locked* if $\rho(F) = \{p/q\}$ for some p $\in \mathbb{Z}^2$, q $\in \mathbb{N}$, and *partially mode-locked* if it is mode-locked, but not fully so.

Figure 9 shows an example of some orbits of a partially mode-locked map. Its rotation set satisfies the relation $\rho_1 - 2\rho_2 = -1$. To see this, the period 5 orbits are found to have rotation vector (3,4)/5. The invariant circle has homotopy type (2,1). Lifting to the plane, the invariant circle gives an infinite family of curves, dividing the plane into strips of average slope (2,1). Under $F^5 T_{(-3,-4)}$ each such strip is mapped to itself. Thus the rotation set ρ' for $F^5 T_{(-3,-4)}$ is contained in the line $\rho_1' - 2\rho_2' = 0$. The rotation set ρ for F is related to ρ' by $\rho' = 5\rho - (3,4)$. Hence the result.

Figure 10 shows an example of some orbits of a fully mode-locked map. It has rotation set $\{(1,1)/2\}$. This is because the periodic orbits visible have rotation vector (1,1)/2, and under $F^2 T_{(-1,-1)}$, the "squares" bounded by saddle manifolds are invariant, so the rotation set for $F^2 T_{(-1,-1)}$ is $\{0\}$.

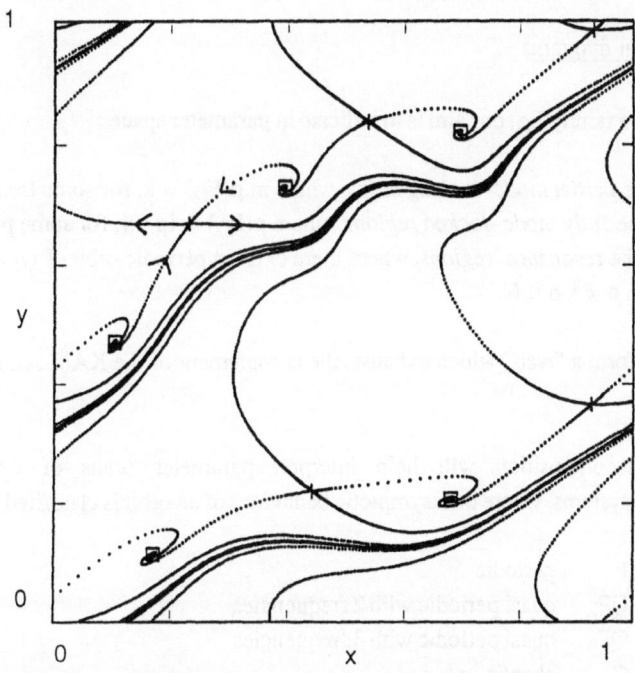

Figure 9. Some orbits of a partially mode-locked map for a = 0.7, Ω = (0.598, 0.8055).

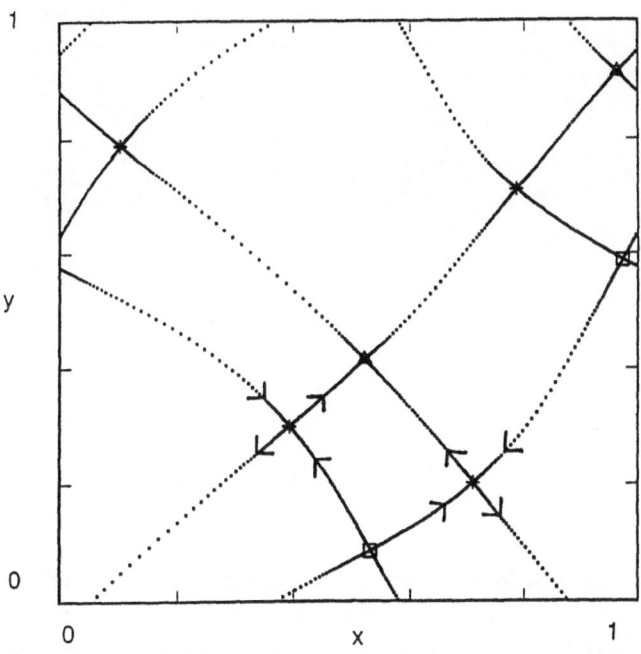

Figure 10. Some orbits of a fully mode-locked map, for a = 0.7, Ω = (0.499, 0.48).

2.5. Bifurcation diagrams

Given a family F_Ω, our aim is to indicate in parameter space :

(i) the *partial mode-locking strips*, where $m.\rho(F_\Omega) = k$, for some $(m,k) \in \mathbf{Z}^2 \times \mathbf{Z} \setminus 0$

(ii) the *fully mode-locked regions*, where $\rho(F_\Omega) = \{p/q\}$, for some $p \in \mathbf{Z}^2$, $q \in \mathbf{N}$

(iii) the *resonance regions*, where there exists a periodic orbit of type (p,q), for some $p \in \mathbf{Z}^2$, $q \in \mathbf{N}$.

These form a "web" which exhausts the complement of the KAM set, apart from a few pathologies.

This decomposition will help interpret parameter scans of experimental and computational systems, where the asymptotic behaviour of an orbit is classified into :

P   periodic
QP_2 quasi periodic with 2 frequencies
QP_3 quasi periodic with 3 frequencies
C   chaotic,

according to some algorithm. See for example, Figures 11 and 12, from [LC] and [BGKM].

Our decomposition will provide a theoretical framework for understanding such pictures and other features of coupled oscillator systems.

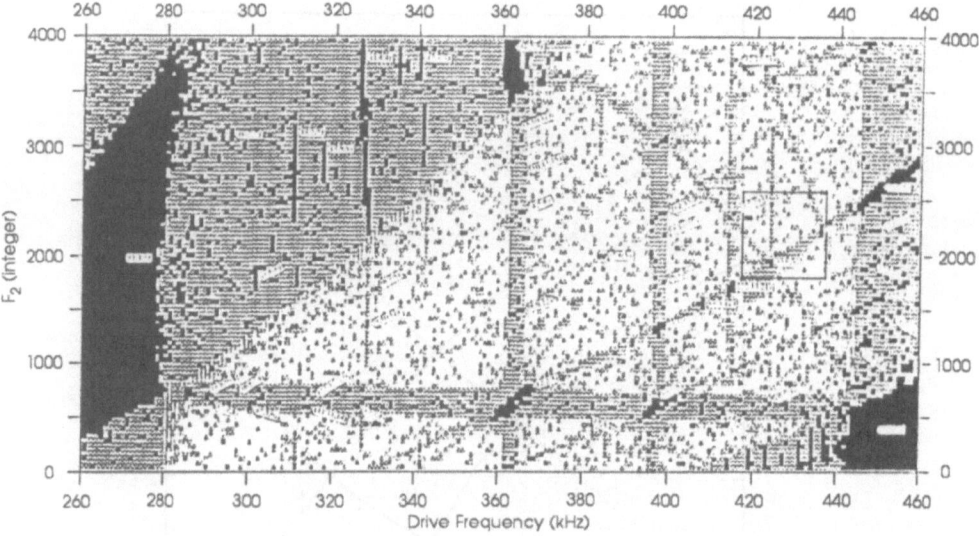

Figure 11. Parameter scan for an electronic circuit, indicating the asymptotic behaviour of some initial condition. Periodicity is plotted in black. Two-frequency quasiperiodicity is represented by open triangles and chaos by open rectangles. Three frequency quasiperiodicity is left blank. All discernable resonances have been labelled with their characteristic triplets (from [LC]).

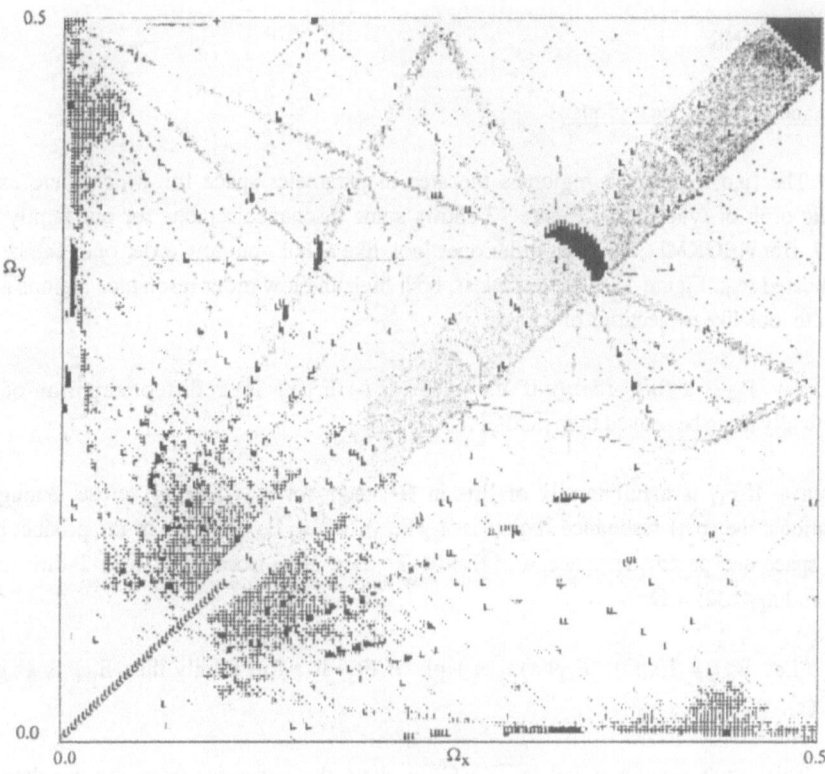

Figure 12. A parameter scan for the family (∗) with a = 0.7, indicating the asymptotic behaviour of one initial condition. Periodicity is (mostly) indicated in black, QP_2 in grey, chaos with black crosses and QP_3 is left blank.

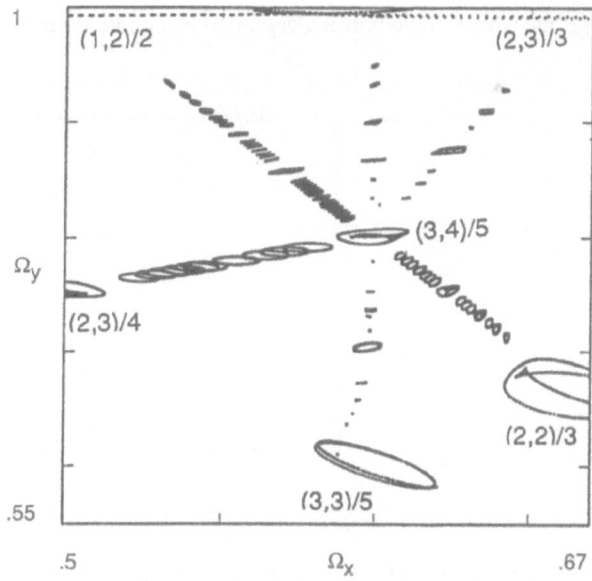

Figure 13. Resonance regions for various rotation vectors for the family (∗) at a = 0.7.

47

3. PATTERNS

3.1. Shape of resonance regions

The (p,q) resonance region is the area in parameter space for which there exists a periodic orbit of type (p,q). Figure 13 shows some resonance regions for our family (∗) at a = 0.7, from [BGKM]. The high order ones look like annuli, but low order ones can be more complicated (e.g. Figure 13). Nevertheless, both high and low order resonance regions always appear to look like projections of a torus.

Say F_Ω is a *full family* if $\forall x \in \mathbb{R}^2$, $\Omega \mapsto F_\Omega(x)$ is a diffeomorphism of $\mathbb{R}^2$. Analytically it can be proved that, [KMG] :

Theorem : If F_Ω is a full family of lifts in $\mathbb{R}^2$, then, for q = 1 or F_Ω close enough to translations, the (p,q) resonance region is $\pi_\Omega(B_{p,q})$, where $B_{p,q}$ is a set in the product of the phase space and parameter space $\{(x,\Omega): x \in \mathbb{T}^2, \Omega \in \mathbb{R}^2\}$, isomorphic to a 2-dimensional torus, and $\pi_\Omega(x,\Omega) = \Omega$.

Proof : Let $B_{p,q} = \{(x,\Omega) : F_\Omega^q(x) = x + p\}$. If F_Ω^q is a full family then $B_{p,q}$ is a graph, $\Omega = \tilde{\Omega}(x)$. □

This result was extended by [KMG] to show that even far from the translations, a resonance region for a "periodic" family (defined in [KMG]) has at least the topology of the torus.

Note that there is not necessarily an attracting periodic orbit for every point of a resonance region. Typically, there are Hopf curves (e.g. Figure 14), across which an attractor turns into a repellor. An invariant circle is also produced or absorbed. It usually dies in a homoclinic wedge, the whole structure being organised by double eigenvalue +1 points (Figure

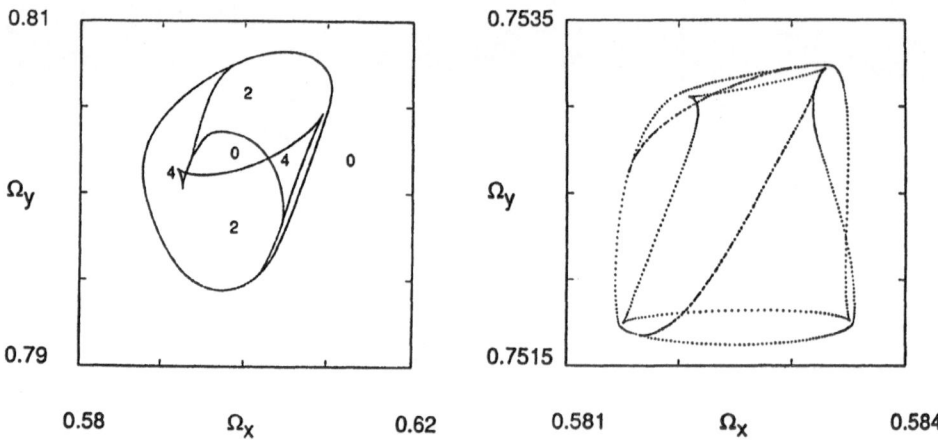

Figure 14. The (3,4)/5 and (7,9)/12 resonance regions for (∗) at a = 0.7, also showing Hopf bifurcation curves.

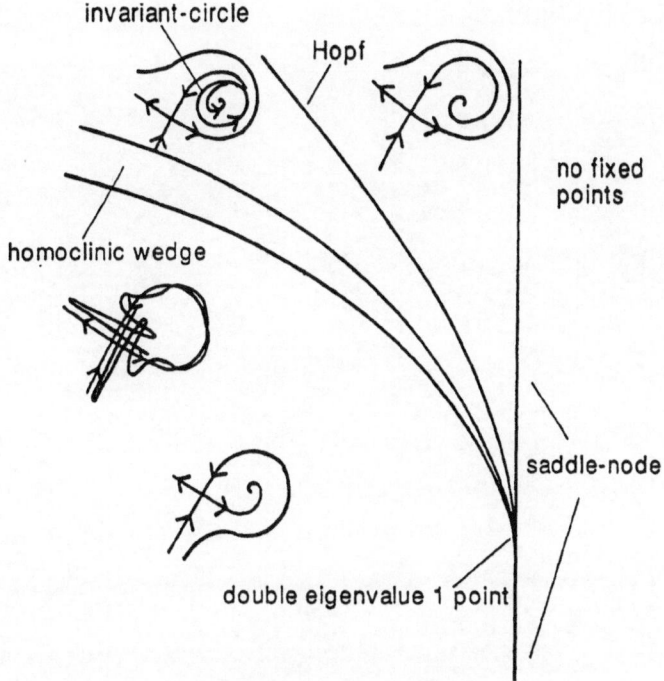

invariant-circle

Hopf

no fixed
points

homoclinic wedge

saddle-node

double eigenvalue 1 point

Figure 15. Unfolding of a double eigenvalue +1 point

15·). Under reasonable conditions we can prove existence of at least two of these points on each curve of saddle-nodes [KMG].

3.2. Relation between mode locking and resonance

How do the regions of full and partial mode-locking relate to the resonance regions?

If F is (p,q) fully mode-locked, then it can be proved that it has a type (p,q) periodic orbit [BGKM]. So the (p,q)-fully mode-locked region is contained in the (p,q) resonance region.

Figure 16 shows a parameter scan of family (∗) with a = 0.7 near the (3,4)/5 resonance, also indicating a region of full mode-locking inside it.

A variety of partial mode-locking strips can be seen converging on the resonance in Figure 16. What they do inside the resonance is not clear from this picture, however, because inside the resonance many of the partial mode-locking strips have no attracting rotational invariant circle, so QP_2 is not observed.

Partial mode-locking strips cannot intersect. So given (p,q) ∈ $\mathbb{Z}^2 \times \mathbb{N}$, most strips with m.p = k.q must terminate, either on a fully mode-locked region or otherwise. The question is how. To answer this, we will study torus maps close to translations, in the neighbourhood of a resonance region, hence close to a rational translation. But these maps can be well approximated by flows, so we shall gain information by studying the bifurcation diagrams of flows on $\mathbb{T}^2$.

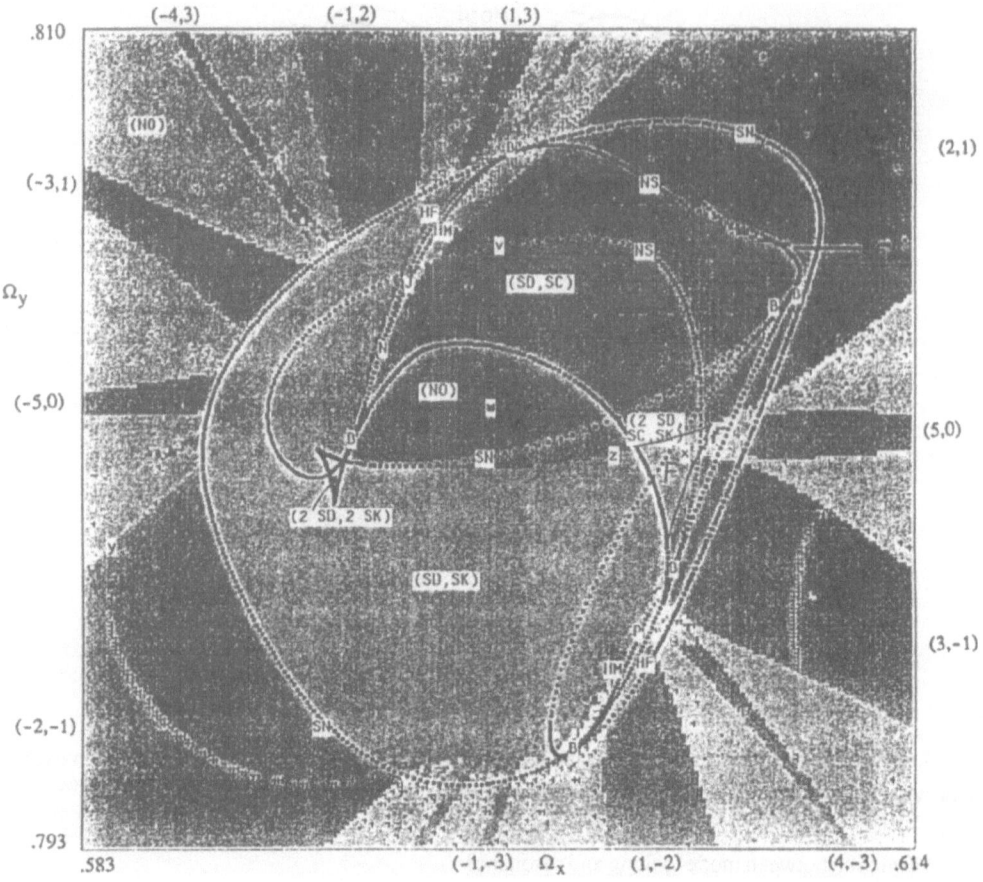

Figure 16. Parameter scan of family (∗) with a = 0.7 near the (3,4)/5 resonance, indicating QP₂ in dark grey, and P and QP₃ in two shades of light grey. The QP₂ strips are labelled by their homotopy type relative to (3,4)/5, i.e. m ∈ $\mathbb{Z}^2$ such that m.ρ = m.(3,4)/5, where ρ is the rotation set. A region of full mode-locking is also indicated (F).

3.3. The flow approximation

Let us sketch why maps near $T_{p/q}$ can be approximated by the time-1 map of a flow on $\mathbb{T}^2$ [BGKM, App A].

<u>Step 1</u> : Use flow equivalence to make p/q = (0,0)/1

Every diffeomorphism F of $\mathbb{T}^2$ isotopic to the identity can be obtained as the return map of the cross-section $\{\theta_0 = 0\}$ of a flow on $\mathbb{T}^3$ (this is the reverse procedure to that of Section 1.2).

There exist (m,k) such that m.p − kq = 1. If F is close enough to $T_{p/q}$ then m.ρ(F) − k > 0. It follows that there exists a cross-section Σ homotopic to $\{(-k,m).\theta = 0\}$ [Fri].

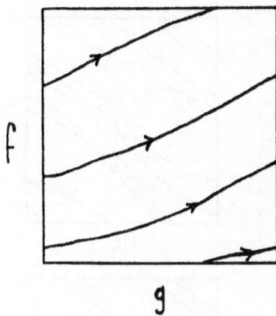

f   ʂ

Figure 17. Flow equivalence of a circle diffeomorphism with an orbit of rotation number 1/3 to one of rotation number 0.

The return map to Σ is said to be *flow equivalent* to F. It is qualitatively equivalent to F in the sense that they are both return maps of the same flow, but there are some changes. In particular, every periodic orbit of type (p',q') transforms to one of period $m.p' - kq'$ (see Figure 17 for the analogue for n = 1). It has the advantage that it can be chosen close to the identity.

<u>Step 2</u> : F within ε of the identity in C^r implies that one can approximate F to $O(\varepsilon^r)$, by the time–1 map of a flow. This is proved by Taylor expansion of the time–1 map of a vector field and comparing coefficients.

3.4. The set of winding ratios for flows on a 2-dimensional torus

Let φ be a lift to $\mathbb{R}^2$ of a flow on $\mathbb{T}^2$. The key tool to its study is the set of winding ratios, W_φ.

<u>Definition</u> : The *winding ratio* of an orbit of φ is a point in the unit circle in $\mathbb{R}^2$, or the origin, defined by :

$$w = \lim_{t\to\infty} \frac{\varphi_t(x)}{|\varphi_t(x)|} \in S^1 \quad \text{if } |\varphi_t(x)| \to \infty$$

or $w = 0$ if $|\varphi_t(x)|$ is bounded.

This is well defined as the limit above is shown to exist by Weil [W]. We ignore the case where $|\varphi_t(x)|$ is unbounded but does not converge to ∞. We will sometimes specify winding ratios by points in $\mathbb{R}^2$ not on S^1, it being understood that one should identify all points in $\mathbb{R}^2$ of the form sw, $s \in \mathbb{R}^+$.

By a theorem of Markley, [Mar], the set of winding ratios always has the form of a subset of $\{w, 0, -w\}$, for some $w \in S^1$. Hence we define the following classes of flows

$$B_w = \{\varphi : W_\varphi = \{w\}\}, \, w \in S^1$$
$$C_w = \{\varphi : W_\varphi = \{0,w\}\}, \, w \in S^1$$
$$D_w = \{\varphi : W_\varphi = \{w,-w\}\}, \, w \in S^1$$
$$E_w = \{\varphi : W_\varphi = \{w,0,-w\}\}, \, w \in S^1$$
$$F \;\; = \{\varphi : W_\varphi = \{0\}\}$$

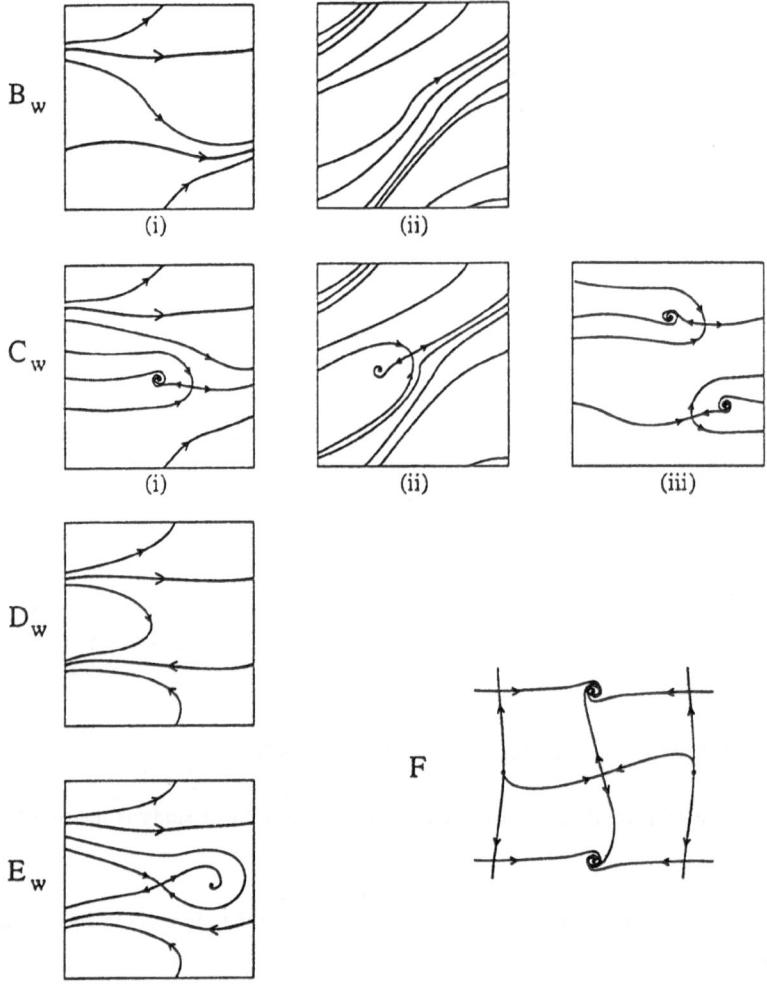

Figure 18. Examples of flows on $\mathbb{T}^2$ of types B, C, D, E and F: (i) w rational, (ii) w irrational, (iii) not a Cherry flow.

Examples are shown in Figure 18.

<u>Definitions:</u> We say that $w \in S^1$ is *rational* if it is of the form $(l,m)/|(l,m)|$, for some $(l,m) \in \mathbb{Z}^2 \backslash 0$. If $\varphi \in D_w \cup E_w$, then w is rational [FM].

We say that φ is *partially mode–locked* if $\varphi \in P = \bigcup_{w \text{ rational}} (B_w \cup C_w \cup D_w \cup E_w)$, and φ is *fully mode–locked* if $\varphi \in F$.

The *homotopy type* of a flow φ in P_w, denoted $[\, w \,]$, is the equivalence class of w under the relation $w_1 \sim w_2$ if $w_1 = \pm w_2$.

Given a family φ_μ, of flows on $\mathbb{T}^2$, we define the *resonance region* to be the set of parameter values for which φ_μ has an equilibrium point i.e. $\varphi_{\mu_t}(x) = x$, $\forall t$. This is equivalent to saying that $0 \in W_{\varphi_\mu}$.

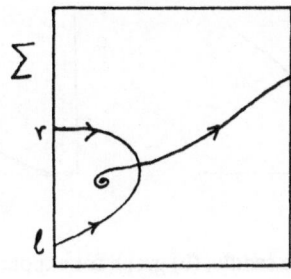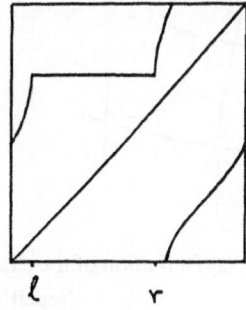

Figure 19. A Cherry flow and associated circle map

3.5. Cherry flows

Type B flows always have a cross–section. Hence their study reduces to that of diffeomorphisms of the circle, with a natural correspondence between winding ratio and rotation number. A large class of type C flows also reduce to circle maps, though with "flats".

Definition : A *Cherry flow* is a type C flow with a rotational transverse section Σ, such that the orbits of a non–empty proper subset $\Sigma' \subset \Sigma$ return to Σ, the induced map g: $\Sigma' \to \Sigma$ is continuous, and $\lim_{x \to l,r} g(x)$ are the same for any gap $[l,r]$ (component of $\Sigma \backslash \Sigma'$) (Figure 19).

This generalises the definition of [Cher,PdM,Boyd]. Examples appear in Figure 18 C(i),(ii).

Note : We say a curve is *rotational* if it is homotopically non–trivial, i.e. not contractible to a point.

Extending g to each gap $[l,r]$ by setting $g(x) = g(l) = g(r)$ for $x \in (l,r)$, we obtain a continuous map of the circle with *flats* (intervals on which g is constant). The rotation number of g is still defined and varies continuously with the flow. For a monotone family g_μ of circle maps with flats we have that (e.g. [Boyd]):

  (i) if ω is irrational then $\{\mu : \rho(g_\mu) = \omega\}$ is a single point,
  (ii) $\{\mu : \rho(g_\mu) = p/q\}$ is a non–trivial interval,
  (iii) $\lambda\{\mu : \rho(g_\mu) = \omega, \omega$ irrational$\} = 0$,

where λ is Lebesgue measure. Hence the behaviour of $\rho(g_\mu)$ is a complete devil's staircase.

The behaviour of g at the ends of flats depends on the exponents of the relevant saddles. The *exponent of a saddle* with eigenvalues v_1 and v_2, with $v_1 < 0 < v_2$, is $\alpha = -v_1/v_2$. For $\alpha = 1$, the saddle is called *neutral*. If $\alpha \neq 1$ then the passage map past the saddle has the asymptotic form $x \mapsto Cx^\alpha$ (Figure 20).

The analogous case of a *backwards Cherry flow* (one whose time–reverse is a Cherry flow) can be treated similarly. The return map now has *jumps* instead of flats, but the above properties hold, by considering the inverse.

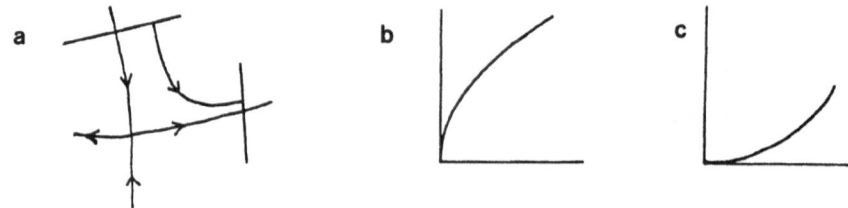

Figure 20. (a) Definition of the passage map past a saddle, (b) graph for exponent $\alpha < 1$ (repelling), (c) $\alpha > 1$ (attracting).

3.6. Some basic bifurcation theory for 2–dimensional flows

A *bifurcation* is a change in qualitative behaviour as parameters pass through an exceptional case. The *codimension* of a bifurcation is the number of independent conditions required to specify the exceptional case. An *unfolding* of a bifurcation is a family which contains the bifurcation in a persistent way.

The codimension–1 bifurcations of 2–dimensional flows were classified by Sotomayor [Soto] (see [GH] for a review) :

(i) Saddle–node of equilibria : A normal form for its unfolding is :

$$\begin{cases} \dot{x} = x^2 - \mu \\ \dot{y} = \lambda y \end{cases}$$

The result is sketched in Figure 21 (for the case $\lambda < 0$).

(ii) Hopf (Poincaré–Andronov) : A normal form for its unfolding (in polar coordinates) is :

$$\begin{cases} \dot{\theta} = \omega \\ \dot{r} = \lambda r (r^2 - \mu) \end{cases}$$

The result is sketched in Figure 22 (for the case $\lambda < 0$).

(iii) Saddle–node of periodic orbits : Taking a local transverse section of a periodic orbit with Floquet multiplier equal to 1, we obtain the following normal form for the unfolding of the return map :

$$x' = x + x^2 - \mu$$

The result is sketched in Figure 23.

(iv) Saddle connection : The unfolding is sketched in Figure 24. A special case is when the saddles are the same : *homoclinic bifurcation*. Then an attracting or repelling periodic orbit is produced, according as the exponent α of the saddle is greater or less than 1 (Figures 25,26).

$\mu < 0$ $\mu = 0$ $\mu > 0$

an equilibrium with   a saddle and a node

an eigenvalue 0

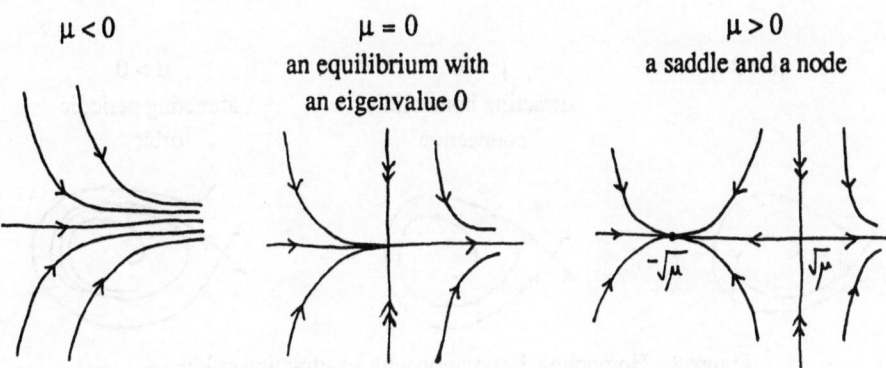

Figure 21. Saddle–node bifurcation for a two-dimensional flow.

$\mu < 0$ $\mu = 0$ $\mu > 0$

an equilibrium with   periodic orbit $r = \mu^{1/2}$

a complex conjugate pair of

eigenvalues on the imaginary axis

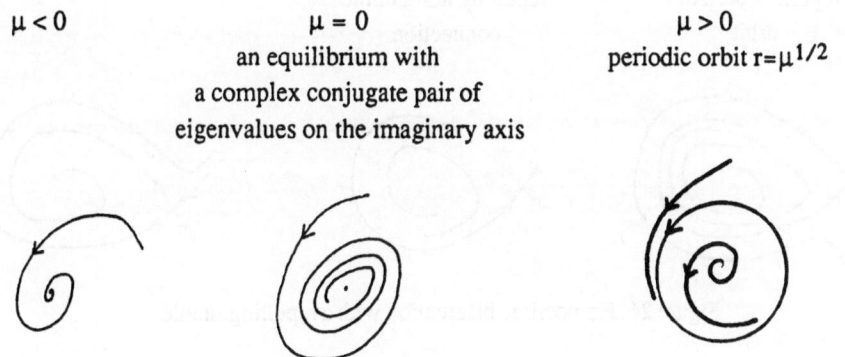

Figure 22. Hopf bifurcation for a 2-D flow

$\mu < 0$ $\mu = 0$ $\mu > 0$

two periodic orbits

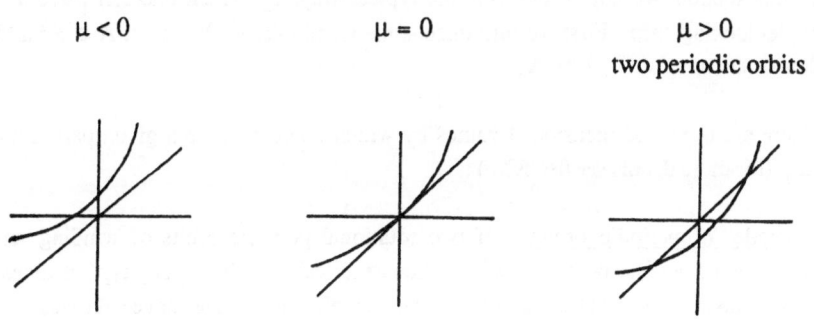

Figure 23. Return maps for a saddle–node bifurcation of periodic orbits for a 2D flow.

$\mu < 0$ $\mu = 0$ $\mu > 0$

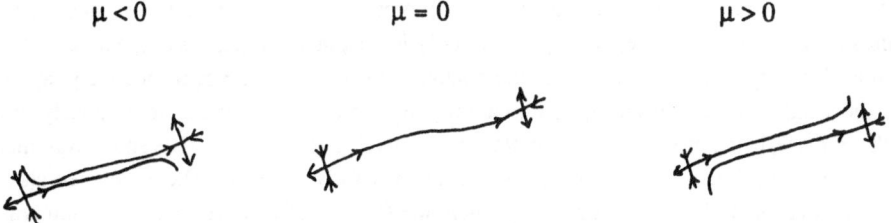

Figure 24. Unfolding of a saddle connection.

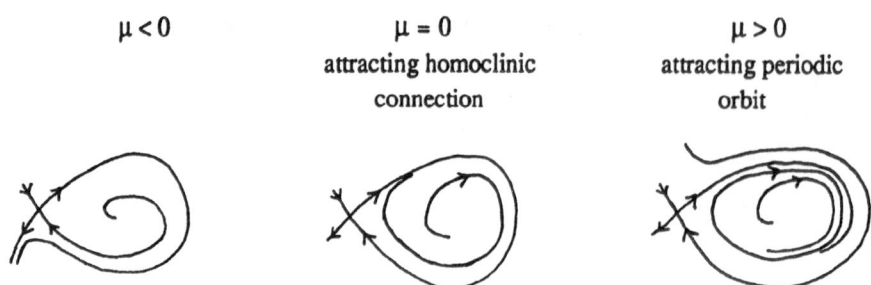

Figure 25. Homoclinic bifurcation with an attracting saddle.

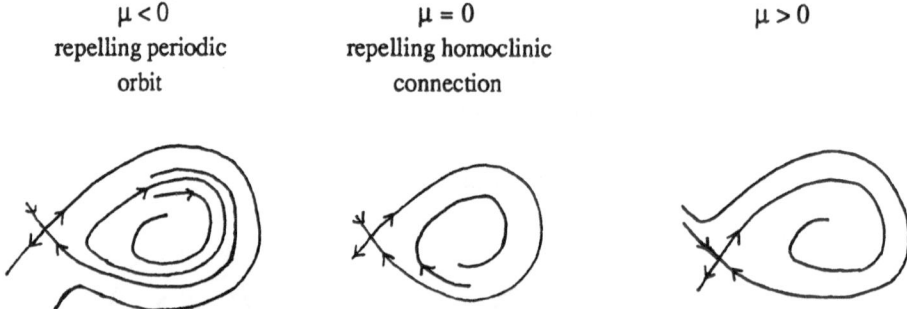

Figure 26. Homoclinic bifurcation with a repelling saddle.

3.7. Boundary of partial mode-locking

In this section, we aim to describe the typical ways by which one can leave a given partial mode-locking strip. First we introduce some terminology. For a set $A \subset \mathbb{R}^2$ and $p \in \mathbb{Z}^2$, we denote the translate $T_p A$ by A_p.

There are three codimension-1 routes by which one can leave a given partial mode-locked strip (for more details see [BGKM]) :

(i) <u>Saddle-node of periodic orbits</u> : If two rotational periodic orbits of winding ratio w coalesce, leaving none of homotopy type [w], then generically the homotopy type changes or φ becomes fully mode-locked. There are two subcases, which themselves have subcases.

(a) <u>with a connecting orbit</u> (see Figure 27): If there is a orbit which is asymptotic to the saddle-node orbit γ in the past and to its translate $\gamma_{0,1}$ in the future, then one can construct a rotational transverse section and hence obtain a monotone circle map. It may have flats and/or jumps, however. If it has none, or only flats or only jumps, then the same will (generically) be true for all nearby parameter values, so the results of Section 2.2 or 3.5, respectively, apply. Thus the winding ratio will change on breaking the saddle-node, varying continuously in a devil's staircase, being incomplete or complete, respectively. If it has both flats and jumps, then the situation is much more complicated and remains to be fully analysed. There are intervals of parameter arbitrarily close to the boundary where the flow is fully mode-locked and intervals arbitrarily close where it is partially mode-locked.

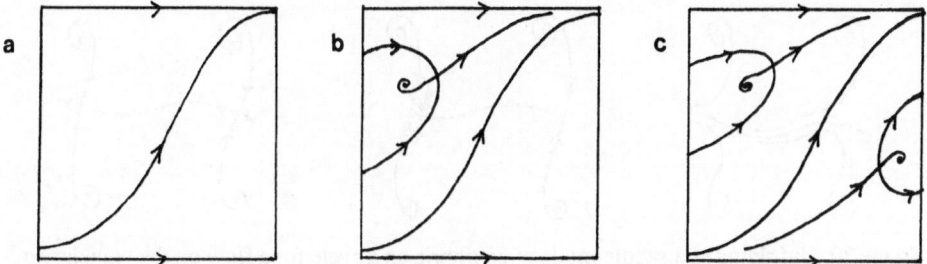

Figure 27. Saddle–node of periodic orbits with a connecting orbit, three case: (a) type B, (b) Cherry flow, (c) flats and jumps.

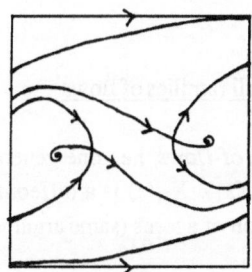

Figure 28. Saddle–node of periodic orbits on the boundary of partial mode-locking, without a connecting orbit.

(b) <u>without a connecting orbit</u>: In this case, the flow becomes fully mode-locked on leaving the partial mode-locking region. An example is shown in Figure 28.

(ii) <u>Rotational homoclinic bifurcation</u> : This is similar to (i) but with a rotational homoclinic cycle instead of a saddle–node. There are again two subcases, with subcases (Figure 29).

(a) <u>with a connecting orbit</u>: A rotational transverse section can be constructed. The return map is a monotone circle map with at least one flat or jump. If it has only flats or only jumps, then we obtain a complete devil's staircase for the winding ratio. If it has both, then there are intervals of partial mode-locking arbitrarily close and intervals of full mode-locking arbitrarily close.

(b) <u>without a connecting orbit</u>: The flow becomes fully mode-locked.

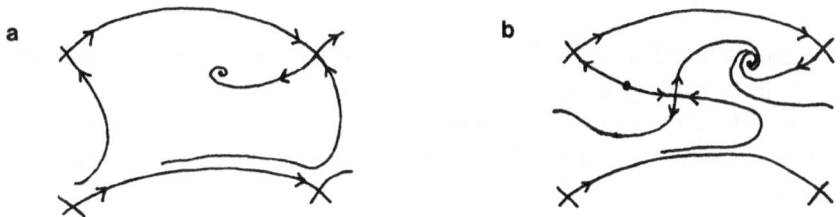

Figure 29. Examples of rotational homoclinic connection on the boundary of partial mode-locking, (a) with a connecting orbit, (b) without.

Figure 30. Unfolding of a saddle–node of equilibria on a cycle for a flow on the boundary of partial mode–locking.

(iii) <u>saddle–node of equilibria on a cycle</u> : This bifurcation converts a type C flow to a type F flow (Figure 30).

3.8. <u>Global bifurcation diagram for full families of flows</u>

Let us consider a *full family of flows*, i.e., one generated by a family of vector fields $\dot{x} = X_\Omega(x)$, such that for all $x \in \mathbb{R}^2$, $\Omega \mapsto X_\Omega(x)$ is a diffeomorphism. The resonance region for such a family is always a projection of a torus (same argument as in Section 3.1).

Since a full family is a two-parameter family, we should expect to see codimension-2 bifurcations. There is a huge variety of possible codimension-2 bifurcations for two-dimensional flows. They can be classed under the following headings:

B   Takens–Bogdanov (an equilibrium with double eigenvalue 0)
D   Degenerate Hopf (an equilibrium with an imaginary pair of eigenvalues and whose first Birkhoff invariant is zero)
Q,T,U Heteroclinic cycle (two saddles s, t, with saddle connections s → t and t → s)
N,P   Double homoclinic (a saddle with two homoclinic orbits, using all four branches of invariant manifold)
J,K   Neutral homoclinic (a saddle with a homoclinic orbit and exponent = 1)
Y,Z   Saddle–node loop (a saddle–node whose 1-D unstable manifold coincides with one branch of strong stable manifold [or vice versa])
V,W,G,H,S,M,L   Others

The letters are symbols we use to denote various varieties of these bifurcations in our Figures. For a review of all but the heteroclinic cycle and the "others", see [DG]. For the remainder, see [BGKM]. No doubt, this list is not complete!

Far enough from $\Omega = 0$, the flow is of type B and the parameter space divides up into partial mode-locking strips with irrational flow in between. As one makes one large circuit in parameter space the winding ratio w makes one complete revolution. Hence there exists $\Omega \in \mathbb{R}^2$ where w is either undefined or discontinuous. What happens to all the partial mode-locking strips in the centre?

There are many ways of generating lots of partial mode-locking tongues. We shall study in detail three of them, all codimension-2 bifurcations.

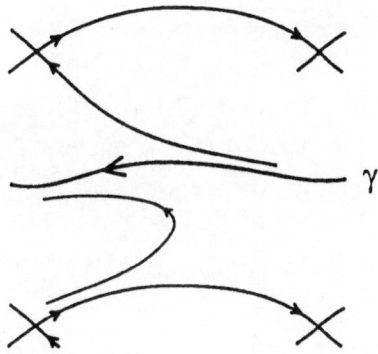

Figure 31. Phase portrait for the organising centre of a type 2 half–plane fan

(i) <u>Half plane fan (type 2)</u> : Suppose we have a saddle–node of periodic orbits and a repelling rotational homoclinic cycle, in opposite directions, with connecting orbits, like Figure 31.

Let ε be a parameter that breaks the rotational homoclinic cycle, and η a parameter that removes the saddle–node. With appropriate choice of signs for ε and η, we obtain phase portraits as in Figure 32. Three quadrants are partially mode–locked. The positive quadrant is the interesting one. Orbits can get through the places where the homoclinic connection and the saddle–node were. But they go a long way to the right past the saddles and then a long way to the left through the region of the saddle–node, the amounts depending on ε and η respectively.

If we suppose no obstructions, then all the flows in the positive quadrant are Cherry flows. The winding ratio varies continuously from $(-1,0)$ to $(1,0)$. Hence we obtain a bifurcation diagram like Figure 32. Other varieties of half–plane fan are possible.

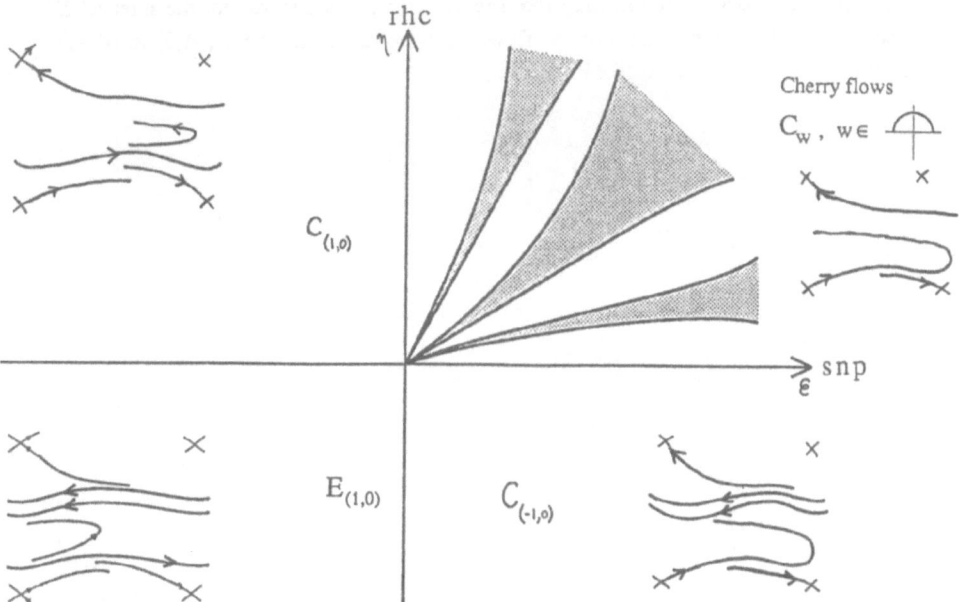

Figure 32. Bifurcation diagram for the type 2 half–plane fan, with some phase portraits.

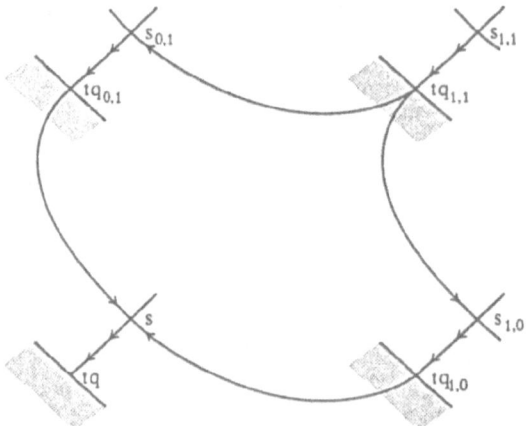

Figure 33. Phase portrait for a saddle-node fan point.

(ii) <u>Saddle-node fan</u> : This occurs in the boundary of full mode-locking. We have a saddle-node of equilibria, tq, whose stable manifold coincides with one branch of the unstable manifold of a saddle, s. And also, both branches of the stable manifold of s lie in the unstable half-plane of tq, like Figure 33.

Let μ be a parameter which unfolds the saddle-node, and ε one which breaks the connection $s \to t$. For $\mu = 0$ the flow is fully mode-locked (because there is an invariant fundamental domain). For $\mu > 0$ (tq becomes a saddle t plus a source q), the flow can be seen to remain fully mode-locked (see Figure 34).

For $\mu < 0$, however, it becomes a Cherry flow. To see this, take a cross-section Σ as indicated in Figure 35(a). There is a small interval Σ' on $\Sigma_{1,0}$ which maps onto Σ, the rest never reaching Σ. Thus we obtain a return map like Figure 35(b). As ε is varied, the interval Σ' moves across 0 and the winding ratio of the flow varies continuously from $(-1,0)$ to $(0,-1)$. Hence the bifurcation diagram looks like Figure 34.

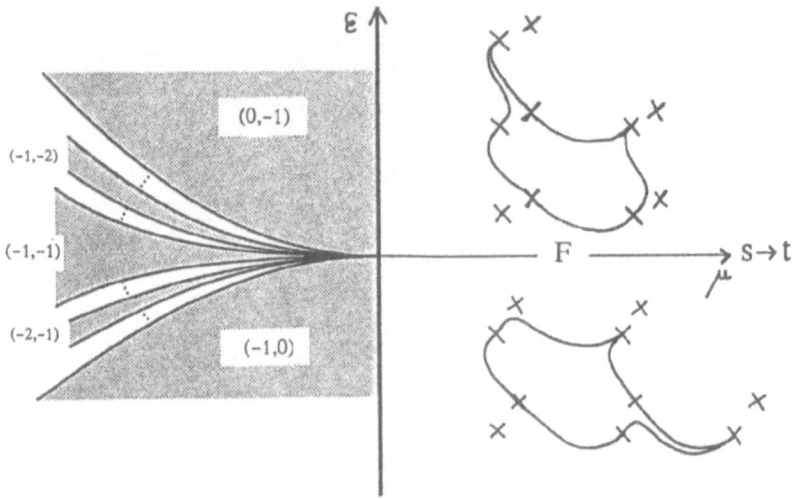

Figure 34. Bifurcation diagram for the saddle-node fan, with some phase portraits.

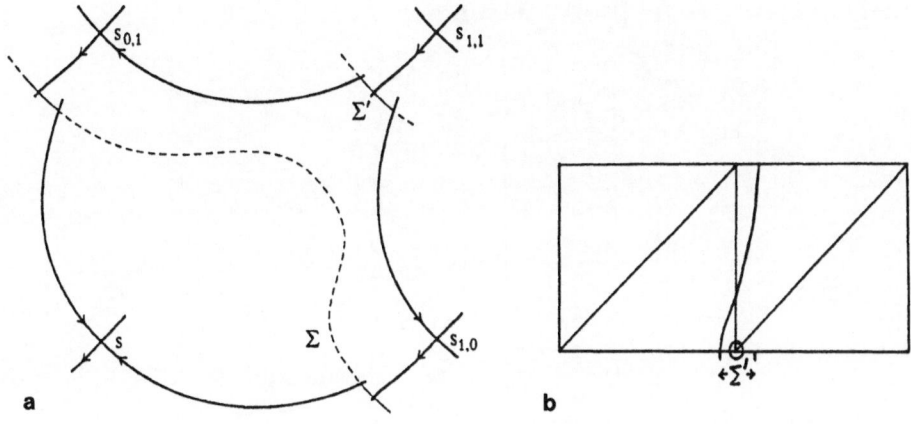

Figure 35. Saddle–node fan for $\mu < 0$: (a) transverse section, and (b) return map.

(iii) <u>Fractal boundary</u> : Instead of generating a fan of partial mode-locking strips, the third codimension-2 bifurcation that we will describe generates a bifurcation set with a region of full mode-locking having a fractal boundary and an infinite set of partial mode-locked strips leaving from it. We give here a brief description of the simplest case. For more details see [BGKM].

Suppose we have an attracting homoclinic connection γ between a saddle, s, and its translation, $s_{1,0}$, a heteroclinic connection between $s_{1,0}$ and another saddle t, and a connecting orbit $t \to \gamma_{0,-1}$ as in Figure 36(a). Using the unfolding parameters η for the homoclinic connection and ε for the heteroclinic one as indicated in Figure 36(b), we obtain a skeleton for the bifurcation diagram, shown in Figure 37.

There is more structure in the lower right quadrant, however. In each band between curves of $t \to s_{n,-1}$ and $t \to s_{n+1,-1}$ connections, one finds the structure of Figure 38.

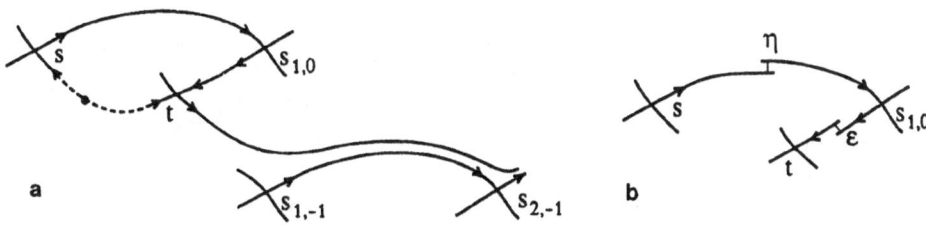

Figure 36. (a) Organising centre for the fractal boundary, (b) unfolding parameters.

61

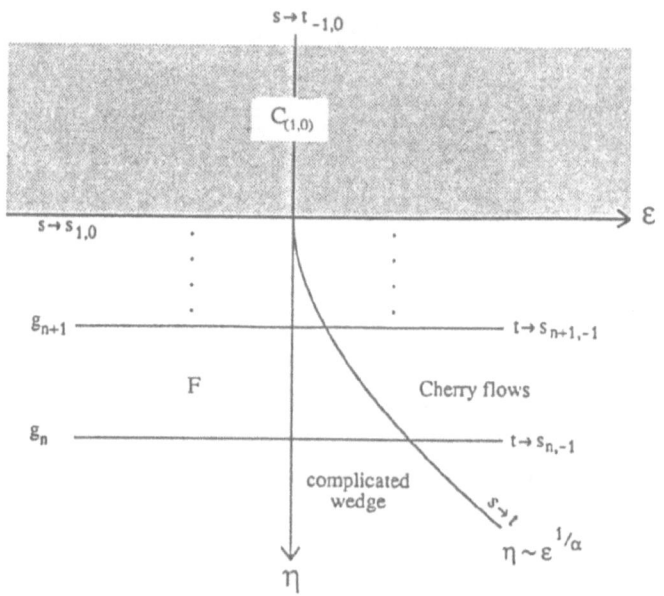

Figure 37. Skeleton of the bifurcation diagram for the fractal boundary.

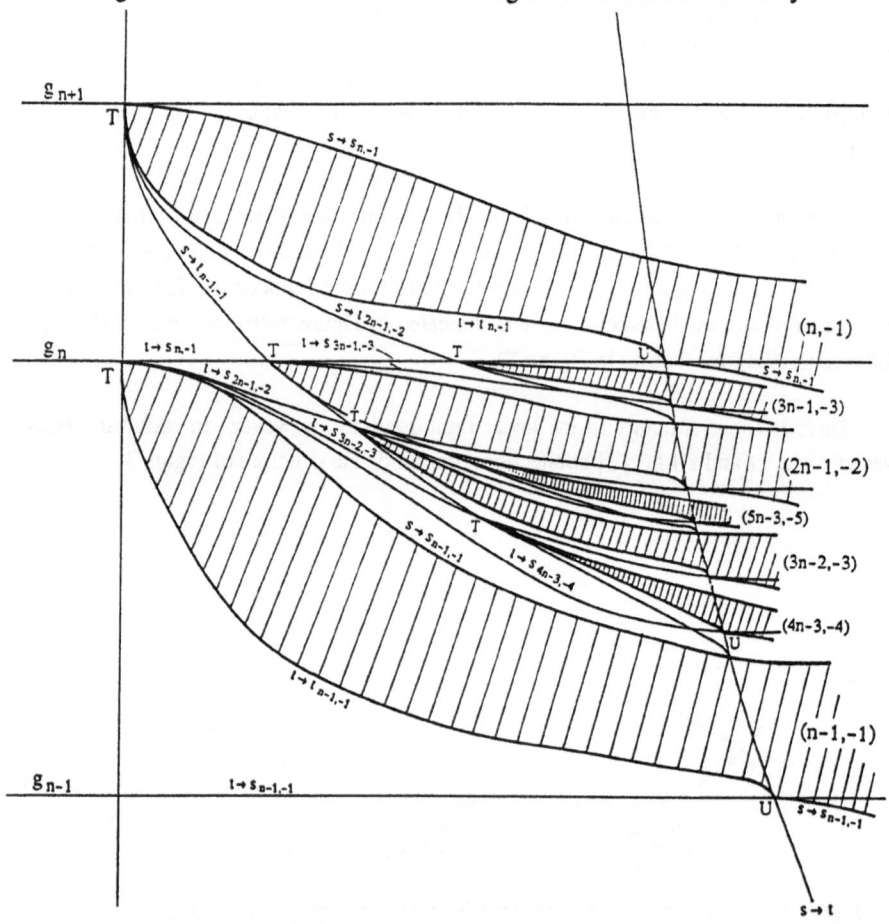

Figure 38. Structure in the complicated wedge of Figure 36.

3.9. Implications

The theory of sections 3.7 and 3.8 allows us to conjecture a lot of features of the global bifurcation diagram for full families of flows on $\mathbb{T}^2$ from a relatively small amount of information.

As an example, we present conjectures for the global bifurcation diagram of the flow approximation to the family of maps (∗) in the neighbourhood of the (3,4)/5 resonance, at a = 0.7, from [BGKM]. Note that there is not a unique flow approximation; what we mean is a family of flows whose time-1 maps are closest, in some norm, to the family of diffeomorphisms (after flow equivalence to turn (3,4)/5 into (0,0)/1).

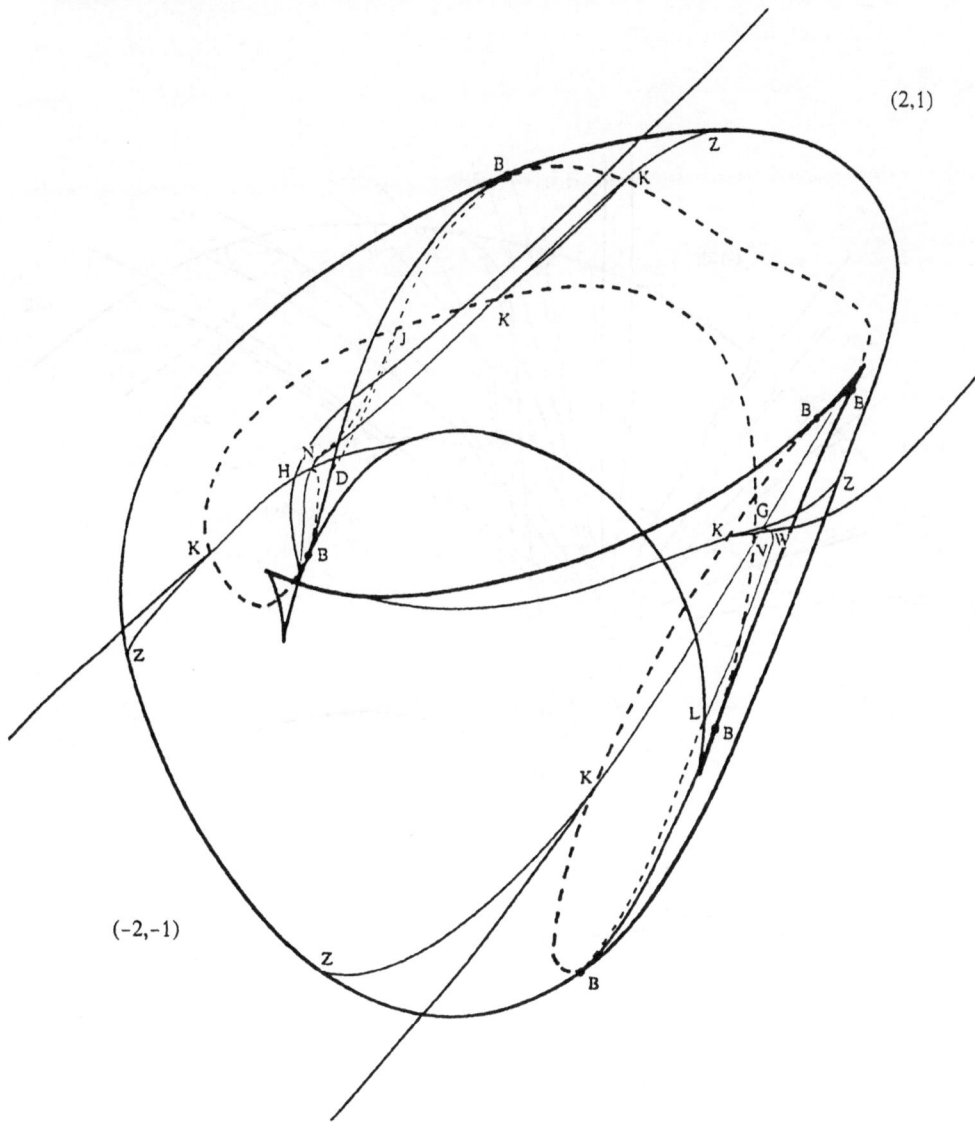

Figure 39. Conjectured basic features of the bifurcation diagram for the flow approximation for the (3,4)/5 resonance of (∗) at a = 0.7.

Figure 39 shows basic features which can mostly be deduced from Figure 16. The most important things to note are that there must be a half-plane fan point H on the left, and a saddle-node fan point S and a fractal boundary (associated with the point G) on the right. This leads to the conjectured bifurcation diagrams of Figures 40 and 41, on the left and right respectively.

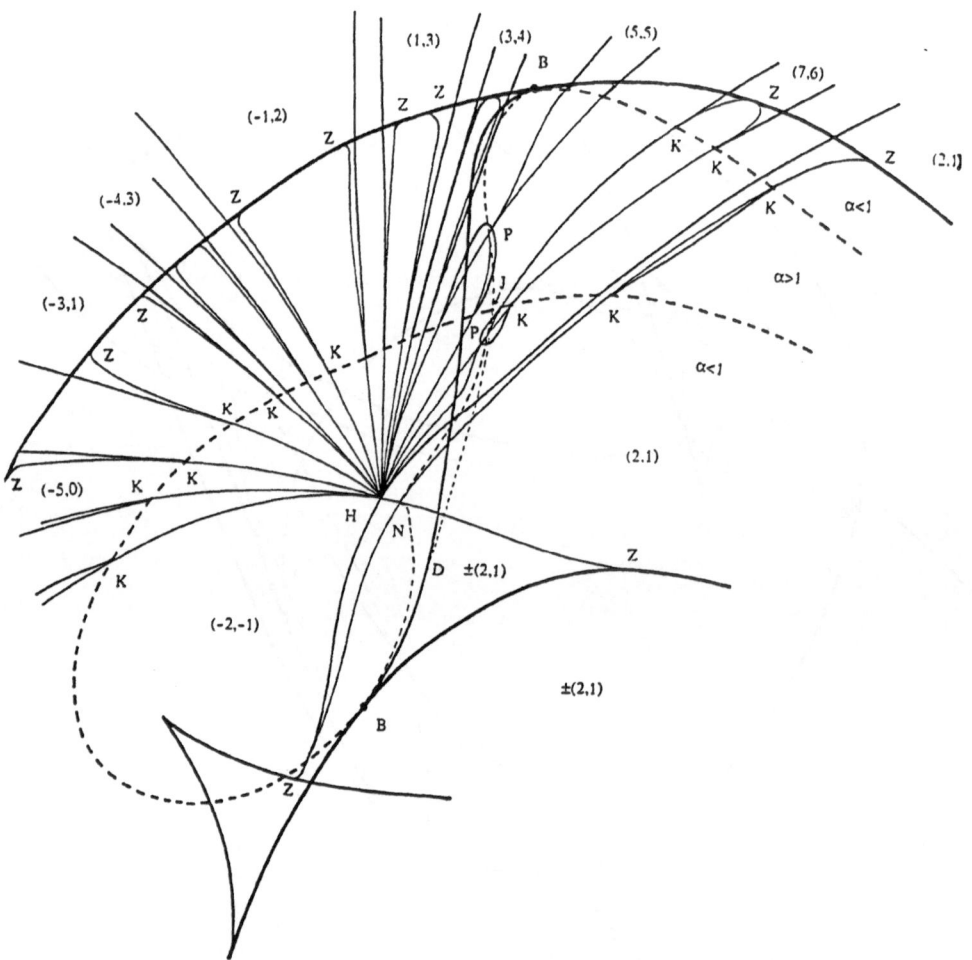

Figure 40. Conjectured structure in the upper left of Figure 38

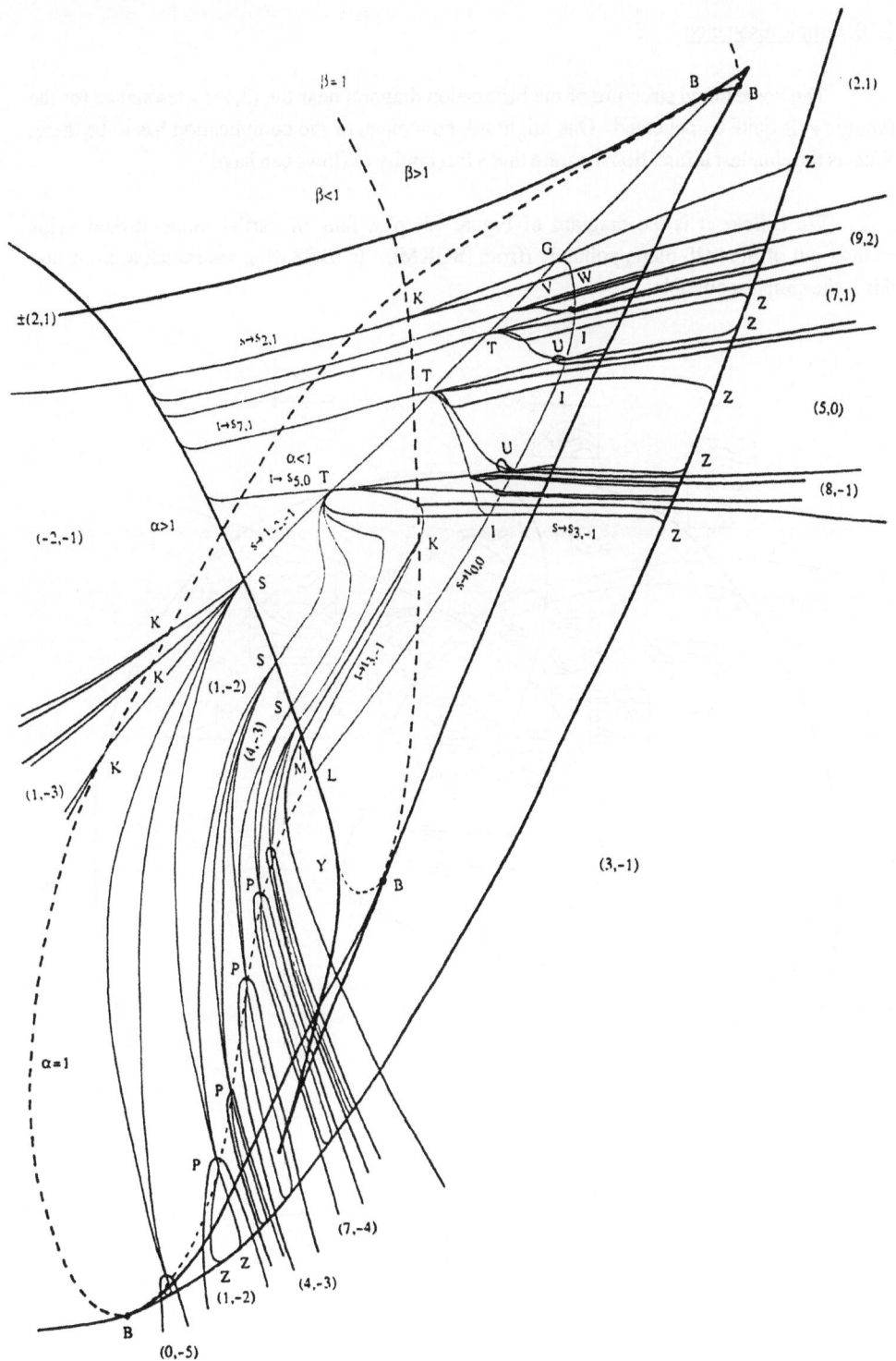

Figure 41. Conjectured structure on the right of Figure 38.

3.10. Simple resonance

The conjectured structure of the bifurcation diagram near the (3,4)/5 resonance for the family (∗) is quite complicated. One might ask how much of the complication has to be there. What is the simplest bifurcation diagram that a full family of flows can have ?

We believe it is the diagram of Figure 42, plus fans of partial mode-locked strips coming out of the half-plane points H (from [BGKM]). In [BGKM2], we sketch a proof that this is the simplest possible.

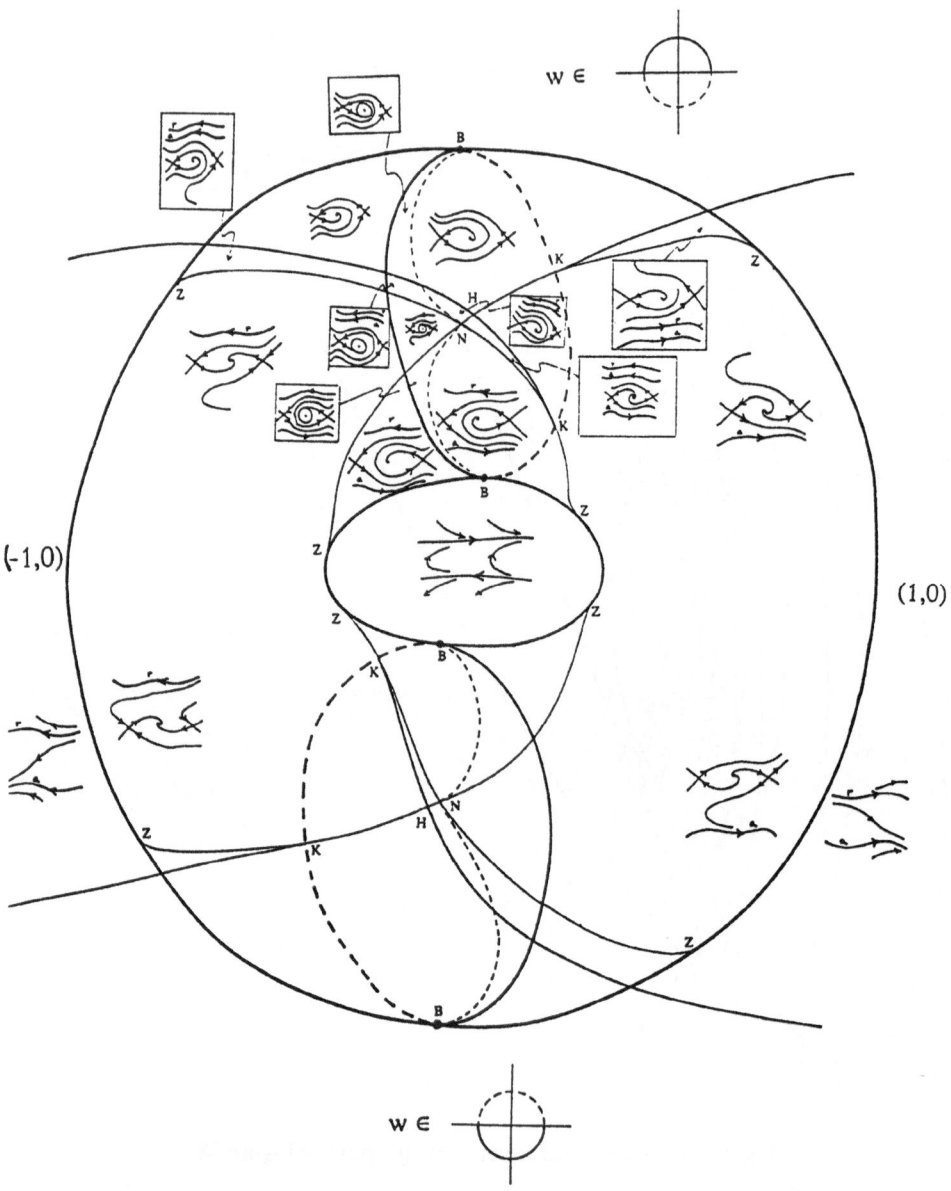

Figure 42. Conjectured simplest bifurcation diagram for a full family of flows.

4. CHAOS

Diffeomorphisms of a torus can have transverse homoclinic orbits, horseshoes and chaos. Close to the translations it occurs only in very small sets in the product of parameter and phase space, however, and the chaos is very weak, because the flow approximation is good. Numerical results show that even a = 0.7 in our family counts as small, e.g. Figure 43. Nevertheless, we shall show that chaos is typically present in every family of perturbations of the translations, and it can be large.

4.1. How to define chaos ?

We say a map f has *topological chaos* if it has positive topological entropy.

Definition : Let X be a compact metric space with metric d, and f a continuous map of (X,d) to itself. Given $\varepsilon > 0$, $n \geq 0$, say two points x, y in X have ε-distinguishable orbit segments of length n if $d(f^i(x), f^i(y)) > \varepsilon$ for some $0 \leq i \leq n$. Let $N(n,\varepsilon) = \max\{\#S\}$, over sets S of points with ε-distinguishable orbit segments of length n (#S stands for the number of elements in S). $N(n,\varepsilon)$ is finite by compactness.

Let $\gamma(\varepsilon) = \overline{\lim_{n\to\infty}} \frac{1}{n} \log N(n,\varepsilon)$ be the growth rate of $N(n,\varepsilon)$.

The *topological entropy*, $h_{top}(f)$, of the map f is :

$$h_{top}(f) = \lim_{\varepsilon\to 0} \gamma(\varepsilon)$$

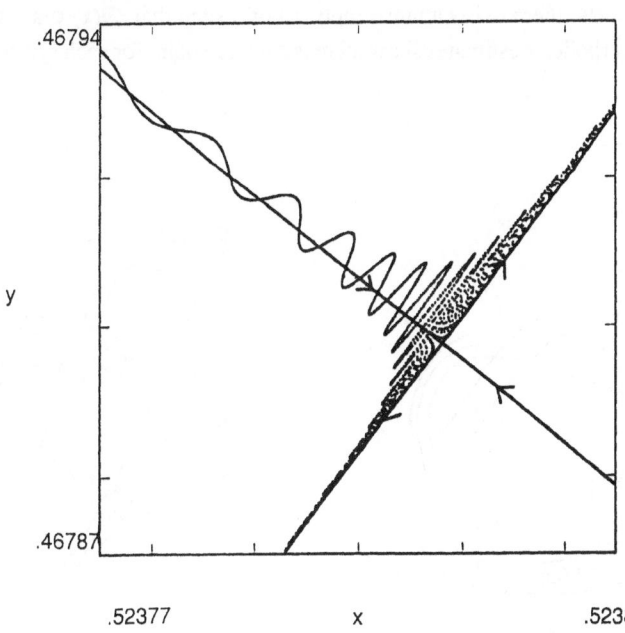

Figure 43. A blowup of transverse homoclinic oscillations near a saddle for (∗) at a = 0.7, Ω = (0.603, 0.801258051).

In other words, f has topological chaos if and only if for ε small enough, the number of ε-distinguishable orbit segments of length n grows exponentially with n.

Caution : A map may have topological chaos without having a chaotic attractor in any sense. The chaos can be transient. The advantage of using this definition of chaos is that one can prove things about topological chaos.

4.2. Characterization of topological chaos

Theorem [K] : If f is a $C^{1+\varepsilon}$ 2-dimensional diffeomorphism then it is topologically chaotic if and only if f^m has a horseshoe for some $n \in \mathbb{N}$.

Definition : A *horseshoe* is an invariant set Λ on which f is hyperbolic and conjugate to the shift σ on $2^{\mathbb{Z}}$ = {doubly infinite sequences of 0's and 1's}, with topology given by saying two sequences a, b are close if $\exists$ large N such that $a_n = b_n$, $\forall \mid n \mid \leq N$. The shift is defined by $[\sigma(a)]_n = a_{n+1}$.

Birkhoff and Smale (e.g. [GH]) showed that f has a horseshoe if it has a hyperbolic fixed point, p, with a transverse homoclinic intersection $q \neq p$ of its stable and unstable manifolds (the converse is also true, using the equivalence to the shift).

We sketch the proof (see Figure 44). Choose a "rectangle" Q as indicated. After some number N of iterations, $f^N(Q)$ intersects Q. If Q is thin enough then the intersection has two components, which we call 0 and 1. Let $F = f^N$. Given any finite sequence $a_{-m} .... a_n$ of 0's and 1's there is a rectangle of points x such that $F^j(x) \in a_j$ for $j = -m,..., n$. Hence, given any infinite sequence $a \in 2^{\mathbb{Z}}$, there exists a non-empty set of x such that $F^j(x) \in a_j$, $\forall j \in \mathbb{Z}$. For N large enough, hyperbolicity estimates show there is a unique point for each symbol sequence.

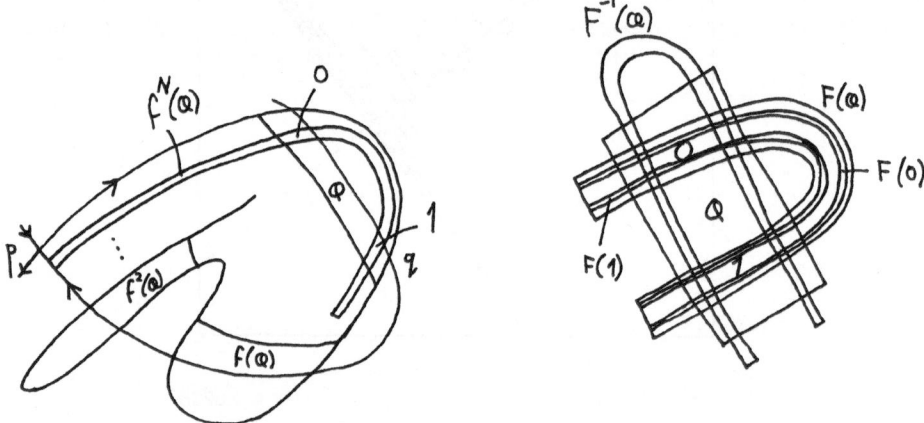

Figure 44. Construction of a horseshoe for a map with a transverse homoclinic orbit.

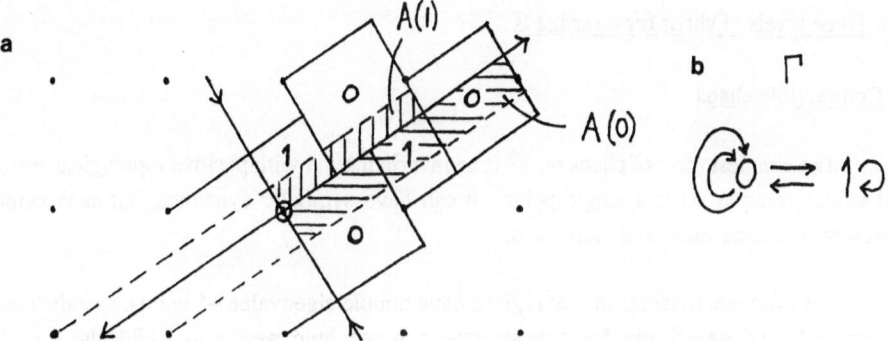

Figure 45. (a) Markov partition for A, and (b) associated graph Γ

Generalisation 1: Markov shift

Let Γ be a finite, irreducible and aperiodic directed graph and Σ be the set of doubly infinite paths in Γ, with the analogous topology to $2^{\mathbf{Z}}$. One specifies the paths by edge sequences e. The dynamics is again given by the shift σ on Σ, defined by $[\sigma(e)]_n = e_{n+1}$. An alternative definition is given in [GH].

Generalisation 2 : Quotient of a Markov shift

This is a Markov shift where some symbol sequences are identified. For example, the torus diffeomorphism A given by the matrix :

$$\begin{bmatrix} 2 & 1 \\ 1 & 1 \end{bmatrix}$$

(which is not isotopic to the identity), is equivalent to the shift on the graph of Figure 45 (b) after certain identifications are made.

To see this, consider the "Markov partition" shown in Figure 45 (a) (for a definition see [GH]). The images of the two regions 0 and 1 are also shown. We see that the allowed transitions are represented by the graph Γ, with two routes from 0 to itself because A(0) intersects two copies of 0. So for every orbit there is an associated path. Conversely, it can be shown that for every path in Γ there is an orbit of A which follows that path. However, some points, namely those whose orbit at some time falls on a partition boundary, can be represented by more than one path. To preserve a 1–1 correspondence between orbits of A and paths in Γ, we need to identify certain paths. So the dynamics is equivalent to a quotient of a Markov shift. It is important to note, however, that the identifications are small : they are at most 4–1, only finitely many periodic paths are identified, and the set of points on $\mathbb{T}^2$ with ambiguous path has measure zero.

We say a map has *symbolic dynamics* if it has a subset Λ on which f is semiconjugate to a Markov shift or a quotient with "small" identifications in some sense, which needs to be made precise according to the context.

4.3. Three levels of chaos for torus maps

(i) Contractible chaos

The simplest form of chaos on $\mathbb{T}^2$ is an invariant set Λ with positive topological entropy and whose rotation set is a single point. It can have symbolic dynamics, but each symbol represents the same amount of translation.

For example, most resonance regions have double eigenvalue +1 points, as indicated in Section 3.1, and generically from these issue a homoclinic wedge in which the map has contractible chaos of the form shown in Figure 15.

(ii) Rotational chaos

This is characterised by an invariant set Λ, whose rotation set is an interval I on a commensurate line, with a semi-conjugacy to a quotient of Markov shift, containing periodic orbits for each rational point of I, with distinct symbol sequences. For example, the situation of Figure 46 leads to rotational chaos.

We obtain a set $\tilde{\Lambda} \subset Q$ with symbolic coding a $\in 2^{\mathbf{Z}}$ such that :

$$f^i(x) \text{ returns to } Q \text{ in } \quad N \text{ steps} \quad \text{if} \quad a_{i+1} = 0$$
$$\text{in} \quad N+1 \text{ steps} \quad \text{if} \quad a_{i+1} = 1$$

Let $\Lambda = \bigcup_i f^i \tilde{\Lambda}$, hence the rotation set is the interval $[\,1/N+1, 1/N\,]$ (cf [ACHM]).

In fact, by allowing sequences a $\in \{N, N+1, N+2, ...\}^{\mathbf{Z}}$, we have that for each such sequence a there is a point in Q whose orbit takes a_1 steps for the first return to Q, then a_2 steps for the second, and so on. Hence we obtain a set whose rotation set is $[\,0, 1/N\,]$.

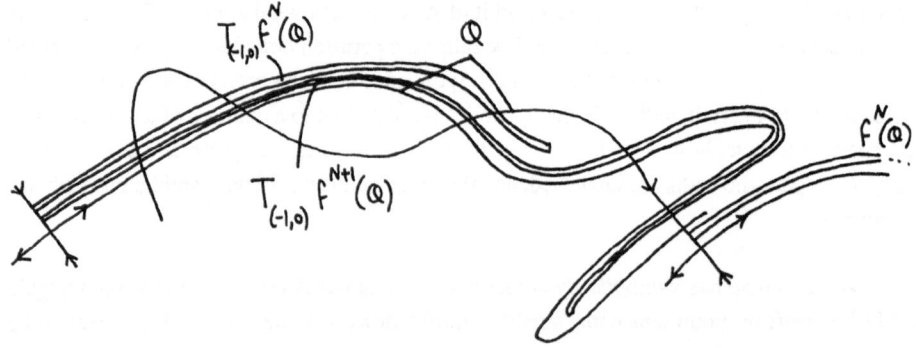

Figure 46. Transverse rotational homoclinic orbit leads to rotational chaos.

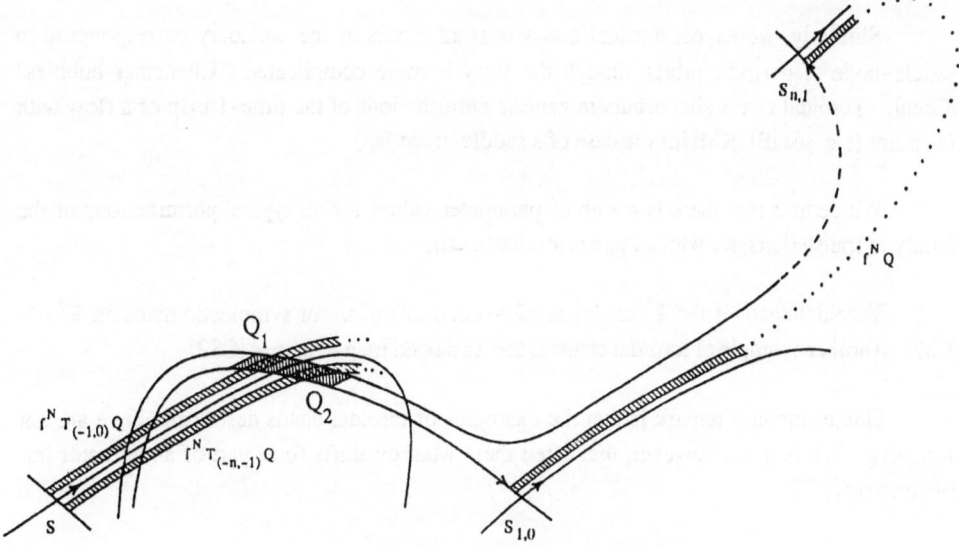

Figure 47. A situation leading to toroidal chaos.

(iii) Toroidal chaos

This level is characterised by an invariant set Λ with $\text{Int}(\rho(\Lambda,f)) \neq \emptyset$, and a semi-conjugacy to a quotient of a Markov shift with periodic points for every rational point of $\rho(\Lambda,f)$, with distinct symbol sequences.

For example, the situation of Figure 47 leads to toroidal chaos. The rectangle $Q = Q_1 \cup Q_2$ is mapped into a long thin strip after a few iterations, which when translated back by $T_{(-1,0)}$ or by $T_{(-n,-1)}$, intersects Q again, after N_1, N_2 iterations, respectively. From this we obtain a set Λ with toroidal chaos. Its rotation set is shown in Figure 48.

This situation arises for all generic perturbations of the time-1 maps of flows on the part of the boundary of partial mode-locking corresponding to rotational homoclinic connections. So all these parts of the boundary of partial mode-locking for the flow approximation generically fatten into strips of toroidal chaos.

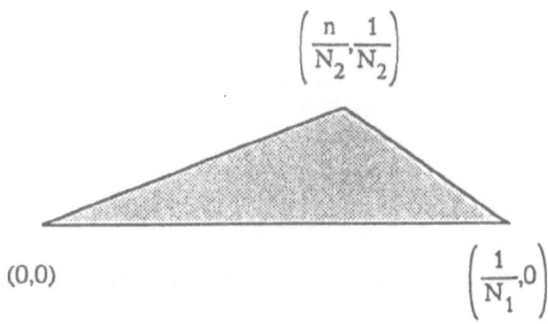

Figure 48. The rotation set for the set Λ arising from Figure 46.

Similarly, we expect toroidal chaos near all points of the boundary corresponding to saddle-node of periodic orbits, though the story is more complicated ("Chenciner bubbles" [Chen]). Toroidal chaos also occurs in generic perturbations of the time-1 map of a flow with fan point (e.g. see [BGKM] for the case of a saddle-node fan).

We deduce that there is a web of parameter values for all typical perturbations of the family of translations, for which there is toroidal chaos.

Toroidal chaos is the $\mathbb{T}^2$ analogue of Arnol'd diffusion for symplectic maps on $\mathbb{T}^2 \times \mathbb{R}^2$ [Chir]. Another example of toroidal chaos is the Zaslavski map with q=4 [CSZ].

One cautionary remark is that the examples of toroidal chaos described above are not attracting. We suspect, however, that often these Markov shifts form part of an attractor (cf. Hénon map [BC]).

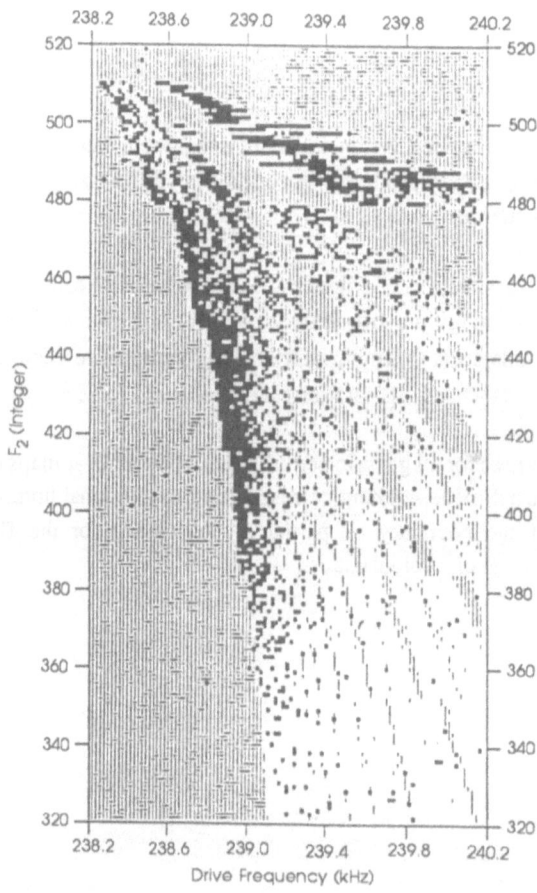

Figure 49. A parameter scan for an electronic circuit. Periodicity is indicated by horizontal lines, QP_2 by vertical lines, and chaos by black squares. QP_3 is left blank (from [LC]).

We believe toroidal chaos is a very significant form of chaos for 3-frequency systems. We conjecture, for example, that the strips of chaos observed by [LC] in Figure 49 are toroidal chaos, and the fan could be a saddle-node or half-plane fan.

4.4. Int $\rho(F) \neq \emptyset \Longleftrightarrow$ toroidal chaos

We close by discussing a nice criterion for toroidal chaos.

Theorem : Int $\rho(F) \neq \emptyset \Longleftrightarrow$ toroidal chaos

Step 1 : Int $\rho(F) \neq \emptyset \Rightarrow \exists$ periodic points for each rational $p/q \in$ Int $\rho(F)$, [Fra].

The proof uses the idea of chain recurrence :

Say x is *chain recurrent* if $\forall \, \varepsilon > 0 \, \exists \, n > 0$ and sequence $x = x_0,...,x_n = x$ such that

$$d(x_{k+1}, F(x_k)) < \varepsilon , \quad k = 0,...,n-1$$

lemma : If F is a lift of annulus or torus map and has a chain recurrent point then it has a fixed point.

To prove step 1, apply this lemma to $F^q T_{-p}$

Step 2 [LM]: Choose 3 periodic orbits of types (p_i, q_i), $i = 1, 2, 3$, whose rotation vectors p_i/q_i are not collinear, with (p_i, q_i) in lowest terms.

Remove the 3 periodic orbits, obtaining a torus with punctures, which we denote by $\mathbb{T}^2_P$, and restrict f to $\mathbb{T}^2_P$ obtaining a map $f_P : \mathbb{T}^2_P \mapsto \mathbb{T}^2_P$. Using the Nielsen-Thurston classification of surface homeomorphisms up to isotopy [FLP], one can show that f_P is isotopic to a pseudo-Anosov map φ. Let us not state the definition, but the important properties are :

(i) φ is conjugate to a quotient of a Markov shift.
(ii) f_P isotopic to φ implies f_P has an invariant subset Λ on which it is semi-conjugate to φ.

From step 1 and convexity of $\rho(F)$, φ has periodic orbits of all rotation vectors in the convex interior of $\{p_i/q_i : i = 1, 2, 3\}$. Hence f has toroidal chaos.

5. POSTSCRIPT

In these lectures, we have glimpsed the richness of behaviour of families of torus maps. Our theoretical framework allows us to understand diagrams like Figures 11, 12, and 49, and to make predictions of the features one should expect to see for systems of three weakly coupled oscillators.

There are many questions remaining, however, like unfolding the flows on the boundary of partial mode-locking when the circle map has flats and jumps, classifying and unfolding all the codimension-2 bifurcations for flows on $\mathbb{T}^2$, understanding the way the bifurcation diagram changes on perturbing the time-1 map of a saddle-node of periodic orbits to a generic diffeomorphism, understanding the large-scale structure of the mode-locking web for diffeomorphisms, and proving existence of toroidally chaotic attractors. It is a fertile field for further research.

The applications to physics and other sciences are also wide open. While in my opinion, it is unlikely that flows on $\mathbb{T}^3$ have much to do with the transition to turbulence (see [Mac]), there are many situations involving coupled modes of oscillation to which this theory could be relevant. One cautionary remark, however, is that torus break-up needs adding into the set of possible behaviours, and this has been found to happen before much appreciable chaos is observed [BGOY]. The phenomena that can result are already very complicated for two coupled oscillators, e.g. [LeC, Cas].

A major challenge for the future is to extend the theory to higher dimensions. Many of the concepts (such as the rotation set ρ, mode-locking, levels of chaos) generalise to diffeomorphisms of $\mathbb{T}^d$, $d > 2$, but many of the results do not (e.g. convexity of ρ, periodic points for $p/q \in \text{Int } \rho$, chaos for $\text{Int } \rho \neq \emptyset$). The potential applications are vast, e.g. networks of oscillators, synchronisation of power stations, instabilities in spatially extended systems, and the workings of the brain.

Acknowledgements

I would like to acknowledge fruitful collaborations on this subject with Claude Baesens, John Guckenheimer, Seunghwan Kim and Jaume Llibre, and the UK Science and Engineering Research Council for making these collaborations possible. I would also like to thank Paul Linsay, Jerry Gollub and Albert Libchaber for permission to reproduce their figures. Lastly I would like to thank Anamaria Oliveira for her assistance in writing up these notes, and Claude Baesens for a critical reading of the manuscript.

REFERENCES

[Arn]   Arnol'd VI, Geometric methods in the theory of Ordinary Differential Equations, Springer (1983).

[ACHM] Aronson DG, Chory MA, Hall GR, McGehee RP, Bifurcations from an invariant circle for two-parameter families of maps of the plane: a computer assisted study, Commun Math Phys 83 (1982) 303-354.

[BGKM] Baesens C, Guckenheimer J, Kim S, MacKay RS, Three coupled oscillators: mode-locking, global bifurcation and toroidal chaos, Physica D, to appear.

[BGKM2] Baesens C, Guckenheimer J, Kim S, MacKay RS, Simple resonance regions of torus diffeomorphisms, IMA Conf Proc series (Springer), to appear.

[BGOY]   Battelino PM, Grebogi C, Ott E, Yorke JA, Chaotic attractors on a 3-torus and torus breakup, Physica D 39 (1989) 299-314.

[BC]   Benedicks M, Carleson L, The dynamics of the Hénon map, preprint

[BH]   Biswas DJ, Harrison RG, Experimental evidence of three-mode quasiperiodicity and chaos in a single longitudinal, multi-transverse mode cw CO_2 laser, Phys Rev A 32 (1985) 3835-7.

[BMS]   Bogoljubov NN, Mitropolskii JuA, Samoilenko AM, Methods of accelerated convergence in nonlinear mechanics (Springer, 1976).

[Boyd]   Boyd C, On the structure of the family of Cherry fields on the torus, Ergod Th Dyn Sys 5 (1985) 27-46.

[Cas]   Casdagli MC, Periodic orbits for dissipative twist maps, Erg Th Dyn Sys 7 (1987) 165-173.

[Chen]   Chenciner A, Bifurcations de points fixes elliptiques III.-Orbites périodiques de petites périodes et élimination résonnante des couples de courbes invariantes, Publ Math IHES 66 (1988) 5-91.

[Cher]   Cherry TM, Analytic quasi-periodic curves of discontinuous type on a torus, Proc Lond Math Soc 44 (1938) 175-215.

[Chir]   Chirikov BV, A universal instability of many oscillator systems, Phys Repts 52 (1979) 265-379.

[DG]   Dangelmayr G, Guckenheimer J, On a four parameter family of planar vector fields, Arch Rat Mech Anal 97 (1987) 321-352.

[FLP]   Fathi A, Laudenbach F, Poénaru V, Travaux de Thurston sur les surfaces, Astérisque 66-67 (1979).

[Fra]   Franks J, Realising rotation vectors for torus homeomorphisms, Trans Am Math Soc 311 (1989) 107-115.

[FM]   Franks J, Misiurewicz M, Rotation sets of toral flows, preprint.

[Fri]   Fried D, The geometry of cross sections to flows, Topology 21 (1982) 353-371.

[GB]   Gollub J, Benson SV, Many routes to turbulent convection, J Fluid Mech 100 (1980) 449-470.

[GH]   Guckenheimer J, Holmes PJ, Nonlinear Oscillations, Dynamical Systems, and Bifurcations of Vector Fields, Springer (1983).

[H]   Herman MR, Sur la conjugation differentiable des difféomorphismes du cercle à des rotations, Publ Math IHES 49 (1979) 5-234.

[HJ]   Hollinger F, Jung C, Single longitudinal-mode laser as a discrete dynamical system, J Opt Soc Am B 2 (1985) 218-225.

[K]   Katok A, Lyapunov exponents, entropy and periodic orbits for diffeomorphisms, Publ Math IHES 51 (1980) 137-173.

[KMG]   Kim S, MacKay RS, Guckenheimer J, Resonance regions for families of torus maps, Nonlinearity 2 (1989) 391-404.

[IL]   Iooss G, Los J, Quasi-genericity of bifurcations to high dimensional invariant tori for maps, Commun Math Phys 119 (1988) 453-500.

[Lang]   Langford WF, Periodic and steady-state mode interactions lead to tori, SIAM J Appl Math 37 (1979) 22-48.

[LeC]   Le Calvez P, Propriétés des attracteurs de Birkhoff, Ergod Th Dyn Sys 8 (1987) 241-310.

[LC]   Linsay PS, Cumming AW, Three-frequency quasiperiodic, phase-locking and the onset of chaos, Physica D 40 (1989) 196–217.

[LM]   Llibre J, MacKay RS, Rotation vectors and entropy for homeomorphisms of the torus isotopic to the identity, Erg Th Dyn Sys, to appear.

[Mac]   MacKay RS, An appraisal of the Ruelle–Takens route to turbulence, in The Global geometry of Turbulence, ed. Jimenez J (Plenum), to appear.

[MT]   MacKay RS, Tresser C, Transition to topological chaos for circle maps, Physica D 19 (1986) 206–237; and Erratum, Physica D 29 (1988) 427.

[Mar]   Markley NG, The Poincaré–Bendixson theorem for the Klein bottle, Trans Am Math Soc 135 (1969) 159–165.

[ML]   Maurer J, Libchaber A, Effect of the Prandtl number on the onset pof turbulence in liquid ^{4}He, J Physique Lett 41 (1980) L515–8.

[MZ]   Misiurewicz M, Ziemian K, Rotation sets for maps of tori, preprint.

[NRT]   Newhouse SE, Ruelle D, Takens F, Occurrence of strange Axiom-A attractors near quasiperiodic flows on $\mathbb{T}^m$, $m \geq 3$, Commun Math Phys 64 (1978) 35–40.

[Ox]   Oxtoby JC, Ergodic sets, Bull Am Math Soc 58 (1952) 116–136.

[PdM]   Palis J, de Melo W, Geometric theory of dynamical systems, Springer (1982).

[RhT]   Rhodes F, Thompson CL, Topologies and rotation numbers for families of monotone functions on the circle, preprint.

[Ru]   Ruelle D, Chaotic evolution and strange attractors (CUP, 1989).

[RT]   Ruelle D, Takens F, On the nature of turbulence, Commun Math Phys 20 (1971) 167–192; and 23 (1971) 343–4.

[Sell]   Sell GR, Resonance and bifurcation in Hopf-Landau Dynamical Systems, in Nonlinear Dynamics and Turbulence, ed Barenblatt GI, Iooss G, Joseph DD, Pitman (1985).

[Shub]   Shub M, Global Stability of dynamical systems, Springer, 1987.

[Soto]   Sotomayor J, Generic bifurcations of dynamical systems, in Dynamical Systems, ed Peixoto MM (1973) 549–560.

[Wal]   Walters P, An introduction to ergodic theory, Springer, 1982.

[W]   Weil A, Les familles de courbes sur le tore, Mat Sb 43 (1936) 779–781.

Quasiperiodicity, Mode-Locking, and Universal Scaling in Rayleigh-Bénard Convection

Robert E. Ecke

Physics Division and Center for Nonlinear Studies
Los Alamos National Laboratory, Los Alamos, NM 87545

Abstract

Quasiperiodicity, mode-locking and universal scaling dynamics are described for Rayleigh-Bénard convection in a dilute solution of ^{3}He in superfluid ^{4}He. Examples from experimental data are used to illustrate analysis techniques of nonlinear dynamics: power spectra, phase space reconstruction, Poincaré sections, transients and stability eigenvalues, return maps, multifractal $f(\alpha)$ analysis, and scaling function dynamics. Using these tools we show that the route to chaos in this system of two intrinsic oscillatory modes has the same universality as the sine circle map.

Introduction

These lectures on quasiperiodicity and mode-locking in a model fluid system, Rayleigh-Bénard convection, will review the theory of quasiperiodicity derived from 1-D circle maps, discuss how this perspective fits with the classical view of mode-locking, and illustrate many techniques of dynamical systems, applied in a real experiment to obtain a detailed understanding of experimental data. In particular, Sec. 1 deals with simple physical models of mode-locking and with the theory of the generic 1-D map model of mode locking, the sine circle map. An historical perspective is presented on experiments on quasiperiodicity, mode-locking and chaos with an emphasis on fluid experiments. In Sec. 2, the experimental work and basic analysis techniques of convection in a small-aspect-ratio cell using a ^{3}He-superfluid-^{4}He mixture are discussed. Comparisons of fundamental ideas of mode locking with experiment will then be presented: phase angles, convective amplitudes, 1-D return maps and demonstration of the tangent bifurcation to mode locking. Another interesting and promising tool in experimental dynamical systems, the study of transients, will be illustrated. Section 3 will concentrate on analysis tools for making comparisons between the theoretical predictions of the universal theory of quasiperiodicity and experimental data. This will focus on understanding the dynamics of the attractor in the language of multifractal thermodynamics and the scaling function formalism. This "statistical mechanical" approach yields effective algorithms for analyzing data.

Chaos, Order, and Patterns, Edited by R. Artuso *et al.*
Plenum Press, New York, 1991

1 Quasiperiodicity and Mode Locking

Frequency entrainment of two nonlinear oscillators was first observed by Christian Huyghens in 1665 when he noticed that clocks set on a shelf would synchronize their motion [1]. The simple example of pendulum-style clocks is an intuitive way to understand the basic elements of frequency entrainment [2]. Consider two uncoupled clocks, each composed of a pendulum motion and an energy input to overcome dissipation. In Fig. 1, if the upper plate is rigidly supported then the motions of the two clocks are uncoupled. The crucial element of the oscillation is that it is nonlinear, i.e., that the oscillation frequency depends on the amplitude of the motion. In the case of a clock the nonlinearity comes primarily from the impulses of the rachet mechanism which sustains the motion. The individual oscillators are in limit cycle states. The phase space of the two oscillators is composed of the amplitude and phase of each oscillator and is thus four dimensional. When the amplitudes are only weakly perturbed by the coupling (i.e. slaved to the phase) it is adequate to consider the motions of the system as being confined to the surface of a torus imbedded in a 3D phase space, Fig. 2. In the absence of coupling one motion winds around the large amplitude (major) axis with amplitude A and phase $\phi_A = \omega_A t$. The other motion winds around the minor axis with amplitude $B < A$ and with $\phi_B = \omega_B t$. We denote Ω as the ratio of the frequencies of the uncoupled oscillators $\Omega \equiv \omega_B/\omega_A$. For almost all (measure one) values of Ω the motion on the torus is ergodic, i.e. the trajectory of the motion eventually passes through all points on the torus. An equivalent statement is that the frequencies are incommensurate (not the ratio of two integers) and that the motion is not periodic but quasiperiodic.

The two oscillators can now be coupled together by, for example, a platform that is allowed to move, Fig. 1. The theorem of Peixoto [3] guarantees that for any type and strength of coupling the two oscillators will mode lock, the locking will consist of q pairs of stable and unstable fixed points for a period q cycle, and the mode-locked interval is structurally stable. Mode locking (entrainment, resonance) means that a particular periodic trajectory with frequency ratio p/q exists over a finite interval in Ω whereas in the uncoupled case the periodic state only exists for the precise value $\Omega = p/q$. The widths of mode-locked intervals are very small for small coupling, however, and Peixoto's theorem does not give much in the way of physical intuition. In order to develop some physical intuition we consider a simpler model of mode-locking consisting of one clock driven by a periodic forcing, Fig. 3. For simplicity we confine the discussion to entrainment at the drive frequency, i.e., the 1:1 resonance. When the drive is precisely tuned to the natural frequency ω_o of the clock there are three periodic points to this dynamical system. The clock without forcing has an unstable fixed point corresponding to no motion (you have to give it a little kick to get it going). This state becomes a periodic repellor with small oscillation amplitude as forcing is applied. There are two other periodic states associated with entrainment of the limit cycle. The attracting periodic point corresponds to in-phase motion of the drive and the pendulum, while the unstable point is associated with out-of-phase motion. For weak forcing the in-phase drive slightly increases the amplitude of the pendulum while the out-of-phase motion decreases the amplitude a little. The stability boundary of the locking can be understood by considering the frequency response curve of the oscillator at fixed drive amplitude, Fig. 4. As the drive is detuned away from ω_o, less amplitude is gained by the in-phase oscillation and less is lost in the out-of-phase mode. At some detuning boundary the oscillator gains (or loses) no net amplitude relative to the undriven oscillation. This marks the end of the mode-locked interval. Another thing to note is that the phase difference between the oscillators changes as

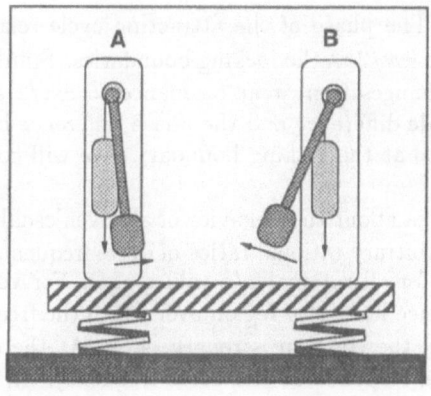

Figure 1. Schematic illustration of coupled pendulum clocks.

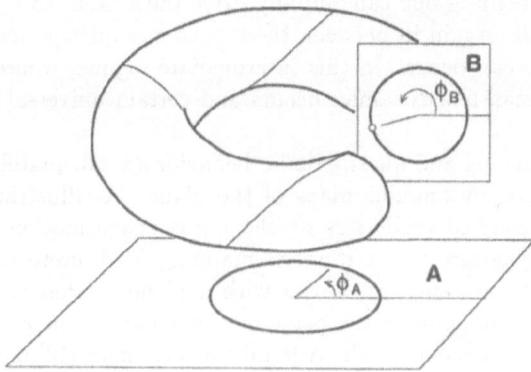

Figure 2. Schematic illustration of the phase space of weakly coupled oscillators. Trajectories lie on a 2-D torus.

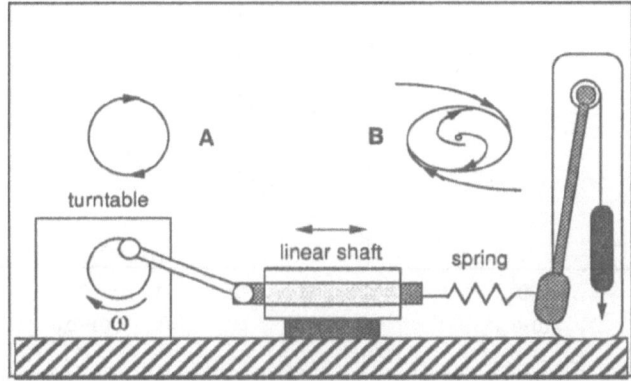

Figure 3. Schematic illustration of a driven pendulum clock. Phase portraits of the drive (A) and the clock with no coupling (B) are shown.

the system is detuned. The phase of the attracting cycle relative to the drive goes from zero at resonance to $\pm\pi/2$ at the locking boundaries. Similarly the relative phase for the unstable cycle changes from π at resonance to $\pm\pi/2$ at the boundaries. In other words the amplitude difference and the phase difference between the stable and unstable cycles go to zero at the locking boundary. We will come back to this point later in section 2.

We now have some idea about the behavior of a driven oscillator at low amplitude. Similar ideas apply for arbitrary rational ratios of drive frequency to natural frequency so that resonances exist for all rational p/q values of Ω. For very small coupling, the width of a q-cycle resonance horn (the region over which the frequencies are entrained) varies as K^q where K is the coupling strength [4,5]. At the other extreme of very strong forcing everything responds at the drive frequency. In between there can be complex dynamics and complicated structure of resonance horns. The whole range of such behavior has been demonstrated by numerical studies of forced ODEs [6,7] and in experiments on forced magnetic oscillators [8]. In the case of two intrinsic oscillators the situation is more complex. Recent numerical work on coupled chemical reactions [9] shows that for large coupling one or the other oscillators becomes dominant while in certain parameter regimes one can actually drive the system to a fixed point, that is no oscillations at all. Again in between these extremes quasiperiodic, mode-locked and chaotic dynamics can occur. In this intermediate regime, where the coupling is not too great, circle-map-like dynamics occurs and certain universal scaling behavior is expected.

In general mode-locking and quasiperiodic behavior are adequately modeled, from a dynamics perspective, by annulus maps of the plane. As illustrated in Fig. 2 we can describe weakly coupled oscillators as the phase space motion on a torus. To convert continuous dynamics to an iterated mapping in discrete time, one studies the intersections of the system trajectories with a plane (a Poincaré section). The resultant points form a mapping of the plane onto itself or, in the case of small radial perturbations, of an annulus onto itself. A family of such maps [10] is $P_{\Omega,K,\lambda}(r,\theta)$ given by

$$r_{n+1} = 1 + \lambda(r_n - 1) - \frac{K}{2\pi}\sin(2\pi\theta_n),$$

$$\theta_{n+1} = \theta_n + \Omega + \lambda(r_n - 1) - \frac{K}{2\pi}\sin(2\pi\theta_n).$$

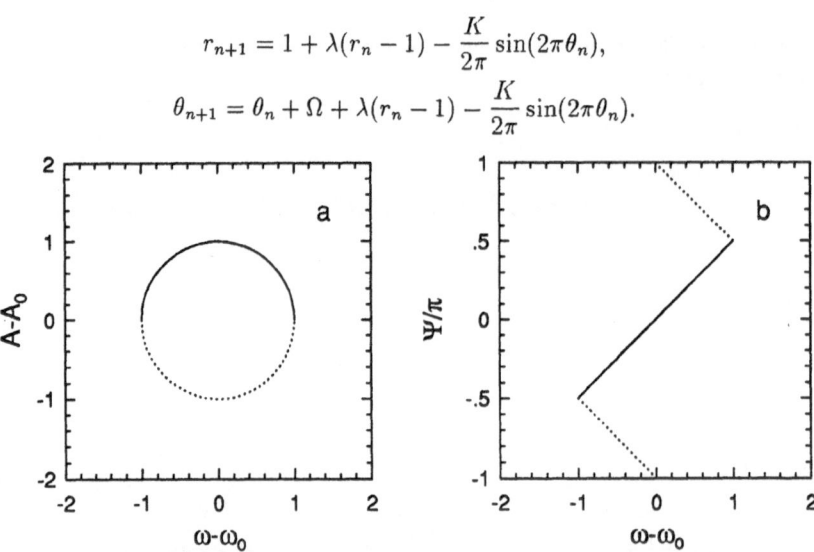

Figure 4. Schematic frequency response curves for the a) amplitude deviation A-A$_0$ (A$_0$ is the unperturbed amplitude) and b) phase difference (normalized by π) of the driven clock plotted vs. detuning parameter $\omega - \omega_o$. The solid curve is for the attracting cycle and the dashed curve for the unstable cycle.

In the case that the radial contraction is infinite (or very large) then $\lambda = 0$ and the equations reduce to the extensively studied sine circle map:

$$\theta_{n+1} = \theta_n + \Omega - \frac{K}{2\pi}\sin(2\pi\theta_n), \, mod\, 1. \qquad (1)$$

This mapping has two control variables: Ω the bare winding number and K which represents the nonlinearity/coupling. Many excellent papers describe the fractal structure of the mode-lockings [11,12,13] the renormalization group theory of universal scalings [10,14,15], and other features of circle map behavior [16]. The basic features are presented here.

The circle map (1) is represented graphically by plotting θ_{n+1} versus θ_n as in Fig. 5. For $K < 1$ this is an invertible mapping; the mapping is single valued for forward and backward iterations. As K is increased from zero at fixed Ω the ratio of the two motions, the rotation number, defined as

$$\rho \equiv \lim_{n\to\infty}\left[\frac{\theta_n - \theta_o}{n}\right],$$

deviates from Ω. In the case of mode locking, ρ takes on the value of some rational ratio p/q, where p,q are integer, over some finite interval in Ω. The condition for mode locking to a p/q resonance in some general 1-D circle map, $\theta_{n+1} = f(\theta_n)$, is that $f^q(\Omega_q^p, x_q) = p+x_q$, where x_q is a periodic point of a q-cycle, Ω_q^p is a value of Ω for which $\rho = p/q$, and f^q is the qth iterate of the mapping function f. One can see graphically what this condition means by plotting the qth iterate map, θ_{n+q} versus θ_n, see Fig. 5. At either locking boundary the mapping becomes tangent to the $\theta_{n+q} = \theta_n$ line at q points. This is called a tangent bifurcation and is the transition between quasiperiodic and mode-locked states. For values of Ω between the boundaries there are q stable periodic points and q unstable periodic points, the stability depending on the slope of the curve through the $\theta_{n+q} = \theta_n$ line.

There are two main features of the circle map in the K, Ω parameter space. One is the tongue structure of resonance horns as K is varied, Fig. 6. At $K = 0$ the irrationals have measure one on the unit interval and the rationals, which make up the resonance intervals, have zero width. The measure of irrational holes, μ decreases, Fig. 7, as the resonance widths increase until at $K = 1$ the mode-locked intervals form a devil's staircase, Fig. 8, with fractal dimension d_F=0.870 [11,12]. In the subcritical interval $0 < K < 1$ the set of p:q resonances is not a "thin" fractal but instead obeys

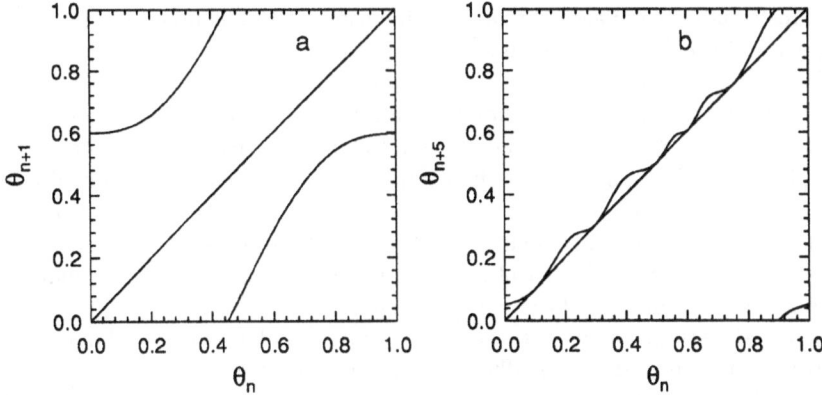

Figure 5. Discrete mappings of the circle from the sine circle map with K=0.95 and p/q=3/5 at the tongue boundary: a) θ_{n+1} vs. θ_n and b) θ_{n+5} vs. θ_n.

so-called fat fractal scaling [17,18]. A fat fractal has finite measure but still has scaling at all sizes given by $\mu(\epsilon) = \mu(0) + A\epsilon^\beta$ where β is the fat fractal exponent and ϵ the coarse graining size. From numerical experiments it is conjectured that $\beta = 2/3$ for all subcritical staircases [5]. The transition to fractal scaling at K=1 comes from the development of a cubic inflection point in the circle map. This inflection point also produces universal scaling dynamics at special irrational winding numbers. These special irrationals can be written as an infinite continued-fraction expansion of the form

$$\rho = \cfrac{1}{n + \cfrac{1}{m + \cfrac{1}{r + \dots}}}, \tag{2}$$

where $n, m, r, \dots$ are integers. This series can be represented by a sequence of numbers written as $[n, m, r, \dots]$. The simplest such expansion is for all the terms to be one, the sequence $[1, 1, 1, \dots]$. One can sum this series and obtain the so called "golden mean" winding number $\rho_g \equiv (\sqrt{5} - 1)/2$. The theoretical results for this winding number are the ones most often compared with experiment. Any winding number which has the same asymptotic expansion of ones should also obey the predicted scalings but convergence will be slower.

Experiments on nonlinear dynamical systems prompted the development of many of the theoretical models described here. We briefly review those original experiments and discuss the more recent experiments on forced nonlinear oscillators. Many of these results are descirbed in more detail in reference [19]. Mode-locking was first observed quantitatively in fluid experiments where a sequence of Hopf bifurcations resulted in states with two incommensurate frequencies [20,21,22]. Subsequent theoretical work attempting to understand elements of mode-locking in these experiments resulted in the universal theory of golden-mean scalings and the fractal "devil's" staircase. Many of the universal predictions, however, require the simultaneous tuning of two experimental parameters to obtain both criticality and a rotation number at the golden mean. The systems originally studied did not have multiple control parameters which were readily accessible. As a result researchers adopted the approach of producing a stable limit cycle in a nonlinear system and forcing it externally with a periodic signal of adjustable amplitude and frequency. Such a system is more likely to be a close approximation to the circle map model because radial contraction to a stable limit cycle can be tuned by varying the control parameter of the nonlinear system. A multitude of such experiments ensued with systems of fluid convection [23,24,25], electronic materials [26,27,28,29,30, 31], open fluid flows [32], laser oscillations [33], and recently thermoacoustic oscillations [34,35]. Most studies compared the fractal dimension of the locking intervals and the power spectra of the dynamics [10,36]. The scaling exponent δ [14] is somewhat harder to obtain experimentally but was measured quite accurately in the convection work of Stavans et. al. [24,25]. Another characterization of the universal circle map scaling which focuses on the dynamics instead of the resonance widths is the multifractal $f(\alpha)$ analysis [37]. The $f(\alpha)$ analysis was first applied to circle map data from Rayleigh-Bénard experiments with great success [38]. All forcing experiments [39] report good agreement with the predictions of circle map scaling. From these experiments one can conclude that the critical behavior of many externally-forced nonlinear oscillators has the same universal scaling as the sine circle map.

The originally studied experiments of two intrinsic oscillators received much less attention owing to the experimental difficulties. The case for two independent oscillations is more complex, in principle, because each oscillator can adjust to the dynamics of the other; an additional degree of freedom is available. In addition, the size of the radial

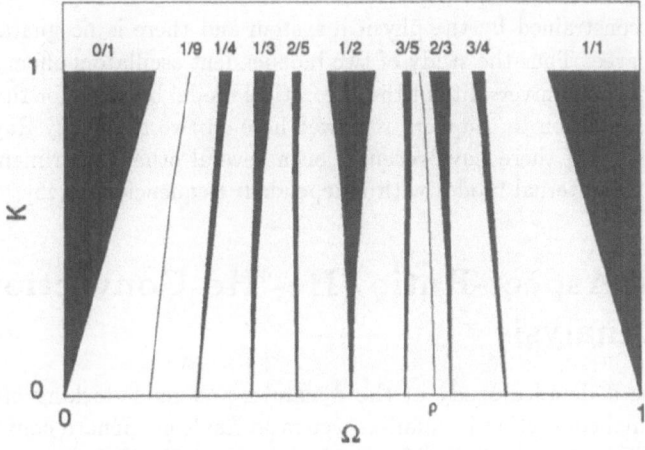

Figure 6. Regions of mode-locking for the sine circle map in the K, Ω space; only large lockings are shown. The golden mean line is labeled ρ.

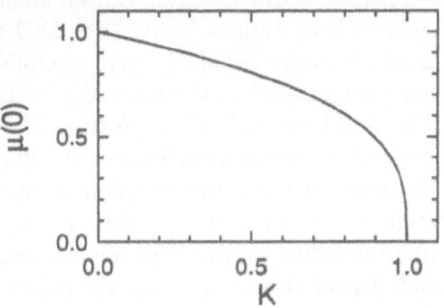

Figure 7. Irrational measure of holes $\mu(0)$ vs. nonlinearity parameter K.

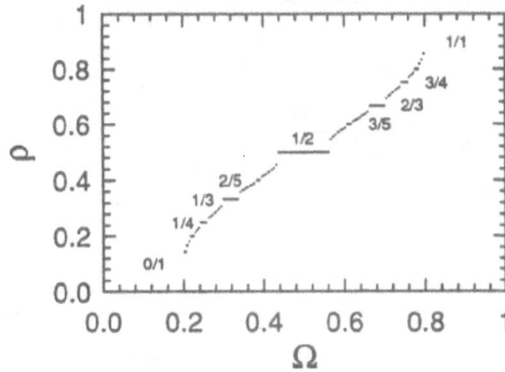

Figure 8. Winding number ρ vs. bare winding number Ω showing a theoretical devil's staircase of mode-locked intervals. Large lockings are labeled.

dissipation is constrained by the physical system and there is no guarantee that the dissipation is large. Thus the study of two independent oscillators offers a more severe test for not only the universality of the theoretical model but also for the genericity of the results. In addition to the work reviewed here on two frequency Rayleigh-Bénard convection [41,42,43] there have recently been several other experimental studies of systems with two internal modes with independent frequencies [44,45].

2   Small-Aspect-Ratio ^{3}He-^{4}He Convection: Data and Analysis

We have made a detailed study of the dynamics and mode-locking of two intrinsic modes of thermal convection in small-aspect-ratio Rayleigh-Bénard convection in a dilute solution of ^{3}He in superfluid ^{4}He [41,42,43,46,47,48,49,50]. In the parameter space of Rayleigh number R, a dimensionless control parameter proportional to the temperature difference across the fluid layer, and Prandtl number Pr, another dimensionless parameter which is a strong function of the mean cell temperature, many regions of different convective behavior exist, Fig. 9. The superfluid nature of the mixture does not play an important role here and we can consider it to be a classical single component fluid. Details and references to this assertion can be found in references [51,52]. For small R, the fluid conducts heat diffusively. At R $\simeq$ 2000 there is a supercritical (continuous) transition to steady convection [51] where the fluid motion is believed to be two convection rolls oriented parallel to the short side of the rectangular convection cell (height 0.8 cm, length 2.1 cm and width 1.6 cm). The actual flow is inferred from room temperature experiments since visualization of helium fluid convection has not yet been realized. The state of the convective flow is partially determined from measurements of the spatially-averaged convective heat transport. Another probe of the state of the system, more sensitive to time dependence, consists of a local temperature transducer. (In small aspect ratio convection the side walls severely constrain the spatial structure of the fluid flow and measurement of the dynamics of the system at a single point is adequate to characterize the dynamical state.) The local probe is

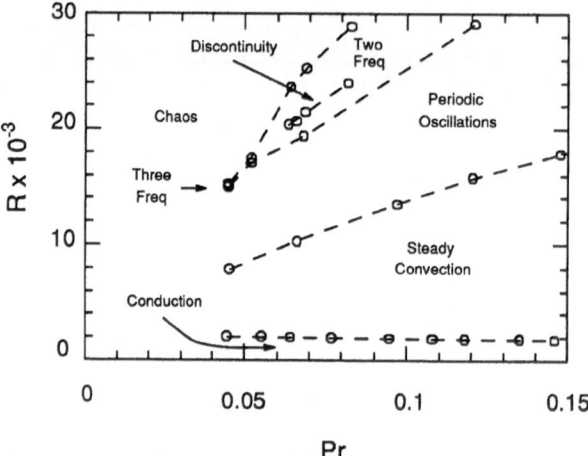

Figure 9. Phase diagram of convective states in parameter space of Rayleigh number and Prandtl number. Regions are described in the text.

a differential thermocouple with high temperature sensitivity, 0.3 $\mu K/\sqrt{Hz}$. Details of the experimental techniques can be found elsewhere [51]. The output of the probe, a voltage proportional to local temperature differences $\delta T(t)$, is digitized to produce a time series. Standard techniques of phase space reconstruction [53,54,55] are used to produce phase space trajectories of the dynamical system. Frequency spectra are also calculated from the time series using an FFT algorithm and standard windowing techniques [56]. The peak frequencies for periodic states are obtained by interpolation of the windowing function, thereby increasing the accuracy of the frequency determination over taking the peak value by as much as three orders of magnitude. Using these techniques we can characterize the evolution of dynamical states of the system as R is increased.

The first time dependence begins at the transition to periodic oscillations of frequency f_1. This transition is a forward Hopf bifurcation [57,58,59] and the onset value depends strongly on Pr. Another Hopf bifurcation at higher R gives rise to a second frequency f_2, incommensurate with the first. This second mode is only weakly interacting with the initial limit cycle mode and not until there is a discontinuous transition to a different second mode does measurable mode-locking occur. The physical mechanism for this transition is not known. Within a region of parameter space above the discontinuous transition, quasiperiodic (incommensurate frequencies), mode-locked and chaotic states exist. As in the theoretical discussions the state of the dynamics is often well represented by the winding number $W \equiv f_2/f_1$, where f_1 and f_2 are the fundamental frequencies determined from spectral analysis of time series data. (W is related to ρ in the theory as discussed below.) In Fig. 10 is a typical quasiperiodic time series and power spectrum in which many distinct peaks are observed. Each peak can be written as a sum, difference, or harmonic of the two fundamental peaks. In table 1 the fundamental frequencies and some of the other peaks are compared with the decomposition into combination frequencies. The mode-locking structure know as the "devil's" staircase is constructed from a sequence of such spectra as a function of R at fixed Pr. This structure is illustrated in Fig. 11 for experimental data and is qualitatively similar to the circle map version shown in Fig. 8. (Note that the range of experimentally accessible winding numbers, $1/8 < W < 2/11$, is substantially less than one.) By making a series of such measurements at different values of Pr the locking regions in the {R,Pr} parameter space are determined, Fig. 12. The tongues are seen

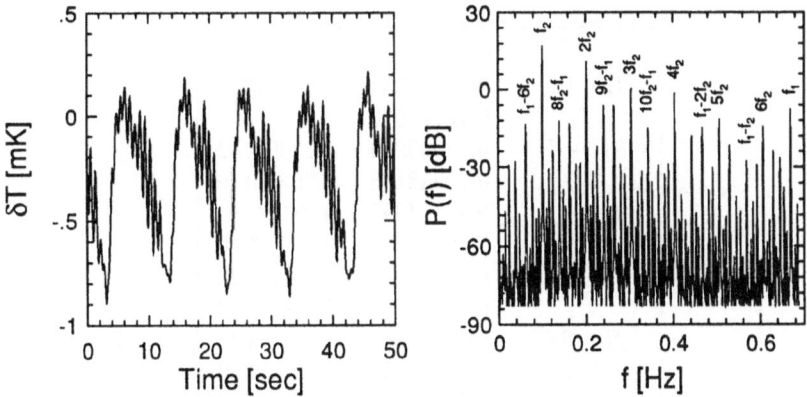

Figure 10: Quasiperiodic a) time series of temperature oscillations and b) corresponding power spectrum. Combination frequencies are labeled in the spectrum.

to broaden as Pr decreases (this is why we plot 1/Pr in Fig. 12) while the value of the winding number is controlled primarily by changing R. A simple mapping of the control variables R and Pr onto the circle map variables of K and Ω assigns R the role of Ω and Pr the part of $1/K$. In general one is not guaranteed such a direct relationship but it works well here.

The states which border the region where quasiperiodic and mode-locked states are observed are higher dimensional attractors for large R and simple quasiperiodic state with no mode-locking at small R [41,47,46,48,49,50]. In this review we will concentrate on the dynamical states that are not higher dimensional and that are well described by circle map models. An important characteristic of the theory is the critical line where critical scaling behavior is expected. In the more complex experimental parameter space of R and Pr one needs a way to define and locate such a line if it exists. Below we discuss how we accomplish this experimentally. First we describe simple quasiperiodic and mode-locked dynamics that occur in a region of parameter space in our fluid system that is far below criticality.

In our simple picture of mode locking and frequency entrainment developed in Sec. 1, there are several ideas that can be demonstrated with experimental data. One is the concept that there is energy pumping between modes as a mechanism for frequency locking. In Fig. 13 are three plots that illustrate this nicely. On the top is the winding number as a function of R/R_c showing locking to a W=2/13 resonance. In the second and third figures the amplitudes A_2 and A_1 of the fundamental frequencies f_2 and f_1 are shown. In the locking interval A_2 is boosted slightly relative to a baseline trend (shown as a solid line in the figure) while A_1 is sharply reduced; the motion of mode 2 is being increased at the expense of mode 1. The other idea is the importance of the phase of one oscillator relative to the other. Recall that in-phase motion on resonance is associated with the attracting state whereas out of phase motion is characteristic of the unstable cycle. It is straightforward to obtain the absolute phase at a particular frequency from a Fourier transform of the time series. When the resonance is not one to one, however, the phase difference is a little difficult to define. In order to

Table 1. Quasiperiodic frequency decomposition showing the deduced combination of the fundamental frequencies f_1 and f_2, the measured frequency and the calculated frequency for a given combination.

combination	f measured	f calculated
$f_1 - 6f_2$	0.06257355	0.06257358
f_2	0.10124070	——
$8f_2 - f_1$	0.13990820	0.13990782
$2f_2$	0.20248148	0.20248140
$9f_2 - f_1$	0.24114892	0.24114852
$3f_2$	0.30372200	0.30372210
$10f_2 - f_1$	0.34239028	0.34238922
$4f_2$	0.40496285	0.40496280
$f_1 - 2f_2$	0.46753636	0.46753638
$5f_2$	0.50620357	0.50620350
$f_1 - f_2$	0.56877589	0.56877708
$6f_2$	0.60744363	0.60744420
f_1	0.67001778	——

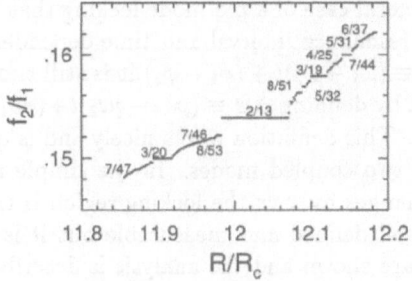

Figure 11. Frequency ratio f_2/f_1 vs. R/R_c showing experimental devil's staircase of mode-locked intervals for $1/Pr=14.9$, slightly above criticality. Prominent lockings are indicated.

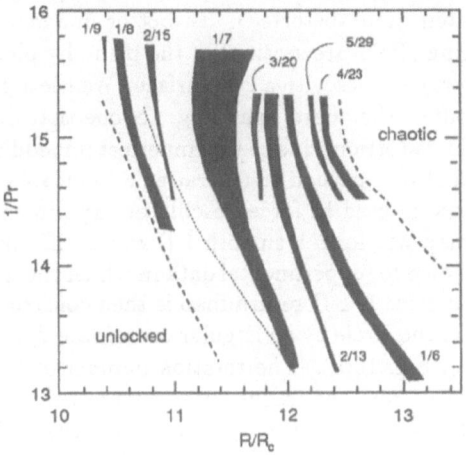

Figure 12. Experimental regions of mode locking in the $1/Pr$ and R/R_c parameter space. The dashed lines represent the hysteretic discontinuity at low R/R_c, increasing R/R_c (– – – –) and decreasing (---------), and the transition to high dimensional chaos at large R/R_c. For $1/Pr>14.6$ there is structure in the tongues not shown on this global plot.

understand this in the general case, we will first consider two oscillators with motions $\cos(\omega_2 t + \phi_2)$ and $\cos(\omega_1 t + \phi_1)$ where the absolute phases of the oscillations represent the uncertainty in the time origin. If the frequencies are equal then the phase difference is fixed, does not change with time and is a physically relevant quantity. This is true only so long as the frequencies are exactly equal since any small frequency difference will cause the phases to evolve according to $\Delta \phi = (\omega_2 - \omega_1) t$. We want to define a phase difference in the general case of a p:q mode locking that has the same properties of being fixed inside the resonance interval and time dependent outside. If we define the phase difference $\Delta \Psi \equiv (\omega_1 - \omega_2) t + (\phi_1 - \phi_2)$ it is still time dependent even if the state is mode-locked. But by defining $\Delta \Psi \equiv (p\omega_1 - q\omega_2)t + (p\phi_1 - q\phi_2)$ the coefficient of t is zero for a p:q locking. This definition works nicely and is quite useful in analyzing the interactions between two coupled modes. In the simple model described in Sec. 1, the phase difference changes by π as the locking region is traversed. In experiment this phase difference is well defined and measurable but it is not always equal to π. In [41] several examples are shown and the analysis is described in more detail. Here we illustrate the phase angle dependence of the data of Fig. 13. In Fig. 14 the phase difference is shown over the mode-locked region and spans a total range of about $\pi/2$. No theory is presently available that explains quantitative features of this phase angle difference. Phase differences constructed for the sine circle map also show deviations from a total change of π over the resonance region [41].

We now consider quasiperiodicity and mode locking from a dynamical systems perspective. One can construct the dynamical state of a system composed of two nonlinear, interacting modes of oscillation in a three dimensional phase space consisting of delay coordinates of the local probe temperature oscillations. The trajectories in this phase space lie on a manifold which is topologically equivalent to a 2-D torus (in practice the torus is often quite distorted). To better understand the dynamics one constructs a 2-D mapping (Poincaré section) of the plane by plotting the interpolated intersections of system trajectories with a 2-D surface. We use a planar surface oriented in such a way as to produce the "best" mapping, i.e. one that avoids crossings, kinks, etc. In this way we avoid distortions due to the imperfect imbedding of delay coordinate schemes [53,55]. Note that our method of generating a Poincaré section is more general than the strobing technique used in forced oscillator experiments. In the subcritical regime, this Poincaré map will form a smooth 1-D curve diffeomorphic to a circle. In Fig. 15 an example is shown for experimental data in which the dynamical state is very close to a period-2 resonance. A 1-D return map is then constructed by parametrizing the iteration of points on the circle by an angular coordinate θ, producing a circle map, θ_{n+1} as a function of θ_n, Fig. 16(a). The rotation number of this Poincaré section is different from W obtained from the frequency spectrum because of the ambiguity in choosing the plane of the Poincaré section that cuts the torus. It is operationally easier to produce a section by cutting through the minor axis of the torus thereby producing a rotation number ρ which is related to W by $\rho = 1/W$, mod 1. The state represented in Figs. 15 and 16 has $W \approx 2/13$ corresponding to $\rho \approx 1/2$. In this and all future Poincaré sections we shall refer to ρ as the appropriate winding number. To illustrate the tangent bifurcation structure of mode locking we plot the second iterate map for the data in Figs. 15 and 16 which are close to a period-2 locking. (For a period-q locking a tangent bifurcation will occur in the qth iterate map). This is illustrated in Fig. 16(b) for the experimental data.

Having demonstrated aspects of the circle map from experimental data we proceed to develop a more quantitative comparison based on the critical properties of the mapping. The first problem is how to define the critical line in the real experimental parameter space. In the circle map criticality comes about when the map becomes

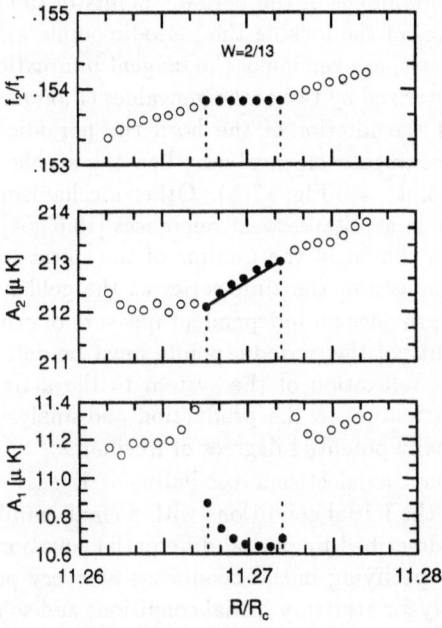

Figure 13. Convective mode amplitudes A_2 and A_1 and winding number f_2/f_1 vs. R/R_c for experimental data around the W=2/13 locking. Data inside (•) and outside (o) the locking region are labeled and the boundaries of the locking indicated (- - - - -).

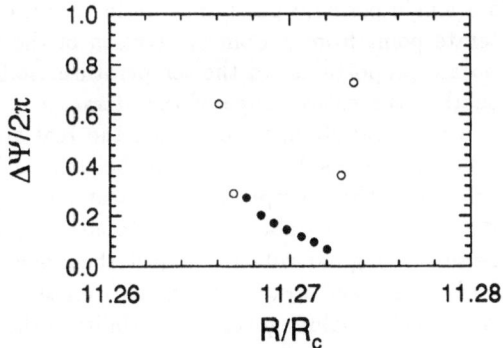

Figure 14. Phase angle difference $\Delta\Psi$ normalized by 2π vs. R/R_c for experimental data with points inside (•) and outside (o) the locking interval labeled.

noninvertible at a cubic inflection point. In the more general case of a 2-D map one can think of criticality as resulting from a loss of smoothness or differentiability of the invariant circle as, in general, the return map will not have a cubic inflection point on this critical line [60,61]. A theoretical approach [62,63] was based on the structure of the stable and unstable manifolds of the periodic points in the interior of a resonance horn [10,64]. At the edges of the locking the periodic points arise from a saddle-node bifurcation (the 2-D phase-space version of the tangent bifurcation) and thus the stable manifold must be characterized by two real eigenvalues (Flouquet multipliers), see Fig. 17(a). If somewhere in the interior of the horn the periodic points take on spiral stability, i.e., complex conjugate eigenvalues, then the stable manifold is no longer differentiable at those points, see Fig. 17(b). Other mechanisms can also lead to loss of smoothness of the circle as discussed in references [10,62,64] but for simplicity we define our critical line to consist of the minima of the spiral stability regions. Other criteria based on the qualities of the time series at the golden mean are most often used but this technique provides an independent measure of criticality. To realize this method the linear stability of the periodic points must be determined. This requires the analysis of transient relaxation of the system to the attracting periodic points. We therefore turn our attention to the production and analysis of transients in real physical systems with many potential degrees of freedom.

In simple systems such as electronic oscillators it is easy to produce transients because one can specify the initial conditions with a small number of parameters. On the other hand systems described by partial differential equations have spatial degrees of freedom and require specifying initial conditions at every point in space. This is impossible experimentally for arbitrary initial conditions and so another stable state of the system is used as the initial state and transients are produced by making sudden changes in a control parameter. Even if the long time dynamics of the system is low dimensional, as in our experiments on quasiperiodic convection, a sudden change in control parameter is likely to induce dynamics in a much larger dimensional phase space. These higher dimensional motions will damp out more quickly than the low dimensional transients of interest but there is often a mixing of time scales which complicates interpretation of the dynamics. With that caveat we describe cases where the interpretation is straightforward and quantitative results obtainable. Consider transient relaxation to a single periodic point in a period-2 cycle. This means that we take every other iterate point from a Poincaré section of the mode-locked state. (The axes of the section are proportional to the temperature oscillations of the local probe but are scaled so that the entire range of the attractor is ±1.) The data in Fig. 18 are for a state with spiral stability for which the real and imaginary parts of the complex conjugate eigenvalues λ_R and λ_I can be defined by considering the position of the points relative to the asymptotic stable state: $\vec{r}(r, \theta, n) = \vec{r}_0 e^{-\lambda_R n + i\lambda_I n}$ where n is the discrete-time iterate variable. λ_R is obtained by plotting $\ln r$ versus n as in Fig. 19(a). Likewise the angular rotation is just the slope of θ versus n, Fig. 19(b). We get $\lambda_R = 0.18$/iterate and $\lambda_I = 1.7$ radians/iterate. In the region of the resonance horn where two real eigenvalues govern the stability of the periodic points one eigenvalue is typically much larger than the other and only the smaller one is accessible to experimental measurement. An example of pure radial relaxation is shown in Fig. 20 where we get $\lambda_R = 0.3$/iterate. By making a series of such measurements across the 2:13 resonance we can specify the stability of the periodic point inside that tongue. In Fig. 21 λ_R and λ_I are shown for $Pr^{-1} = 14.77$, where spirals exist over most of the horn, and for $Pr^{-1} = 14.56$ where there are no spirals. The radial eigenvalue decreases

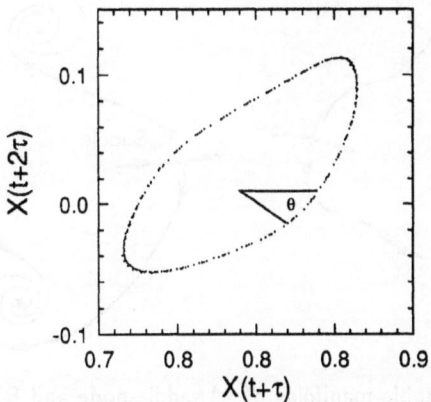

Figure 15. Experimental Poincaré section in delay coordinate phase space (units normalized to size of attractor). Angle θ indicates parameter for generation of return map.

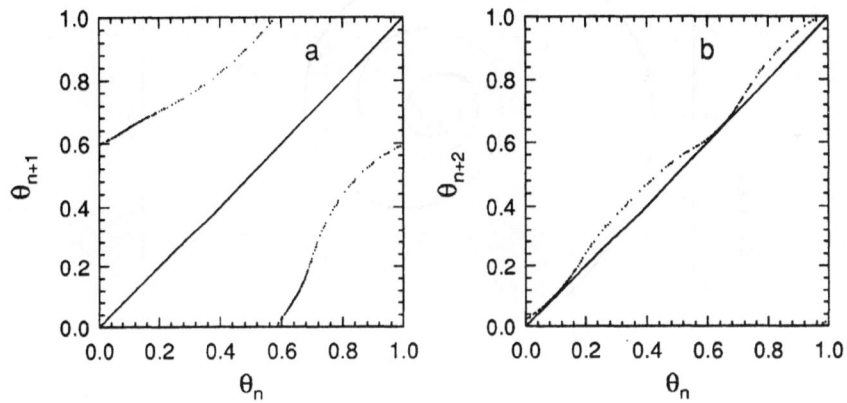

Figure 16. Return maps for a) first iterate θ_{n+1} vs. θ_n and b) second iterate θ_{n+2} vs. θ_n for experimental data near locking to a period-2 resonance.

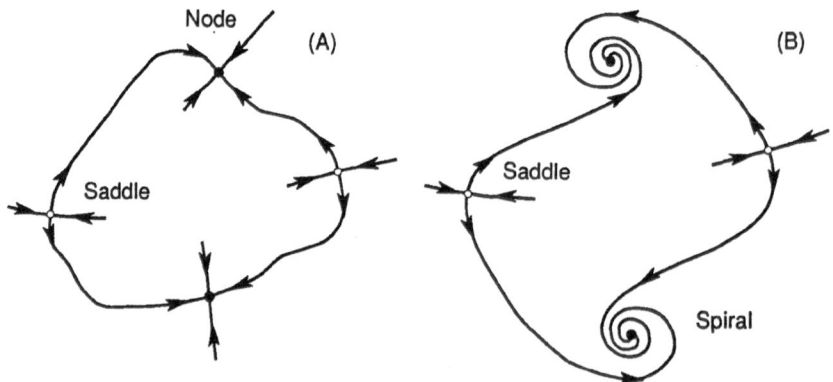

Figure 17. Stable and unstable manifolds of A) saddle-node and B) spiral periodic states. Saddles are donoted by (o) and attracting periodic points by (•)

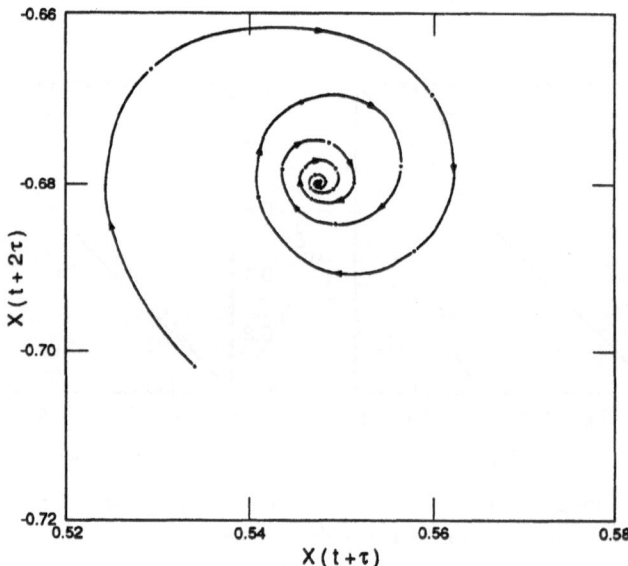

Figure 18. Transient Poincaré section showing spiral approach to a periodic point in the 2/13 resonance horn; $R/R_c=12.025$ and $1/Pr=14.77$. Solid curves are a guide to the eye and not system trajectories.

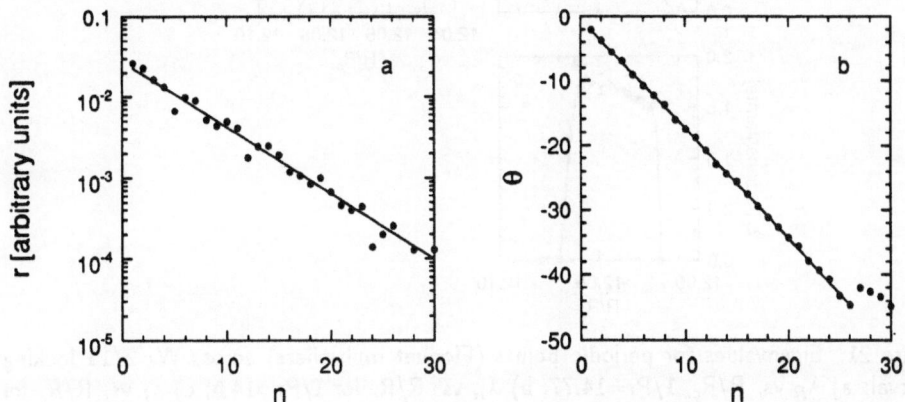

Figure 19. Plots of a) $\log_{10} r$ and b) θ vs. iterate number n for data from Fig. 18. Solid lines are straight line least squares fits to the data where the slope determines the eigenvalues λ_R and λ_I.

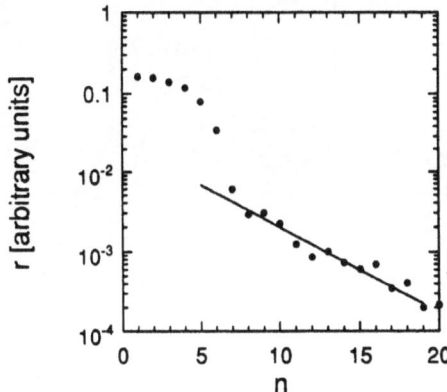

Figure 20. $\text{Log}_{10} r$ vs. n for experimental data with pure real eigenvalues. Only the smaller eigenvalue is measured by the fit to the data (———).

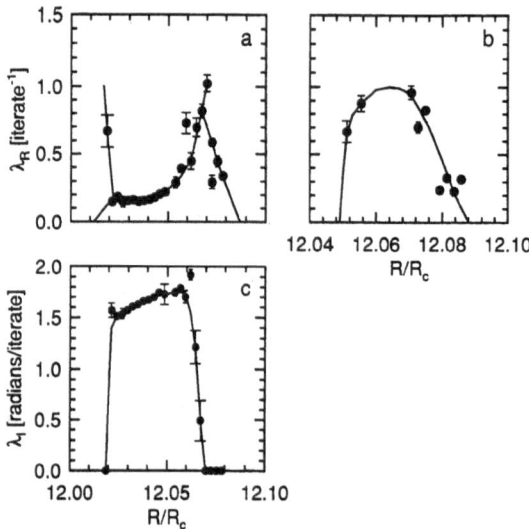

Figure 21. Eigenvalues for periodic points (Floquet multipliers) across W=2/13 locking interval: a) λ_R vs. R/R_c, 1/Pr=14.77; b) λ_R vs. R/R_c for 1/Pr=14.6; c) λ_I vs. R/R_c for 1/Pr=14.77. Lines are guide to eye and error bars reflect statistical uncertainty in the fits.

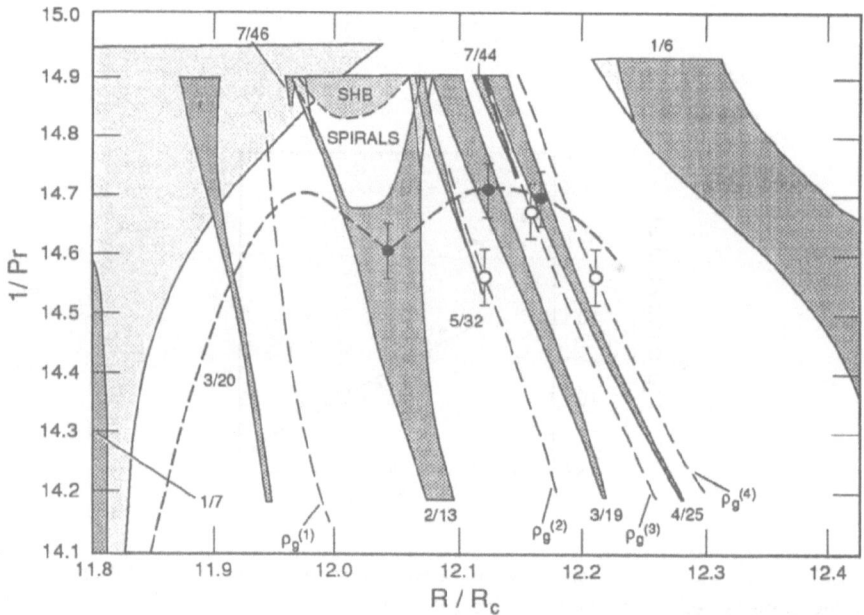

Figure 22. Resonance horns in parameter space of inverse Prandtl number 1/Pr and normalized Rayleigh number R/R_c. Dashed vertical lines trace golden-mean lines with winding numbers $\rho_g^{(1)}$, $\rho_g^{(2)}$, $\rho_g^{(3)}$, and $\rho_g^{(4)}$. Lightly shaded regions indicate hysteresis; the interior structure of the 2/13 resonance consists of stable spiral periodic cycles and secondary Hopf bifurcations (SHB) of the periodic cycles. Also shown are estimated location of critical line (- - - -) from analysis of spiral periodic cycles (•) and of $f(\alpha)$ spectra (o).

towards the middle of the locking for the spiral case but is peaked for the non-spiral data. From this information the spiral stability boundaries of the 2:13 resonance horn are identified. The approximate threshold for spirals is determined in the same way for the 3/19 and 4/25 resonances. The accuracy with which this lower boundary can be determined decreases as the cycle length increases. Another limitation is that as the two real eigenvalues become close in magnitude, as they must before the transition to spiral stability, it becomes difficult to resolve the spiral component before the transient has decayed to the noise threshold. The critical line, determined by this constraint and by the condition that the critical line is below any tongue hysteresis, is placed slightly below the locus of apparent spirals. The tongue diagram, the spiral region of the 2:13 resonance, and the critical line are shown in Fig. 22. The general shape of the critical line follows reference [62]. Also illustrated are lines of constant winding number with asymptotic golden mean expansions. These are defined as $\rho_g^{(n)} = [n, 1, 1, 1, \ldots]$ according to (2). It is at the intersections of these constant golden mean lines and the critical line that universal predictions of the sine circle map model should apply. In the next section we describe those predictions and the analysis tools for making quantitative comparisons between theory and experiment.

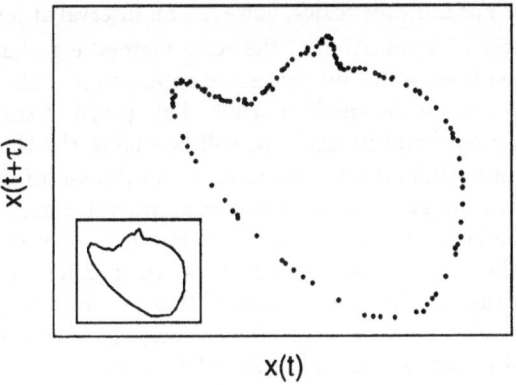

Figure 23. Poincaré section for experimental data with $R/R_c = 12.22$, $1/Pr=14.57$, and ρ within 50 ppm of $\rho_g^{(4)}$. Inset: connected segments for p/q=28/129 approximation to data points.

3 Universal Scaling Dynamics: Experimental Analysis

There are a number of quantitative comparisons that can be made between experimental data and predictions of universal scaling theory. Calculation of the fractal dimension of the staircase, evaluation of the contraction exponent δ for interval widths around the golden mean, the self similar structure of the power spectrum of the data at criticality are all useful and well described analysis techniques. These are reviewed in detail elsewhere [19] and are difficult to apply for our system. Instead this section will focus on analysis of the dynamics of the system represented by a Poincaré section of the phase space trajectories. As described earlier the phase space motion, for states below criticality, is constrained to a 2-D torus. The Poincaré section of a torus is a circle, Fig. 12, and even at criticality the section is very line like, i.e., a 1-D curve. In Fig. 23 the Poincaré section is shown for a state with $\rho = \rho_g^{(4)}$ to within 0.005%. The inset shows a high order approximation to the data with a 28:129 cycle. This curve is used to compute the arc lengths that define the segments discussed below. Figure 24 is the corresponding 1-D return map. Before discussing this data in detail the general analysis techniques of multifractal $f(\alpha)$ spectra and scaling functions [65] need to be presented.

A simple fractal structure is one that can be generated by successive application of a single scale factor. The center-third Cantor set is such a set. Many fractal sets, however, have multiple scalings and cannot be adequately described by a single number, the fractal dimension of the set. The concepts of generalized dimension D_q and $f(\alpha)$ spectrum were created to provide a framework for analyzing such fractals. Many aspects of multifractals are reviewed in reference [66]. To understand the $f(\alpha)$ and scaling function techniques it is useful to consider the dynamics of a model system which generates a Cantor set by successively dividing up the unit interval: the zeroth level consists of the unit interval, the first level has two segments, the second has four, etc. In general there are Q_n segments labeled $\Delta_k^{(n)}, k = 0, \ldots, Q_n - 1$ at the nth level. Consider a Cantor set where the two intervals at the first level are divided into four different intervals on the second level by application of four scale factors, $\{\sigma_i\}$. Successive application of these scalings produces a fractal set. In the case of the center-third Cantor set there is only a single scale so $\sigma = \sigma_i = 1/3$ so that on the nth level all intervals have size $(1/3)^n$. For multiple scales, however, an interval at level n has a size that depends on the history of application of the scale factors, e.g., $\left|\Delta^{(n)}\right| = \sigma_1\sigma_2\sigma_1\sigma_3\ldots$. The sizes are bounded from above by successive application of the largest scale factor and similarly from below by the smallest scale. This range of scales is an important element in characterizing a multifractal and will be one of the things extracted in an $f(\alpha)$ analysis. The other important parameter is the Haussdorf dimension of the set which corresponds in a rough sense to the average fractal dimension. The $f(\alpha)$ formalism is a technique for extracting the information from a set of intervals. A better characterization of this fractal is the set of scales which generate the intervals. The scaling function approach is the identification of these scale factors.

A useful analogy exists between some of the averaging over scales in a multifractal and the averaging of energy states in statistical mechanics [67]. In this picture the statistical averaging over intervals gives the thermodynamics whereas elucidation of the individual scale factors is a more microscopic description of the fractal as derived from statistical mechanics. A "thermodynamic" description of these intervals can be

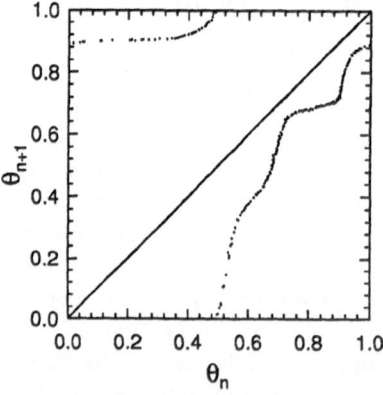

Figure 24. Return map θ_{n+1} vs. θ_n for data in Fig. 23.

defined by a "partition function"

$$Q_n^{-P(\beta)} = \sum_{k=0}^{Q_n-1} |\Delta_k^{(n)}|^\beta,$$

from which follows the definition of the thermodynamic "pressure":

$$P(\beta) = -\lim_{n\to\infty} \frac{\ln \sum_{k=0}^{Q_n-1} |\Delta_k^{(n)}|^\beta}{\ln Q_n}. \tag{3}$$

(Here we follow Feigenbaum [68] in defining a canonical formulation of multifractal thermodynamics.) The "internal energy" is defined as

$$u(\beta) = \frac{\partial P(\beta)}{\partial \beta} = -\lim_{n\to\infty} \frac{\sum_k |\Delta_k^{(n)}|^\beta \ln |\Delta_k^{(n)}|}{\ln Q_n \sum_k |\Delta_k^{(n)}|^\beta}, \tag{4}$$

from which follows relationships for α, $f(\alpha)$, D_q, etc., from the microconanical formulation [37] in terms of $P(\beta)$ and $u(\beta)$:

$$\begin{aligned} \alpha &= 1/u(\beta), & \tau &= -\beta, & f(\alpha) &= \beta - \frac{P(\beta)}{u(\beta)}, \\ D_q &= \frac{\beta}{1 + P(\beta)}, & q &= -P(\beta). \end{aligned} \tag{5}$$

Applying these definitions to the 1/3 Cantor set with $Q_n = 2^n$ and $|\Delta_k^{(n)}| = 3^{-n}$ one gets $\alpha = f(\alpha) = \ln 2/\ln 3$ as expected. Another way to obtain this result uses a statistical mechanical transfer matrix formalism that reveals more about the scaling dynamics than the averages computed from the thermodynamic approach [69,70]. We define the transfer matrix as operating on two segments Δ_0 and Δ_1 with scale factors $\sigma_1, \sigma_2, \sigma_3, \sigma_4$. Consider a two-by-two matrix:

$$\mathbf{T}^\beta = \begin{pmatrix} \sigma_1^\beta & \sigma_2^\beta \\ \sigma_3^\beta & \sigma_4^\beta \end{pmatrix} \tag{6}$$

where the scale factors $\sigma_i < 1$ and we use $\mathbf{T}^\beta$ to indicate each element to the power β and not the matrix. $\mathbf{T}$ operates on two intervals Δ_0 and Δ_1 to produce sums of intervals composed with the dynamical scale factors. From the eigenvalues of $\mathbf{T}^\beta$ one can get the thermodynamics directly. The two eigenvalues λ_1 and λ_2 of $\mathbf{T}^\beta$ are obtained from the secular equation $|\mathbf{T}^\beta - \mathbf{I}\lambda| = 0$. The pressure $P(\beta)$ can be written in terms of these eigenvalues:

$$P(\beta) = -\lim_{n\to\infty} \frac{\ln \left(c_1 \lambda_1^{n-1} + c_2 \lambda_2^{n-1} \right)}{n \ln 2}$$

which reduces in the limit to $P(\beta) = -\ln \lambda_1/\ln 2$ where $\lambda_1 > \lambda_2$. The convergence to this limit is given by the ratio λ_1/λ_2 and by the nonuniversal constants c_1 and c_2. For the 1/3 Cantor set $\sigma_i = 1/3$, $\lambda_1 = 2[3]^\beta$, and $\lambda_2 = 0$. Results for α and $f(\alpha)$ follow from (4) and (5).

The more interesting example for generalized Cantor sets is one where the dynamic scale factors are not the same. Solving the secular equation of $\mathbf{T}^\beta$ in 6 for arbitrary σ_i yields an equation for the largest eigenvalue

$$\lambda_{max}(\beta) = \frac{\sigma_1^\beta + \sigma_4^\beta}{2} + \left[\left(\frac{\sigma_1^\beta - \sigma_4^\beta}{2} \right)^2 + (\sigma_2\sigma_3)^\beta \right]^{1/2}$$

from which follows $P(\beta)$, α, and $f(\alpha)$. The scale factors σ_2 and σ_3 only appear as a product and thus the thermodynamics of the four scale Cantor set is infinitely degenerate over internal scales in that there is a 1-D family of sets with the same product $\sigma_2\sigma_3$. A more direct approach in determining the scale factors is to measure ratios of intervals. In the four scale example, Fig. 25, the relationships are $\sigma_1 = \left|\Delta_1^{(2)}\right|/\left|\Delta_1^{(1)}\right|$, $\sigma_2 = \left|\Delta_2^{(2)}\right|/\left|\Delta_1^{(1)}\right|$, $\sigma_3 = \left|\Delta_3^{(2)}\right|/\left|\Delta_2^{(1)}\right|$, and $\sigma_4 = \left|\Delta_4^{(2)}\right|/\left|\Delta_2^{(1)}\right|$. With suitable relabeling these relationships hold at every level although for this example the results are the same. In real dynamical systems, however, scalings usually apply asymptotically for small intervals and so the smallest size compatible with noise considerations should be used. The advantage of this direct technique, an unambiguous determination of the scaling dynamics, is balanced by the practical disadvantage of sensitivity to experimental (or numerical) errors. The thermodynamic approach on the other hand should be more robust since it averages over many intervals. These problems are considered next in the analysis of experimental data.

The dynamics of the circle map produces an ordering of points on the circle which divides up the interval 0 to 2π into segments, each segment consisting of the arc length between points on the circle. For a winding number that is a rational approximant to the golden mean there is a generation of segments on different levels similar to the Cantor set example. The rational approximants for the sequence $[1,1,1,1,\ldots]$ are $1/1$, $1/2$, $2/3$, $3/5$, $5/8$, $8/13$, $\ldots$ with cycles that are the Fibonacci numbers $F_i = \{1,2,3,5,8,13,\ldots\}$. Successively higher approximants correspond to higher cycles, a closer approximation to the golden mean, and more levels in the scaling tree. In Fig. 26 this tree is illustrated for an 8/13 cycle which generates 3 levels and either 3 or 5 scale lengths depending on whether one takes the transfer matrix $\mathbf{T}$ as operating on segments $\Delta_0^{(1)}$ and $\Delta_1^{(1)}$ or on the three segments $\Delta_i^{(2)}$ at level 2. The dynamics of the circle map excludes certain intervals and thus constrains certain elements in $\mathbf{T}$ to be zero, thereby reducing the possible four scales for a 2×2 matrix to three and the possible nine scales for a 3×3 matrix to five. The 2×2 transfer matrix is

$$\mathbf{T} = \begin{pmatrix} \sigma_0 & \sigma_1 \\ \sigma_2 & 0 \end{pmatrix} \tag{7}$$

and the 3×3 transfer matrix is

$$\mathbf{T} = \begin{pmatrix} \sigma_0 & \sigma_1 & 0 \\ 0 & 0 & \sigma_2 \\ \sigma_3 & \sigma_4 & 0 \end{pmatrix}. \tag{8}$$

In the limit as the golden mean is approached the number of levels becomes large and an $N \times N$ matrix is needed to describe all the scalings. This limit can be thought of as describing a function $\mathbf{T}(\mathbf{x})$ whose argument is just the index of the scale factor normalized by the total number of scales. The magnitude of $\mathbf{T}(\mathbf{x})$ is the value of the scaling factor at a particular value of $\mathbf{x}$. Approximations to this function consist of increasingly bigger matrices starting with 2×2, then 3×3, etc. As in (7) and (8) this corresponds to 3-scale and 5-scale approximations of the continuous function. The scale factors are only asymptotically convergent to universal values. In Fig. 27 we show a 5-scale approximation to the scaling function derived from the data shown in Fig. 23. Three of the scales show good agreement with the predictions of the theory whereas the higher order scalings are not at all close to the theory. Some of the discrepancies can be understood to arise from the sensitivity of the segments to noise and drift in the data: the higher order scales σ_3 and σ_4 involve smaller segments mapped more times around the circle and thus are more sensitive to experimental imperfections. This dependence can be avoided by performing thermodynamic averages which produce the $f(\alpha)$ spectrum. Before discussing the relative advantages and disadvantages of

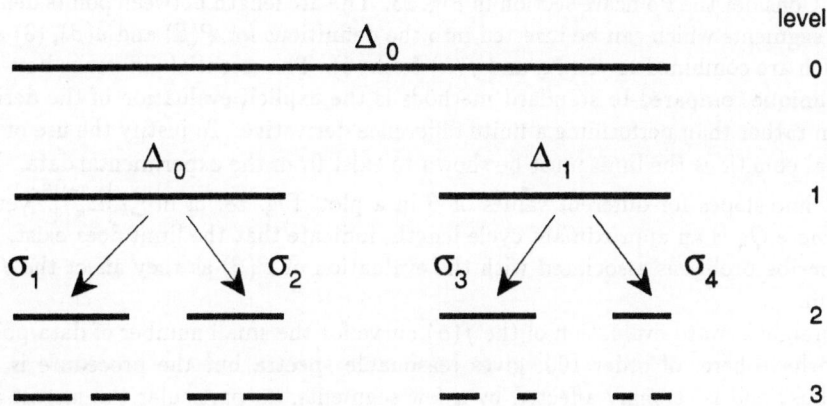

Figure 25. Four-scale Cantor set construction.

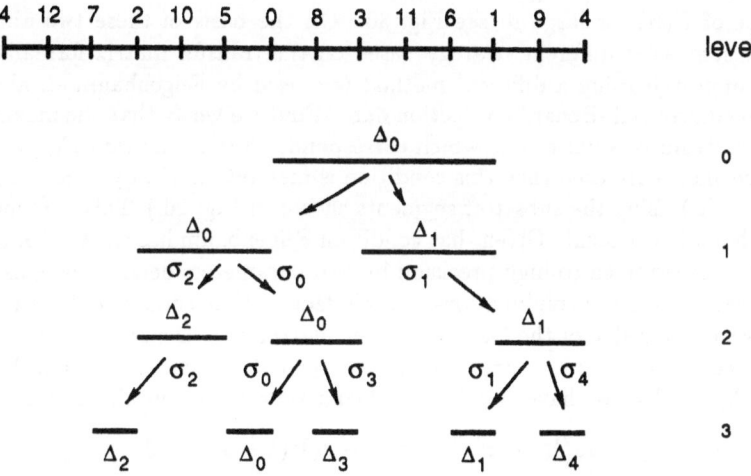

Figure 26. Universal scaling tree for circle map segments with p/q=8/13 and scale factors $\{\sigma_i\}$.

the techniques we describe our methods for constructing the thermodynamic averages. Other results on the determination of a scaling function from experimental data are reported in reference [71].

Standard techniques for calculating $f(\alpha)$ spectra suffer from an arbitrariness in picking scaling regions. Using the canonical approach of Feigenbaum [68] these problems do not arise and the statistics required to obtain reasonable results is much reduced [42,72]. Consider the Poincaré section in Fig. 23. The arc length between points defines a set of segments which can be inserted into the definitions for $P(\beta)$ and $u(\beta)$, (3) and (4), which are combined to yield α and $f(\alpha)$ from (5). The essential difference between this technique compared to standard methods is the explicit evaluation of the derivative sum rather than performing a finite difference derivative. To justify the use of the canonical equations the limit must be shown to exist from the experimental data. The straight line slopes for different values of β in a plot, Fig. 28, of $\ln \sum_k \left| \Delta_k^{(n)} \right|^{\beta}$ versus $\ln Q_n$ where Q_n is an approximant cycle length, indicate that the limit does exist. We now describe problems associated with the evaluation of $P(\beta)$ as they affect the $f(\alpha)$ spectrum.

A straightforward evaluation of the $f(\alpha)$ curve for the small number of data points that we have here, of order 100, gives reasonable spectra but the procedure is not very robust and is strongly affected by a few segments. In particular the largest and smallest segments dominate the determination of the range of scales, i.e., the values of α_{max} and α_{min}. This is because these quantities depend on the interval sizes raised to the power β and thus the largest segment dominates the sum for large positive β whereas the smallest segment is the important one for large negative β. In Fig. 29 we illustrate this for large segments by plotting $P(\beta)$ for all the data and comparing it with $P(\beta)$ for only the largest segment; for $\beta > 8$ there is no difference between the two. This indicates serious problems since it is necessary to have good values of $P(\beta)$ for $\beta > 30$ to accurately obtain α_{max}. There is the additional problem of slow convergence of $P(\beta)$ for large β, see Fig. 30. On the basis of these two difficulties we adopt a somewhat different strategy based on the transfer matrix formalism. An equivalent approach using a different method was used by Feigenbaum et. al. [73] to analyze forced Rayleigh-Bénard convection data. First we verify that the maximum of the $f(\alpha)$ spectrum is equal to one which corresponds to the condition $P(\beta = 1) = 0$. (If all the segments are used then this condition is met automatically so it is necessary to compute $f(\alpha)$ using the subset of segments shown in Fig. 26.) This is found to be true to within a few percent. Given that condition Feigenbaum has shown [69] that the $f(\alpha)$ curve is determined to high precision by two parameters denoted here as s_1 and s_2. These parameters are combinations of scale factors that come about from solving for the largest eigenvalue of the 5-scale transfer matrix, (8). Defining $z = 1/\lambda$, we get the secular equation $1 - z\sigma_0^{\beta} - z^2(\sigma_2\sigma_4)^{\beta} - z^3((\sigma_1\sigma_2\sigma_3)^{\beta} - (\sigma_0\sigma_2\sigma_4)^{\beta}) = 0$. With the condition $P(\beta = 1)=0$ we have $z=1$ for $\beta = 1$ and we get the simplified expression:

$$(1 - zs_1^{\beta})(1 - z^2 s_2^{\beta}) - z^3(1 - s_1)^{\beta}(1 - s_2)^{\beta} = 0, \qquad (9)$$

where $s_1 = \sigma_0$ and $s_2 = \sigma_2\sigma_4$. To find s_1 and s_2 from experimental data we solve (9) using a nonlinear least-squares fitting procedure with z_{min} obtained from the experimental $P(\beta)$ by inverting the relation $P(\beta) = \ln z_{min} / \ln \rho_g$. The range of β is small and centered around zero where many segments contribute to the sum and the averaging is effective. The values of $s_1 = 0.47 \pm 0.01$ and $s_2 = 0.61 \pm 0.2$ do not depend sensitively on the range of β chosen. The $f(\alpha)$ spectrum obtained using these parameters is shown in Fig. 31. Also shown for comparison is a subcritical $f(\alpha)$ curve. In principle a subcritical data set is not multifractal and should collapse to a point. Finite size effects,

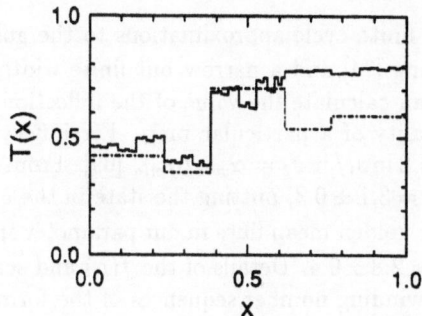

Figure 27. Universal scaling function in the 5-scale approximation of experimental data (–·–·–) of Fig. 23 and for a many-scale theoretical approximation (——)

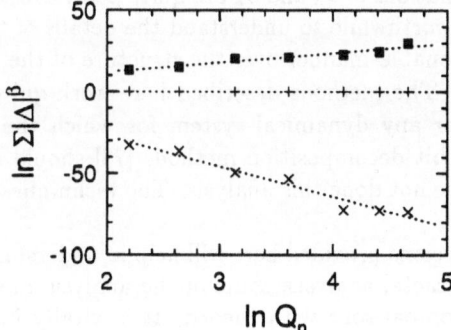

Figure 28. $\text{Log} \sum_k \left| \Delta_k^{(n)} \right|^{\beta}$ vs. $\ln Q_n$ where Q_n is an approximant cycle length with $\beta = 10$ (■), $\beta = 0$ (+), and $\beta = 10$ (×).

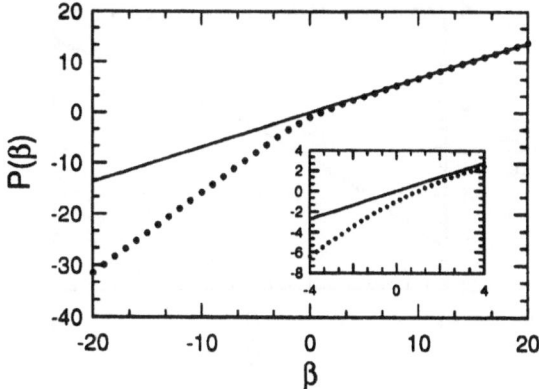

Figure 29. $P(\beta)$ vs. β for experimental data of Fig. 23. For (●) all the data was used and for (—) only the largest segment was used.

however, associated with finite cycle approximations to the golden mean cause some residual multifractal scaling [74] and a narrow but finite width curve [75]. From the quantities s_1 and s_2 we can calculate the value of the inflection point exponent which characterizes the universality of a particular map. For inflection points of the type $\theta \, |2\theta|^{\nu-1}$ we have that $\nu = 2 \ln s_1 / \ln s_2 = \alpha_{max}/\alpha_{min}$ [69]. From the experimental data for $\rho_g^{(4)}$, Fig. 23, we get $\nu = 3.1 \pm 0.3$, putting the state in the universality class of the sine circle map (1). Other golden mean lines in our parameter space, $\rho_g^{(2)}$ and $\rho_g^{(3)}$ yield $\nu^{(2)} = 2.9 \pm 0.3$ and $\nu^{(3)} = 2.8 \pm 0.4$. Details of the $f(\alpha)$ and scaling function analyses and how it is applied to winding number sequences of the form $\rho = [n, 1, 1, 1, \ldots]$ are described in reference [43].

The thermodynamic approach which generates the $f(\alpha)$ spectra provides a robust technique that takes advantage of averaging over the complete fractal structure. On the other hand direct evaluation of the scaling function gives a more complete description of the fractal dynamics. In our experimental scaling function three scale factors σ_0, σ_1, and σ_2 are resolved to within about 10% whereas the thermodynamic approach determines s_1 and s_2 which gives σ_0 and the combination $\sigma_2\sigma_4$. It is interesting that while the direct determinations of σ_3 and σ_4 are quite poor, averaging gives $\sigma_2\sigma_4$ quite accurately. It would be worthwhile to understand the details of the averaging process and to average in a reasonable manner over the structure of the fractal to get a more robust scaling function. The methods described here work quite well for circle map data and presumably for any dynamical system for which the scaling dynamics is understood. Periodic orbit decomposition methods [76] should also be applicable in such systems but we have not done this analysis. The techniques fail when there is no model for the dynamics.

I hope that the discussions provided here will help in understanding the pitfalls and limitations of the multifractal analysis; some of the analysis methods are not robust and can lead to poor comparisons with theory. It is vitally important to establish that the point ordering of the experimental circle map is the same as for the theory and that a scaling function can be constructed with the correct properties. Without such consistency checks the nonrobust features of the $f(\alpha)$ methods suggest reason for caution in comparisons between theory and experiment.

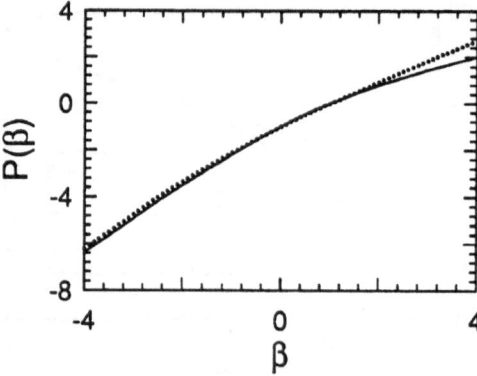

Figure 30. $P(\beta)$ vs. β for experimental data ($\bullet$) of Fig. 23 and fitting curve (———) parameterized by s_1 and s_2. Notice poor convergence of experimental data for positive β.

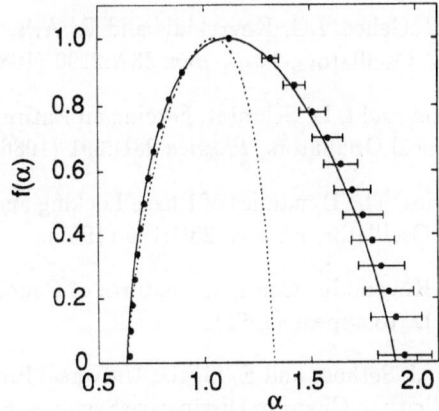

Figure 31. $f(\alpha)$ curves for the experimental data in Fig. 23 (•), for theoretical data from a sine circle map at criticality (——) and for a subcritical experimental data set (- - - - -) with R/R_c=12.375 and $1/Pr$=14.17. The error bars for α that are smaller that 1.1 are of the order of the data points and are not plotted.

Acknowledgements

I would like to acknowledge the people who have contributed to this work: John Wheatley for the original work on ^{3}He-superfluid-^{4}He mixtures, Hans Haucke and Yoshi Maeno who built the experimental apparatus and did much of the early convection work, Ioannis Kevrekidis who taught me about bifurcation theory and helped in understanding the Arnold tongue structure, Doyne Farmer and Dave Umberger for their work on fat fractals, Tim Sullivan for his contributions to the transient analysis and the critical dynamics, and a special thanks to Ronnie Mainieri for teaching me everything I know about $f(\alpha)$ analysis and scaling function dynamics and for helping prepare parts of these notes. This research was sponsored by the U.S. Department of Energy, Basic Energy Sciences, Division of Materials Science.

References

[1] C. Huyghens, letter to his father, dated 26 Feb. 1665, *Ouevres Completes de Christian Huyghens*, (M. Nijhoff, Ed.), The Hague, The Netherlands: Societe Hollandaise des Sciences, 1893, vol. 5, p. 243. ˙

[2] R. Abraham and C. Shaw, 'Dynamics: The Geometry of Motion', *Aerial Press* (1981).

[3] M. Peixoto, Structural Stability on Two Dimensional Manifolds, *Topology* 1:101 (1962).

[4] V.I. Arnold, Loss of Stability of Self-Oscillations Close to Resonance and Versal Deformations of Equivariant Vector Fields, *Func. Anal. Appl.* 11:1 (1977).

[5] R.E. Ecke, J.D. Farmer, and D.K. Umberger, Scaling of the Arnold Tongues, *Nonlinearity* 2:175 (1989).

[6] D.G. Aronson, R.P. McGehee, I.G. Kevrekidis and R. Aris, Entrainment Regions for Periodically Forced Oscillators, *Phys. Rev.* 33A:2190 (1986).

[7] I.G. Kevrekidis, R. Aris, and L.D. Schmidt, Forcing an Entire Bifurcation Diagram: Case Studies in Chemical Oscillators, *Physica* 23D:391 (1986).

[8] P. Bryant and C. Jeffries, The Dynamics of Phase Locking and Points of Resonance in a Forced Magnetic Oscillator, *Physica* 25D:196 (1987).

[9] M.A. Taylor and I.G. Kevrekidis, Common Features of Coupled Oscillatory Reacting Systems, *Physica* D, to appear (1991).

[10] S. Ostlund, D. Rand, J. Sethna, and E. Siggia, Universal Properties of the Transition from Quasiperiodicity to Chaos in Dissipative Systems, *Physica* 8D:303 (1983).

[11] M. Jensen, P. Bak, and T. Bohr, Transition to Chaos by Interaction of Resonances in Dissipative Systems: I. Circle Maps, *Phys. Rev.* A30:1960 (1984).

[12] P. Cvitanovic, M.H. Jensen, L.P. Kadanoff, and I. Procaccia, Renormalization, Unstable Manifolds and the Fractal Structure of Mode Locking, *Phys. Rev. Lett.* 55:343 (1985).

[13] P. Cvitanovic, B. Shraiman, and B. Soderberg, Scaling Laws for Mode Lockings in Circle Maps, *Phys. Scr.* 32:263 (1985).

[14] S.J. Shenker, Scaling Behavior in a Map of a Circle onto Itself: Empirical Results, *Physica* 5D:405 (1982).

[15] M.J. Feigenbaum, L.P. Kadanoff, and S.J. Shenker, Quasiperiodicity in Dissipative Systems: A Renormalization Group Analysis, *Physica* 5D:370 (1982).

[16] K. Kaneko, 'Collapse of Tori and Genesis of Chaos in Dissipative Systems', *World Scientific Pub.*, Singapore (1986).

[17] J.D. Farmer, Sensitive Dependence on Parameters in Nonlinear Dynamics, *Phys. Rev. Lett.* 55:351 (1985).

[18] R. Eykholt and D.K. Umberger, Characterization of Fat Fractals in Nonlinear Dynamical Systems, *Phys. Rev. Lett.* 57:2333 (1986).

[19] J.A. Glazier and A. Libchaber, Quasiperiodicity and Dynamical Systems: An Experimentalist's View, *IEEE Trans. Cir. Sys.* 35:790 (1988).

[20] J. Maurer and A. Libchaber, Rayleigh-Bénard Experiment in Liquid Helium; Frequency Locking and the Onset of Turbulence, *J. Physique Lett. (Paris)* 40:L-419 (1979).

[21] J.P. Gollub and S.V. Benson, Many Routes to Turbulent Convection, *J. Fluid Mech.* 100:449 (1980).

[22] M. Sano and Y. Sawada, Experimental Study on Poincaré Mappings in Rayleigh-Benard Convection, in *Turbulence and Chaotic Phenomena in Fluids*, ed. by T. Tatsumi (North Holland, Amsterdam, 1983).

[23] A.P. Fein, M.S. Heutmaker and J.P. Gollub, Scaling at the Transition from Quasiperiodicity to Chaos in a Hydrodynamic System, *Phys. Scr.* T9:79 (1985).

[24] J. Stavans, F. Heslot, and A. Libchaber, Fixed Winding Number and the Quasiperiodic Route to Chaos in a Convecting Fluid, *Phys. Rev. Lett.* 55:596 (1985).

[25] J. Stavans, Experimental Study of Quasiperiodicity in a Hydrodynamic System, *Phys. Rev.* A35:4314 (1987).

[26] G.A. Held and C. Jeffries, Quasiperiodic Transitions to Chaos of Instabilities in an Electron-Hole Plasma Excited by ac Perturbations at One and Two Frequencies, *Phys. Rev. Lett.* 56:1183 (1986).

[27] S. Martin and W. Martienssen, Circle Maps and Mode Locking in the Driven Electrical Conductivity of Barium Sodium Niobate Crystals, *Phys. Rev. Lett.* 56:1522 (1986).

[28] E. G. Gwinn and R. M. Westervelt, Frequency Locking, Quasiperiodicity, and Chaos in Extrinsic Ge, *Phys. Rev. Lett.* 57:1060 (1986).

[29] E. G. Gwinn and R. M. Westervelt, Scaling Structure of Attractors at the Transition from Quasiperiodicity to Chaos in Electronic Transport in Ge, *Phys. Rev. Lett.* 59:157 (1987).

[30] A. Cummings and P.S. Linsay, Deviations from Universality in the Transition from Quasiperiodicity to Chaos, *Phys. Rev. Lett.* 59:1633 (1987).

[31] Z. Su, R.W. Rollins, and E.R. Hunt, Measurements of $f(\alpha)$ Spectra of Attractors at Transitions to Chaos in Driven Diode Resonator Systems, *Phys. Rev.* A36:3515 (1987).

[32] D. Olinger and K. Sreenivasan, Nonlinear Dynamics of the Wake of an Oscillating Cylinder, *Physical Review Letters* 60:797 (1988).

[33] D. Baums, W. Elsasser and E. Gobel, Farey Tree and Devil's Staircase of a Modulated External Cavity Semiconductor Laser, *Phys. Rev. Lett.* 63:155 (1989).

[34] T. Yazaki, S. Takishima, and F. Mizutani, Complex Quasiperiodic and Chaotic States Observed in Thermally Induced Oscillations of Gas Columns, *Phys. Rev. Lett.* 58:1108 (1987).

[35] T. Yazaki, S. Sugioka, F. Mizutani, and H. Mamada, Nonlinear Dynamics of a Forced Thermoacoustic Oscillation, *Phys. Rev. Lett.* 64:2515 (1990).

[36] B. Shraiman, Transition from Quasiperiodicity to Chaos: A Perturbative Renormalization Group Approach, *Phys. Rev.* A29:3464 (1984).

[37] T.C. Halsey, M.H. Jensen, L.P. Kadanoff, I. Procaccia, and B.I. Shraiman, Fractal Measures and their Singularities: The Characterization of Strange Sets, *Phys. Rev.* A33, 1141 (1986).

[38] M.H. Jensen, L.P. Kadanoff, A. Libchaber, I. Procaccia, and J. Stavans, Global Universality at the Onset of Chaos: Results of a Forced Rayleigh-Bénard Experiment, *Phys. Rev. Lett.* 55:2798 (1985).

[39] Nonuniversal behavior in the fractal dimension of the devils staircase reported in [30] was later shown [40] to be due to the scheme used to compute the fractal dimension; a complete multifractal $f(\alpha)$ analysis yields excellent agreement between experiment and theory.

[40] D. Barkley and A. Cummings, Thermodynamics of the Quasiperiodic Parameter Set at the Borderline of Chaos: Experimental Results, *Phys. Rev. Lett.* 64:327 (1990).

[41] H. Haucke and R. Ecke, Mode Locking and Chaos in Rayleigh-Bénard Convection, *Physica* 25D:307 (1987).

[42] R. Mainieri, T.S. Sullivan, and R.E. Ecke, Two-Parameter Study of the Quasiperiodic Route to Chaos in Convecting ^{3}He-Superfluid-^{4}He Mixtures, *Phys. Rev. Lett.* 63:2357 (1989).

[43] R.E. Ecke, R. Mainieri, and T.S. Sullivan, Universality in Quasiperiodic Rayleigh-Bénard Convection, in preparation.

[44] J. Peinke, J. Parisi, R.P. Huebner, M. Duong-van and P. Keller, Quasiperiodic Behavior of d.c.-Biased Semiconductor Br akdown, *Euro. Phys. Lett.* 12:13 (1990).

[45] M. Bauer, U. Krueger, and W. Martienssen, Experimental Studies of Mode-Locking and Circle Maps in Inductively Shunted Josephson Junctions, *Europhys. Lett.* 9:191 (1989).

[46] H. Haucke, Y. Maeno, R.E. Ecke, and J. Wheatley, Noise-induced Intermittency in Rayleigh-Benard Convection, *Phys. Rev. Lett.* 53:2090 (1984).

[47] H. Haucke, R.E. Ecke, and J.C. Wheatley, Dimension and Entropy for Quasiperiodic and Chaotic Convection, in *Dimension and Entropies in Chaotic Systems*, ed. by G. Mayer-Kress, (Springer Verlag, Berlin, 1986) p. 198.

[48] R.E. Ecke and I.G. Kevrekidis, Interactions of Resonances and Global Bifurcations in Rayleigh-Bénard Convection, *Phys. Lett.* 131A:344 (1988).

[49] R.E. Ecke and H. Haucke, Noise-induced Intermittency in the Quasiperiodic Regime of Rayleigh-Bénard Convection, *J. Stat. Phys.* 54:1153 (1989).

[50] I.G. Kevrekidis and R.E. Ecke Global Bifurcations in Maps of the Plane and in Rayleigh-Bénard Convection, *Cont. Math.* 99:313 (1989).

[51] Y. Maeno, H. Haucke, R.E. Ecke, and J.C. Wheatley, Oscillatory Convection in a Dilute ^{3}He-Superfluid-^{4}He Solution, *J. Low Temp. Phys.* 59:305 (1985).

[52] G. Metcalf and R. Behringer, Convection in ^{3}He-Superfluid-^{4}He Mixtures: Measurement of Superfluid Effects, *Phys. Rev.* A41:5735 (1990).

[53] N.H. Packard, J.P. Crutchfield, J.D. Farmer, and R.S. Shaw, Geometry from a Time Series, *Phys. Rev. Lett.* 45:712 (1980).

[54] A. Fraser and H.L. Swinney, Independent Coordinates for Strange Attractors from Mutual Information, *Phys. Rev.* A33:1134 (1986).

[55] J.-P. Eckmann and D. Ruelle, Ergodic Theory of Chaos and Strange Attractors, *Rev. Mod. Phys.* 57:617 (1985).

[56] F. Harris, On the Use of Windows for Harmonic Analysis with the Discrete Fourier Transform, *Proc. IEEE* 66:51 (1978).

[57] R.E. Ecke, Y. Maeno, H. Haucke, and J.C. Wheatley, Critical Dynamics near the Oscillatory Instability in Rayleigh-Bénard Convection, *Phys. Rev. Lett.* 53:1567 (1984).

[58] R.E. Ecke, H. Haucke, Y. Maeno and J.C. Wheatley, Critical Dynamics at a Hopf Bifurcation to Oscillatory Rayleigh-Bénard Convection, *Phys. Rev.* A33:1870 (1986).

[59] R.J. Deissler, R.E. Ecke, and H. Haucke, Universal Scaling and Transient Behavior of Temporal Modes hear a Hopf Bifurcation: Theory and Experiment, *Phys. Rev.* 36A:4390 (1987).

[60] D. Rand, Universality for the Breakdown of Dissipative Golden Invariant Tori, *Proceedings of the Eighth International Congress of Mathematical Physics*, ed. by M. Mebkhout and R. Seneor, (World Scientific, Singapore, 1987).

[61] X. Wang, R. Mainieri, and J.H. Lowenstein, Circle Map Scaling in a Two-Dimensional Setting, *Phys. Rev.* A40:5382 (1989).

[62] T. Bohr, P. Bak, and M. Jensen, Transition to Chaos by Interaction of Resonances in Dissipative Systems: II Josephson Junctions, Charge-Density Waves, and Standard Maps, *Phys. Rev.* A30:1970 (1984).

[63] T. Bohr, Destruction of Invariant Tori as an Eigenvalue Problem, *Phys. Rev. Lett.* 54:1737 (1985).

[64] D.G. Aronson, M.A. Ghory, G.R. Hall, and R.P. McGehee, Bifurcations from an Invariant Circle for Two-Parameter Families of Maps of the Plane: A Computer-Assisted Study, *Commun. Math. Phys.* 83:303 (1982).

[65] Mitchell J. Feigenbaum, The Transition to Aperiodic Behavior in Turbulent Systems, *Commun. Math. Phys.*, 77:65, 1980.

[66] J.L. McCauley, Introduction to Multifractals in Dynamical Systems Theory and Fully Developed Turbulence, *Phys. Rpts.* 189:225 (1990).

[67] D. Ruelle, 'Thermodynamic Formalism', *Addison-Wesley*, Reading (1978).

[68] M. J. Feigenbaum, Some Characterizations of Strange Sets, *J. Stat. Mech.* 46:919 (1987).

[69] M. J. Feigenbaum, Scaling Spectra and Return Times of Dynamical Systems, *J. Stat. Mech.* 46:925 (1987).

[70] M. J. Feigenbaum, Scaling Function Theory for Circle Maps, *Nonlinearity* 1:577 (1988).

[71] A. Belmonte, M.J. Vinson, J.A. Glazier, G.H. Gunaratne, and B.G. Kenny, Trajectory Scaling Functions at the Onset of Chaos: Experimental Results, *Phys. Rev. Lett.* 61:539 (1988).

[72] A. Chhabra and R.V. Jensen, Direct Determination of the $f(\alpha)$ Singularity Spectrum, *Phys. Rev. Lett.* 62:1327 (1989).

[73] M. J. Feigenbaum, M. Jensen, and I. Procaccia, Time Ordering and the Thermodynamics of Strange Sets: Theory and Experimental Tests, *Phys. Rev. Lett.* 57:1503 (1986).

[74] A. Arneodo and M. Holschneider, Crossover Effect in the $f(\alpha)$ Spectrum for Quasiperiodic Trajectories at the Onset of Chaos, *Phys. Rev. Lett.* 58:2007 (1987).

[75] J. Glazier, G. Gunaratne, and A. Libchaber, $f(\alpha)$ Curves: Experimental Results, *Phys. Rev.* A37:523 (1988).

[76] D. Auerbach, P. Cvitanovic, J.-P. Eckmann, G. Gunaratne, and I. Procaccia, Exploring Chaotic Motion Through Periodic Orbits, *Phys. Rev. Lett.* 58:2387 (1987).

PATTERNS IN CHAOS

Boris V. Chirikov
notes prepared with assistance of Svend E. Rugh

Institute of Nuclear Physics, 630090 Novosibirsk, U.S.S.R.

Classification of chaotic patterns in classical Hamiltonian systems is given as a series
of levels with increasing disorder. Overview of critical phenomena in Hamiltonian dynamics is presented, including the renormalization chaos, based upon the fairly simple
resonant theory. First estimates for the critical structure and related statistical anomalies in arbitrary dimensions are discussed.

Introduction

The main idea I would like to convey here is the inexhaustible diversity and richness
of the dynamical chaos whatever description you choose: trajectories, statistics or,
recently, renormalization. The importance of this relatively new phenomenon - the
dynamical chaos - is in that it presents, even in very simple models to be discussed
below, the surprising complexity of the structures and evolution characteristic of a broad
range of processes in nature, including the highest levels of its organization. Moreover,
dynamical chaos is the only stationary source of any new information and, hence, a
necessary part of creative activity, science included. This is a direct consequence of the
Alekseev-Brudno theorem and Kolmogorov's development of the information theory
(see e.g. refs. [1,2]). The chaos is not always that bad!

Below I restrict myself to the classical mechanics only. The so-called "quantum
chaos" is another story (see e.g. refs. [3,4]). Let me just mention that apart from
very exotic examples, there is no "true" chaos in quantum mechanics, contrary to
common belief. On the other hand, the unavoidable statistical element of quantum
mechanics related to the measurement is very likely associated with the classical chaos
in a measuring apparatus.

With a bit of imagination and fantasy one may even conjecture that any macroscopic
event in this World, which formally is the result of some quantum "measurement", would
be impossible without chaos.

Also, I am not going to consider any dissipative models (very important in practical
applications) because they are not as fundamental as Hamiltonian systems. Besides,

strictly speaking, the dissipative systems are not purely dynamical as the dissipation is inevitably related to some noise.

In what follows I take a physicist's approach to the problem: my presentation will be based on a simple (sometimes even qualitative) theory, combined with the results of extensive numerical (computer) experiments. For a good physical overview of dynamics and chaos see books [5,6].

The principal concept of such a theory is the nonlinear resonance whose phase space picture (quite familiar by now) is depicted in Fig. 4 below. An essential part of this resonance structure is a pair of periodic orbits, the most important being the unstable one as it gives rise to the separatrix and, under almost any perturbation, to the surrounding chaotic layer around. This is precisely the place where chaos dawns.

Again, I have to restrict myself to a simpler case of strong nonlinearity which does not vanish with perturbation. A very interesting weakly nonlinear resonance will be briefly mentioned in section 1.2. below.

These lectures are organized as follows. In section 1 simple models, currently extensively used in the studies of nonlinear phenomena and chaos, are described. They will represent the whole spectrum of complexity classified in section 2. The main sections 3 and 4 are devoted to a detailed description of the so-called critical phenomena in dynamics which reveal the most complicated behavior presently known.

1   Simple models

First, let us consider a number of simple models currently very popular in the studies of dynamical chaos. Most of them are specified by some mappings, or maps, rather than by differential equations. This considerably simplifies both the theoretical analysis and, especially, the computer experiments. In conservative Hamiltonian systems the chaos requires two degrees of freedom, at least. The corresponding Poincaré map is two-dimensional.

1. Strong nonlinearity [7]. Below we shall consider 2D maps of the following form:

$$\overline{y} = y + f(x) \quad ; \quad \overline{x} = x + g(\overline{y}) \tag{1.1}$$

This map is area-preserving, or canonical, reflecting the Hamiltonian nature of the model. The function $f(x)$, periodic in x, describes a perturbation, usually assumed to be small. Hence, y is the unperturbed motion integral. The function $g(y)$, even if it is linear (see eq. (1.6) below), represents the nonlinearity of the x oscillation.

The simplest example of an analytic perturbation is given by

$$f(x) = K \, \sin x \tag{1.2}$$

We shall also consider a smooth perturbation specified by the Fourier series

$$f(x) = \sum_m f_m \, e^{imx} \quad ; \quad f_m \sim K \, |m|^{-\beta} \tag{1.3}$$

where β is the smoothness parameter. The term 'smooth' actually means sufficiently smooth. For $\beta = 2$, for example, the function $f(x)$ is continuous but the first derivative is discontinuous.

I mention two particular forms of nonlinearity. The first one

$$g(y) = \lambda \ \ln |y| \tag{1.4a}$$

models the motion near the separatrix of a nonlinear resonance, so that the map (1.1) with its nonlinearity and perturbation (1.2) describes, in particular, a separatrix chaotic layer [7].

Another form of nonlinearity

$$g(E) = 2\pi\omega(-2 \ E)^{-3/2} \tag{1.4b}$$

corresponds to the Coulomb interaction, or the Kepler law. Here it is convenient to use the unperturbed energy $E < 0$ as a dynamical variable, and ω is the perturbation frequency (see ref. [4]).

The map (1.1) with the nonlinearity (1.4 b) and perturbation (1.2) is called the Kepler map, and it is applied in both celestial mechanics and atomic physics. In the former case the motion of comet Halley driven by Jupiter and Saturn was proved to be chaotic [8]. In atomic physics the Kepler map is a simple model to describe, in particular, a new type of photoelectric effect, the so-called diffusive ionization of Rydberg (highly excited) atoms [9].

The two latter examples show that the map (1.1) can be considered also as a model for time-dependent dynamical systems driven by a periodic perturbation. This is, of course, simply a very convenient approximation in which the feedback from the perturbed degree of freedom to the perturbing one is completely neglected. Then, the model (1.1) can be described by the Hamiltonian

$$H(x, y, t) = G(y) + F(x)\delta_1(t) \ \rightarrow \ G(y) + K \sum_m \cos(x - 2\pi m t) \tag{1.5}$$

where $\delta_1(t)$ is a δ-function of period 1 (one map's iteration), $G'(y) = g(y)$, $F'(x) = -f(x)$, and the last series represents the perturbation (1.2).

The fairly simple map (1.1) can be simplified still further by linearizing the second equation. In this way we arrive, upon appropriate change of the action y, at the so-called standard map

$$\overline{y} = y + K \ \sin x \ ; \quad \overline{x} = x + \overline{y} \tag{1.6}$$

which describes the original model (1.1) locally in y, and which is also very popular now in studies of nonlinear phenomena in Hamiltonian systems. The model (1.6) is completely characterized by a single parameter K. In the Hamiltonian representation (1.5) the 'kinetic energy' for the standard map is $G(y) = y^2/2$. Since x is an angle (phase) variable and y is the angular momentum, the model (1.6) is also called the 'kicked rotator'.

Each term in the series (1.5) describes a particular first order (primary) nonlinear resonance with the 'pendulum' Hamiltonian (for the standard map)

$$H_m = \frac{y^2}{2} + K \ \cos(x - 2\pi m t) \tag{1.7}$$

The resonant value of momentum $y_m = \dot{x}_m = 2\pi m$. In variables $\tilde{x} = x - 2\pi m t$ and $\tilde{y} = y - y_m$ any single resonance is a conservative system. Its motion is strictly bounded in y by the resonance width $\triangle y_m = 4 \sqrt{K}$, due to the nonlinearity, i.e., the dependence of the frequency $\dot{x} = y$ on the momentum y.

2. Weak nonlinearity [10]. The structure of the resonance drastically changes if we add to the Hamiltonian (1.7) the term $\omega_0^2 x^2/2$:

$$H_m = \frac{y^2}{2} + \frac{\omega_0^2 x^2}{2} + K \, \cos(x - 2\pi m t) \qquad (1.8)$$

which breaks down the integrability of the system for any $\omega_0 \neq 0$.

Actually, the model (1.8) is quite different from the model (1.7), as now the variable x is no longer confined to the interval $(0, 2\pi)$, and y is not the angular momentum. One may interpret the Hamiltonian (1.8) as describing a particle-wave interaction. Such models have been studied by many authors in plasma physics (see, e.g., ref. [5]), yet the true understanding has been achieved only recently (see, e.g., ref. [10]). The peculiarity of the model (1.8) is the weak nonlinearity, i.e., the unperturbed ($K = 0$) oscillation is linear (isochronous), which turns out to be a much more difficult problem as compared with strong (unperturbed) nonlinearity (1.4). The resonance is now determined not by initial conditions but by the parameters of the model: $2\pi m = n \, \omega_0$ with any integer $n \neq 0$. In the action-angle variables (I, ϕ) of the harmonic oscillator a single resonance is approximately described by the Hamiltonian

$$H_m \approx K \, J_m(a) \, \cos(m\phi + \frac{\pi m}{2}) \qquad (1.9)$$

where $a = (2 \, I/\omega_0)^{1/2}$ is the oscillation amplitude, and J_m the Bessel function. There are now infinitely many stable and unstable periodic orbits (instead of two for strong nonlinearity) while the separatrices, connecting unstable points, form an unbounded network on the phase plane (I, ϕ). As a result, even a single weakly nonlinear resonance can make the motion completely unstable and unbounded.

2 Levels of Disorder

In this section I shall attempt to 'organize' the great variety of chaos into a series of levels of increasing disorder and complexity.

0. Complete integrability [15] This, zero-th level of the maximal order is characterized by a stable and dynamically predictable motion in terms of individual trajectories. The motion is quasiperiodic, i.e., of a purely discrete spectrum. One may call this a simple dynamics. Yet, in the general theory of dynamical systems this 'simple' motion includes the whole quantum chaos, typically on a finite time scale (see, e.g., ref. [3]). The latter is dynamically equivalent to a many-dimensional linear oscillator which is apparently the simplest model of the quantum chaos [11]. On the other hand, in the formal thermodynamic limit of infinitely many degrees of freedom, this model has provided the foundations of the traditional statistical mechanics, both classical and quantal, for macroscopic systems (for a rigorous theory see, e.g., ref. [12]).

The standard map, as the simplest model, is completely integrable for $K = 0$ only, that is in the unperturbed limit. In this case $y = const$ is the motion integral, and $x = 2\pi r t$ where the quantity

$$r = \frac{\omega}{2\pi} = \frac{\Delta x}{2\pi \, \Delta t} \qquad (2.1)$$

is called the rotation number. This very important parameter of a trajectory is the ratio of motion frequency (ω) to that of the perturbation frequency (2π). Particularly, this ratio determines resonances (with zero perturbation in this limit !) which correspond to rational $r = p/q$. Any resonant trajectory is just q separate points on the phase plane (x, y). For irrational r the trajectory is a continuous straight line $y = 2\pi r$, called the invariant curve.

In spite of great successes in constructing the whole families of completely integrable systems (see, e.g., ref. [13]), they are all exceptional, or non-generic, in the sense that almost any perturbation destroys the integrability.

1. KAM integrability [14] is the generic property of a completely integrable system under sufficiently weak perturbation. The theory of such systems had been initiated by Kolmogorov and was essentially developed by Arnold and Moser (see, e.g., ref. [15]), hence, the abbreviation KAM.

For the standard map this first level of disorder corresponds to a non-zero $K \to 0$. Most invariant (KAM) curves survive weak perturbation, i.e., they are only slightly deformed but remain continuous and, hence, unpenetrable for other trajectories. For this reason the KAM curve is called an absolute barrier (for the motion). This property depends on the rotation number r of the curve which must be sufficiently irrational for the stability against perturbation. Hence, the importance of the parameter r which is used as the label for identification of a given KAM curve at different perturbations.

Curves with resonant $r = p/q$ are all destroyed by any perturbation to form a different structure of the nonlinear resonance (Fig.4). However, the nonintegrable part of this structure is confined to an exponentially narrow chaotic layer only. From a physical point of view such motion can in most cases be considered integrable to a very high accuracy. This is reminiscent of the adiabatic invariance, very important in physics even though it is not exact. Actually, there is a deep relation between the two, and we call KAM integrability the inverse adiabaticity [14,16].

Approximately, the dynamics on this level is as simple as the previous one. Yet, the chaotic component of motion, being of an exponentially small measure, is everywhere dense. As a result, the whole motion structure becomes very complicated. For more than two degrees of freedom the phase space is cut through by a connected network of channels which support global diffusion [7]. Even though the rate of this Arnold diffusion is also exponentially small, it may be important in some special cases. For a weakly nonlinear system the Arnold diffusion is possible even in two degrees of freedom as well (see for instance [10]) in the model (1.8).

2. Complete chaos [20]. Now we turn to the opposite limiting case, the case of fully chaotic motion. In the standard map, as $K \to \infty$, there is a single chaotic component of motion stretched over the whole phase space (cylinder) of the model. The motion spectrum is purely continuous, while a typical individual trajectory is most complicated. The latter means that Kolmogorov's complexity, which is equal to the information associated with the trajectory per unit time, is finite, and equal to the rate of local instability of the motion [1]. Hence, the dynamics on this level is so complicated that the trajectory actually loses its physical meaning.

Nevertheless, the dynamical equations, e.g., map (1.6), can still be applied to describe completely the statistical properties of the unstable motion. Moreover, on this level the statistics turns out to be very simple and already well known from the traditional statistical mechanics. For example, in the standard map it is simply a homogeneous diffusion in y with the rate

$$D_y \equiv \frac{<(\Delta y)^2>}{t} = \frac{K^2}{2} C(K) \to \frac{K^2}{2} \qquad (2.2)$$

where the function $C(K)$ accounts for the dynamical correlation of the phase X, and $C(K) \to 1$ as $K \to \infty$ [17]. For this reason the complexity of the motion on this level is still not the highest one.

3. Critical phenomena: Scale invariance [21]. For a typical (generic) perturbation, neither very weak nor very strong, the whole structure of motion is most complicated because the phase space is generally divided in many separate domains with both regular and chaotic motions. In the standard map, for example, such an intricate behavior

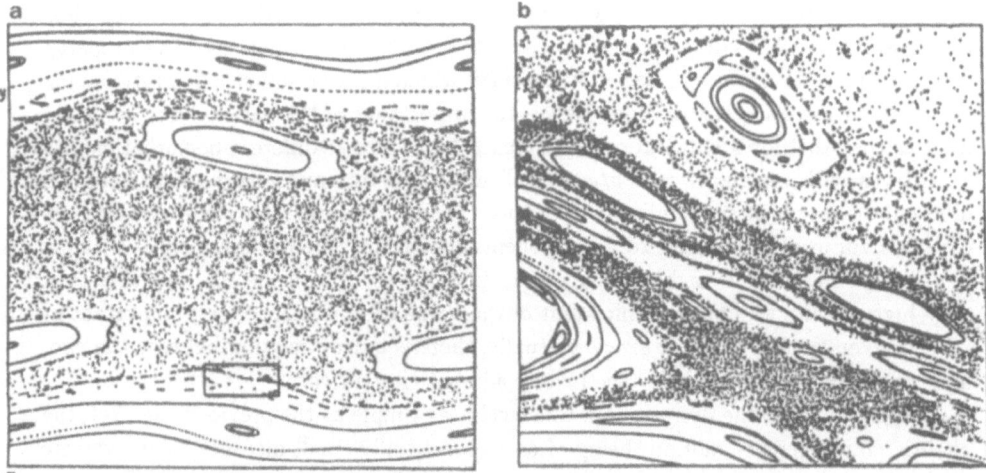

Figure 1. An example of critical structure in the map (3.1) with $\lambda = 5$; Scattered points belong to a single chaotic trajectory: (a) the whole chaotic layer; (b) enlarged part near the chaos border $y \approx -\lambda$ where the motion is described locally by the standard map (1.6) with $K \approx 1$ [19].

corresponds to $K \sim 1$ (see Fig.1), i.e., around the global critical perturbation $K = K_G \approx 1$. The latter is the border between strictly bounded motion for any initial conditions ($K \leq K_G$) and unbounded motion for some initial conditions ($K > K_G$).

In the unbounded chaotic component (for $K > K_G$) the motion is still diffusive with the rate [18]

$$D_y \approx 0.3 (K - K_G)^3 \qquad (2.3)$$

vanishing towards the critical perturbation (cf. eq. (2.2) where the correlation $C(K) \approx 0.6 (K - K_G)^3/K^2$). The main difficulty here is a hierarchical (fractal) structure of the chaotic component. The invariance of the phase space, does not help in this case. The ultimate origin of this complexity is the phase space border between the chaotic and the regular components of motion, which also leads to very peculiar statistical properties of the chaotic motion (see section 4).

The chaos border makes both the individual (chaotic) trajectories as well as the statistical properties of the motion very complicated. Is there any way to simplify the description of such a motion? Or: is it possible to find any order in this mess? Surprisingly it is possible indeed, in some cases, if one compares the critical structure at different scales in the phase plane (section 3.3). Asymptotically, as you enlarge the structure more and more, it repeats itself exactly, with all the dynamical and statistical complexity (see also Fig.5 below)! This peculiar property is called the scale invariance, and it is described by the so-called renormalization group or in brief, renormgroup.

4. Critical phenomena: Renormalization chaos [22]. The variation of the motion structure with the scale in phase space can be considered as a certain abstract dynamics (see section 3.4) which we termed the renormalization dynamics, or renormdynamics [22]. Here the scale plays a role of 'time' and we call it renormalization time. The simplest case of any dynamics is a fixed point (for maps) which in renormalization dynamics corresponds to the scale invariance described above (see also section 3.3). But typically the dynamics is chaotic, and so there must be a sort of renormalization chaos (renormchaos) as well. Guided by this analogy, we have indeed found such chaos [22]!

In this case the renormalization is as complicated as an individual chaotic trajectory of the original dynamical system. Yet, some remnants of order still persist, namely, the universality of renormalization. This means that asymptotically, for a big renormalization time, i.e., for small spatial scales, the critical structure is a universal functional of a single irrational number - the rotation number r_c of the critical curve, e.g., the border curve, in almost any 2D map [21].

Moreover, one can introduce the renormalization statistics ,i.e., statistical description of the renormalization. Then, for almost any r_c, the renormalization statistics is the same, i.e. universal, and it is fairly simple.

5. Critical phenomena: The breakdown of universality [23]. Recently, the first example of still more complicated behavior has been found in ref. [23], where a quasiperiodic driving perturbation was studied. Specifically, the standard map (1.6) was used in numerical experiments with periodically time-dependent parameter $K(t) = K_1 + K_2 \cos(2\pi r_2 t)$, incommensurable with the time step of the map.

To some extent such a model also represents a higher-dimensional behavior. A critical curve is now characterized by the two irrational rotation numbers r_1, r_2. For a particular choice of irrationals r_1, r_2 it was found that the renormalization dynamics depends on the parameters K_1 and K_2. It is thus not clear whether such breakdown is typical. If so, one would expect also a more complicated renormalization statistics.

Here we have come to the frontier of the unknown. Currently, there is no idea what would imply still higher levels of disorder, if such existed.

3 Critical Dynamics

In this section I consider in some detail the two levels of disorder briefly described in section 2 above (levels 3 and 4). This work was done in collaboration with D.L.Shepelyansky.

1. Statistical 'anomalies' in dynamical chaos [24]. We encountered the critical phenomena in studying some statistical properties of motion in a simple map

$$\bar{y} = y + \sin x \quad ; \quad \bar{x} = x + \lambda \ln |\bar{y}| \tag{3.1}$$

of the type described in section 1.1 above. Our studies were stimulated by ref. [25], with the intriguing title 'Numerical Experiments in Stochasticity and Heteroclinic Oscillation'. Actually, the motion in a chaotic separatrix layer had been studied, and we went on with a much simpler model (3.1) (see ref. [7]).

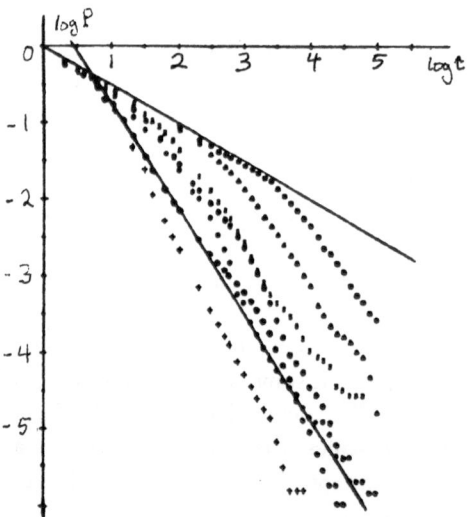

Figure 2. Poincaré recurrences in the chaotic layer of map (3.1) for various $\lambda = 1$ (lower points) through 100 (upper points). The two straight lines are power laws with exponents -0.5 and -1.37, respectively (after ref. [24]).

We studied the statistics of the times t_n when a trajectory crosses the symmetry line $y = 0$. We call the differences $\tau_n = t_{n+1} - t_n$ the Poincaré recurrence times. The same was implicitly done in ref. [25]. Our results are shown in Fig. 2, where $P(\tau)$ is the (integral) probability for $\tau_n > \tau$. The initial part of the distribution is very close to

$$P_f = \frac{1}{\sqrt{\tau}} \quad ; \quad \tau \geq 1 \ , \tag{3.2}$$

and is explained by a free homogeneous diffusion within the chaotic layer before the trajectory reaches the layer border ($y_b \approx \lambda$). This takes the time

$$\tau_f \approx 0.3 \, \lambda^2 \ , \tag{3.3}$$

where the coefficient was estimated from the numerical data.

Curiously, in ref. [25] only this (trivial) part of the distribution $P(\tau)$ was observed. It was the cost for the authors' great concern about the exponential error growth at a chaotic trajectory. To overcome the instability, the computation was performed with the record accuracy of 358 decimal places! As a result, the chaotic trajectory can be followed during a rather short time interval.

Error growth is a serious problem, indeed, as the structural stability of Hamiltonian motion is almost unknown rigorously. Yet, all the numerical experience up to now strongly suggests such stability and, hence, the stability of statistical properties which are of primary interest for chaotic motion. Besides, only structural stability justifies the use of various simple models and approximations.

In our studies of the model (3.1) we checked directly that the distribution $P(\tau)$, which is a statistical property of the motion, does not depend on a particular trajectory within expected statistical fluctuations. The latter noticeably influence the lowest part of the distribution $P(\tau)$ where the number of events per histogram bin is ~ 1 (see Fig.2).

The most interesting is the asymptotics of $P(\tau)$ for $\tau \gg \tau_f$ (3.3). This part characterizes the motion structure of the chaos border at $|y| = y_b \approx \lambda$, or, as we call it, the critical structure.

The following features of the $P(\tau)$ asymptotics seem to be of importance. First, the distribution is a power law and not an exponential:

$$P(\tau) \approx \frac{\tau_f^{p-1/2}}{\tau^p} \; ; \; \tau \geq \tau_f \; ; \; <p> \approx 1.5 < 2 \qquad (3.4)$$

This suggests a hierarchical (fractal) structure of the border. The accuracy of the numerical value for p is not very good, yet we are sure that the important inequality (3.4) always holds.

On the other hand, the distribution $P(\tau)$ behaves as a power law only approximately, in the sense of an average. Irregular oscillations of the local exponent $p(\tau) \equiv d\ln P/d\ln\tau$ clearly show up in Fig.2. These do not depend on the trajectory and, hence, are not statistical fluctuations but characterize the structure of the chaos border. Such structure with a variable exponent $p(\tau)$ is now called multifractality (see,e.g., ref. [26]).

The statistics of the Poincaré recurrences $P(\tau)$ proved to be the most convenient and reliable numerical data to study (cf. ref. [25]). On the other hand, this statistics is directly related to the most important statistical property of time correlation functions [27], such as

$$C_y(\tau) = \frac{\overline{y(t)\,y(t+\tau)}}{\overline{y^2(t)}} \qquad (3.5)$$

which characterizes the 'sticking' of the trajectory near the border. Notice that $\overline{y(t)} = 0$ for the map (3.1).

Indeed, the correlation is proportional to the sticking time, that is (cf. ref. [27])

$$C_y \sim \frac{\tau\,P(\tau)}{<\tau>} \sim \tau^{-p_c} \; ; \; p_c = p - 1 < 1 \qquad (3.6)$$

for $\tau \geq \tau_f$. Here $<\tau> \approx 3\lambda$ is the mean recurrence time, and the latter important inequality follows from eq. (3.4) (see also Fig.3b). For chaotic motion $C_y \to 0$ as $\tau \to \infty$ (mixing property), hence $p_c > 0$, and, for bounded motion, $p > 1$. Notice, also, that due to the ergodicity of the motion $C_y \sim \mu(\tau)$, the measure of the sticking domain (a strip) $\sim |y - y_b|/y_b$, where $y_b \approx \lambda$, is the half-width of the chaotic layer for the map (3.1).

Slow correlation decay due to the sticking of a chaotic trajectory near the chaos border, and especially the inequality (3.6) is responsible for all other statistical 'anomalies' of the motion with a chaos border to be discussed below. A power law decay (3.6) is especially remarkable in view of the strong exponential instability of the motion which is characterized by a positive Lyapunov exponent Λ_+ and the KS-entropy (per map's iteration): $h = \Lambda_+ \approx 0.7$ (see section 6.3 in ref. [7]). The apparent contradiction is explained as follows. The instability rate h is mainly determined by the central part of the chaotic layer while the sticking is a peripheral effect which has a negligible impact

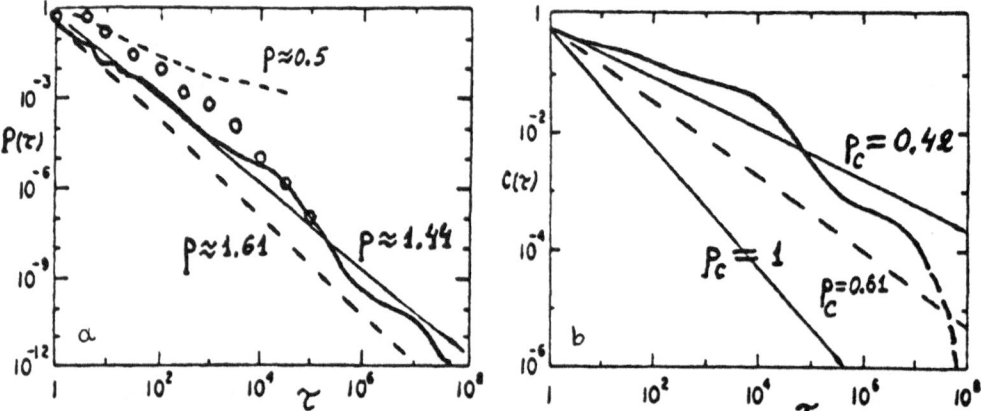

Figure 3. Statistical properties of motion with chaos border: (a) Poincaré recurrences; (b) correlation decay. Solid curves are for the map (3.7) [27] while circles are our data for $\lambda = 3$. Straight lines indicate power laws with the exponents shown. The dashed curve is the effect of noise [22].

on the mean local instability. In other words, the KS-entropy does not discern such statistical anomalies. This can be accomplished using the Renyi entropy K_q which is a generalization of $h = K_1$ (see, e.g., ref. [28]), and which drops to zero for all values of the parameter $q > 1$ in the presence of a chaos border [29].

The critical phenomena at the chaos border and the related statistical anomalies are 'universal' (a very popular word in this field of research!) in that they are approximately the same in any 2D map. In Fig.3(a), for example, our results are compared with those in ref. [27] for a different map on the torus

$$\overline{y} = y + 2(x^2 - a^2) \;\; ; \;\; \overline{x} = x + \overline{y} \;\; , \tag{3.7}$$

with a closed chaos border surrounding the domain of regular motion around the stable fixed point at $y = 0$; $x = -a$ $(0 < a < 1)$. Notice that the two distributions $P(\tau)$ are not identical but rather similar (see below).

2. The resonant theory [30]. To understand the statistical anomalies described above we have developed a resonant theory of critical phenomena in dynamics [30,31]. Let us begin with the simpler problem of an isolated critical KAM curve whose rotation number is some irrational r. According to the KAM theory most invariant curves

are preserved under a sufficiently weak perturbation in the sense that they remain continuous and are only slightly deformed by the perturbation. The theory of critical phenomena follows the transformation of a KAM curve all the way up to the critical perturbation which destroys the curve.

The critical perturbation, e.g. $K_c(r)$ for the standard map, crucially depends on the arithmetic nature of r. Remember that for the everywhere dense set of rationals $r = p/q$, the critical $K_c(p/q) = 0$ (section 2.1). The whole dependence $K_c(r)$ is a fractal function [32].

The physical explanation of this behavior lies in the nature of resonances. Their profound impact on the critical structure is clearly seen in all numerical data (see,

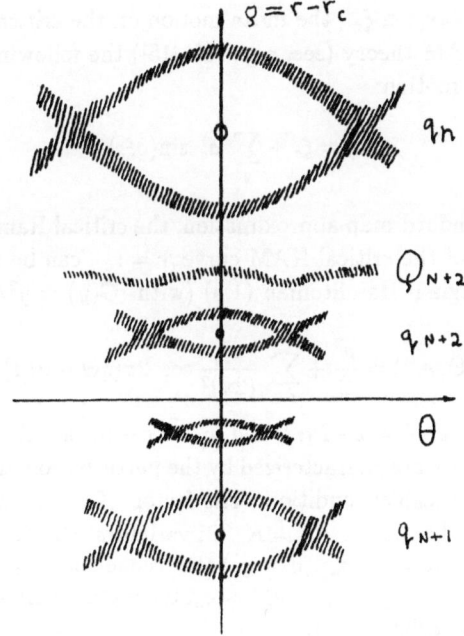

Figure 4. Outline of the critical structure with a few principal resonances, represented by the separatrix chaotic layers (hatched) and stable periodic orbits (circles), and the corresponding scales q_n. Another chaotic layer Q_n is a bottleneck between the scales (section 4.3).

e.g., Fig.1). For irrational r the principal resonances correspond to the best rational approximations of r, i.e., the convergents $r_n = p_n/q_n$ of the infinite continued fraction

$$r = \cfrac{1}{m_1 + \cfrac{1}{m_2 + \ldots}} \equiv [m_1, m_2, \ldots] \; ; \tag{3.8}$$

$$r_n = [m_1, \ldots, m_n] \to r \;\; , \;\; n \to \infty$$

The arithmetic of continued fractions gives for almost any r

$$|\rho_n| \equiv |r_n - r| \sim \frac{1}{q_n^2} \sim |r_{n+1} - r_n| \tag{3.9}$$

From a physical viewpoint, ρ_n is the detuning of the n-th principal resonance with respect to the critical motion. Then, from the resonance overlap criterion [7], the main critical scaling, or the criticality condition, is

$$\Delta \rho_n \sim |\rho_n| \sim \frac{1}{q_n^2} \qquad (3.10)$$

where $\Delta \rho_n$ is the resonance width. These resonances determine the principal scales of the critical structure whose scheme is outlined in Fig.4 (see also Fig.5 below).

To estimate $\Delta \rho_n$, we need the critical Hamiltonian which describes all resonances $r_{pq} = p/q$ $(p, q$ any integers), and not only the primary ones $r_m = m$ from the original Hamiltonian of the type (1.5). Integer resonances $r_m = m$ are obtained in the first approximation $x_c(t) \approx 2\pi r_c t \equiv \xi_c$, the mean motion on the critical KAM curve.

Extrapolating the KAM theory (see, e.g., ref. [15]) the following expression can be assumed for the critical motion:

$$x_c(t) = \xi_c + \sum_q a_q \sin(q\xi_c) \qquad (3.11)$$

Locally in y, in the standard-map approximation, the critical Hamiltonian H_c, which describes some vicinity of the critical KAM curve $r = r_c$, can be written as a natural generalization of the original Hamiltonian (1.5) (with $G(y) = y^2/2$) in the following form

$$H_c(\Theta, \rho, t) = \frac{\rho^2}{2} + \sum_{p,q} \frac{v_{pq}}{(2\pi)^2} \cos 2\pi(q\Theta - v_{pq}t) \qquad (3.12)$$

Here $\rho = r - r_c$; $2\pi\Theta = x - \xi = x - 2\pi rt$ and $v_{pq} = p - qr_c$ are the driving frequencies. Resonances $\rho_{pq} = p/q - r_c$ are characterized by the perturbation amplitudes v_{pq} to be found below from the criticality condition. The factor $(2\pi)^2$ is introduced to recover the original Hamiltonian for which $v_{p1} = K$ (in variables Θ, ρ).

For principal resonances $(p = p_n, q = q_n)$ the frequencies $\nu_n \sim q_n^{-1}$ are minimal, and they determine the time scales $t_n \sim \nu_n^{-1} \sim q_n$, the motion periods at the resonances. We introduce the scaled variables, e.g.,

$$T = \frac{t_n}{q_n} \sim 1 \qquad (3.13)$$

which remain of the same order of magnitude on all scales n.

The width of a principal resonance $\Delta \rho_n \sim v_n^{1/2} \sim q_n^{-2}$ (see eq. (3.10)). Hence $v_n \sim q_n^{-4}$, and another scaled variable

$$V = v_n q_n^4 \sim 1 . \qquad (3.14)$$

Now we can approximately solve the equation

$$\ddot{\Theta} = \dot{\rho} = -\frac{\partial H_c}{\partial \Theta} \approx -\frac{\partial H_c}{\partial \Theta}\bigg|_{\Theta=0} .$$

The latter approximation means that we substitute mean motion ξ for $x_c(t)$. As our original model (3.1) is a map, time t is integer, and we can drop the term pt in the solution $\Theta(t)$ which then takes the form (3.11) with

$$a_q \approx \frac{q}{(2\pi)^3} \sum_p \frac{v_{pq}}{(p-qr)^2} \; ; \; a_n \sim v_n q_n^3 \sim q_n^{-1} \; .$$

Hence, the longitudinal scaled amplitude

$$A = a_n q_n \sim 1 \; , \tag{3.15}$$

and the X scale is q_n^{-1}, which is simply one cell of the resonance chain (see Fig.4).

In the standard-map approximation we have

$$y_c(t) \approx \dot{x}_c(t) = 2\pi r_c + \sum_q b_q \cos(q\xi_c - 2\pi pt)$$

with $b_n = 2\pi a_n \nu_n$, and the transverse scaled amplitude is

$$B = b_n q_n^2 \sim 1 \tag{3.16}$$

Hence, the y scale is q_n^{-2}, the resonance width.

Consider now the periodic orbit at the resonance center (Fig.4). Its stability is determined by the Greene residue [33] (see also ref. [5])

$$R = \sin^2\left(\frac{t_n \omega_n}{2}\right) \sim 1 \tag{3.17}$$

where $t_n = q_n$ is the period, and $\omega_n \approx q_n v_n^{1/2}$ is the small oscillation frequency (see eq. (3.12)). Obviously, R is the scaled variable.

Finally, the scaled rotation number, or rather the scaled detuning

$$D = \rho_n q_n^2 \sim 1 \tag{3.18}$$

determines actually all the other scaled variables.

So far we considered exactly critical conditions that is $K = K_c(r)$. What would be the impact of any deviation $\Delta K = K - K_c(r) \neq 0$? It can be evaluated as follows. The perturbation amplitudes v_n in eq. (3.12) appear in q_n-th order of the perturbation theory and are proportional to $(K/K_c)^q = \exp(q \ln \frac{K}{K_c})$. Hence, for a small deviation from criticality ($\Delta K \to 0$) the amplitude $v_n \sim \exp(Cq_n \Delta K)$ with some $C \sim 1$. At $\Delta K = 0$ the exponential dependence cancels, and only a power law (3.14) remains. Generally,

$$v_n \sim \frac{1}{q_n^4} \exp\left(Cq_n \Delta K\right) \tag{3.19}$$

For $\Delta K > 0$ all scales $q_n \geq (\Delta K)^{-1}$ are destroyed and a chaotic layer of width $\Delta y \sim (\Delta K)^2$ is formed.

From eq. (3.19) the scaled perturbation can be introduced

$$P = q_n \Delta K_n \sim 1 \tag{3.20}$$

which describes the approach to the renormalization limit for a fixed v, for example.

If the original perturbation is non-analytic, i.e., has a power law spectrum $v_q^0 \sim q^{-\beta-1}$ (cf. model (1.3) where $f_m \sim m v_m^0$), the critical conditions are only possible for $\beta > 3$, otherwise $K_c = 0$. Thus, $\beta_c = 3$ is the critical smoothness of the perturbation. I shall come back to this point in section 4.1 below.

3. The renormalization group [21]. This powerful method, well known and widely applied in hydrodynamical turbulence, phase transitions and quantum field theory, was first used in nonlinear dynamics and chaos theory in ref. [33]. Later on, the exact renormalization equations were formulated and studied in ref. [34] for (dissipative) 1D maps, and in ref. [21] for 2D area-preserving (Hamiltonian) maps. The renormalization group equations are an abstract map acting in the space of dynamical maps, and it is based on the arithmetical map for successive convergents $r_n = p_n/q_n$ of the critical rotation number $r_c = (m_n)$:

$$\overline{p} = \overline{m}p + \underline{p} \ ; \ \overline{q} = \overline{m}q + \underline{q} \tag{3.21}$$

where $\overline{p} \equiv p_{n+1}$; $p \equiv p_n$; $\underline{p} \equiv p_{n-1}$ etc. Besides qualitative understanding of critical phenomena (particularly, their universality) this approach provides very efficient numerical algorithms for computing all the parameters of the critical structure. In contrast, our resonant theory, being inherently approximate, allows some analytical estimates.

The resonance overlap criterion, on which the theory is essentially based, can directly provide order-of-magnitude estimates only, for example, for scaled variables (3.14-3.17). However, there exists another group of critical parameters which can be evaluated to a surprising accuracy. Those are the scaling factors, that is the ratios of particular quantities on the neighbouring scales. For example,

$$s_a = \frac{a_n}{a_{n+1}} \ ; \ s_b = \frac{b_n}{b_{n+1}} \ ; \ s_K = \frac{\Delta K_n}{\Delta K_{n+1}}$$

are the renormalization factors for x, y , and the perturbation K , respectively.

The structure of scaled variables shows that all scaling factors are some powers of the main arithmetical factor

$$s_q = \frac{q_n}{q_{n-1}} \tag{3.22}$$

To compare both approaches let us consider the simplest case of a homogeneous continued fraction $r = [m, m, ..., m, ...] \equiv [m^\infty]$. In this case all the scaled variables become asymptotically, as $n \to \infty$, exact invariants of the renormalization group. This is called the scale invariance. For example, $D \to (4 + m^2)^{-1/2}$ (see eq. (3.21)) which is a simple arithmetical property. The other invariants are not yet known except the case of $r = r_G = (1^\infty) = (\sqrt{5} - 1)/2 = 0.618...$ which is called the golden tail (because for asymptotic properties only the tail of the continued fraction matters).

In this particular case, studied in great detail, the renormalization invariants are: $T = 1$ (if, by definition, $t_n = q_n$, see eq. (3.13)); $R = 0.2500888...$; $V \approx (2 \arcsin \sqrt{R})^2 = 1.097052...$; $A \approx 0.167$; $B \approx 2\pi A/\sqrt{5} \approx 0.470$. Notice that from the above relation $R \sim V \sim \Delta\rho_n/\rho_n$ the Greene residue also characterizes the resonance overlap.

Now consider the scaling factor for the area $c_n \sim a_n b_n$ of a resonance cell (the corresponding scaled variable $C = c_n q_n^3 \approx AB \approx 0.0787$):

$$s_c = c_n/c_{n+1} = s_q^3 = 4.236... \tag{3.23}$$

while the exact numerical value via the renormalization group is 4.339... The two numbers are not equal but very close which was a puzzle for the formal renormalization group approach.

A similar situation arises for the perturbation factor: $s_K = s_q = 1.618...$ (resonant theory), and $s_K = 1.627...$ (numerically).

The differences in scaling factors of the two theories can be interpreted as small changes of the exponents of q in scaled variables. For the two examples above we can write:

$$
\begin{aligned}
C &= c_n q_n^{\alpha} \; ; \; \alpha = 3.049960... \\
P &= \triangle K_n q_n^{\beta} \; ; \; \beta = 1.0126966...
\end{aligned}
\tag{3.24}
$$

Other examples will be given below.

The behavior of the asymptotic renormalization invariants A and R is shown in Fig.6 below. Remarkably, the invariant critical structure, which repeats itself on finer

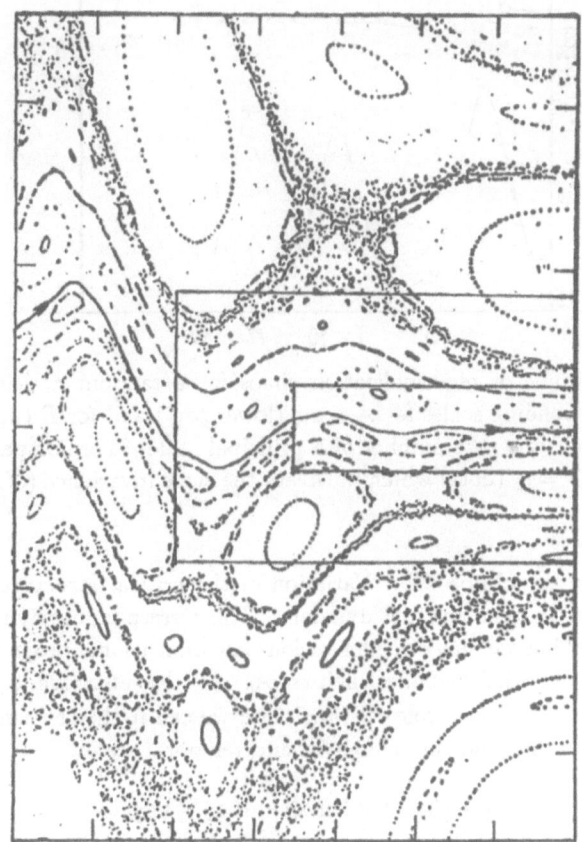

Figure 5. A small part of the critical structure with 3 successive scales shown by rectangles (including the whole picture). The critical curve is indicated by 2 arrows (after ref. [21]).

and finer scales with rapidly increasing precision, is itself of the highest complexity as it contains both chaotic trajectories and an intricate admixture of regular and chaotic components of motion. An example of a tiny part ($\sim 0.01 \times 0.01$) of that structure is shown in Fig.5 [21]. The scale invariance is clearly seen within 3 successively scaled areas indicated by rectangles.

Notice that the scale invariance holds on a particular discrete but infinite set of scales, because the renormalization group equations are based on the arithmetical map (3.21).

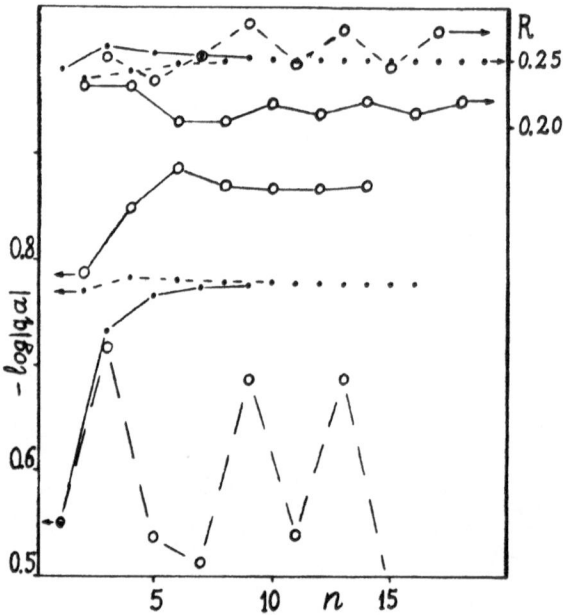

Figure 6. An example of renormalization chaos for a random r_c (circles). Arrows indicate the corresponding scales for $A = a\,q$ (lower part) and for R (upper part); n is the renormalization time (the number of a principal scale). For comparison the same data are given for $r = r_c$ (dots) which illustrate the scale invariance (after ref. [30]).

4. **Renormalization chaos** [22]. Variation of the critical structure from scale to scale can be viewed as some abstract dynamics. The corresponding dynamical space is infinite dimensional but we may consider various few-dimensional projections of that as described by a set of scaled variables such as A_n, R_n, V_n etc. (see, e.g., Fig.6). The serial scale number n plays a role of 'time', and we call it the renormalization time. It is proportional to the logarithm of spatial and temporal scales:

$$n \sim |\ln a_n| \sim |\ln b_n| \sim \ln t_n \sim \ln q_n \qquad (3.25)$$

The renormalization time is discrete as is the renormalization dynamics based on the arithmetical map (3.21).

The scale invariance described in the previous section is the simplest type of renormalization dynamics, namely, a fixed point of the renormalization map. The dynamical interpretation of renormalization suggests other, more complicated, scalings up to a chaotic one which would be the opposite limiting case. Guided by this heuristic approach we conjectured a new type a chaotic behavior - the renormalization chaos [22], and presented an example of the latter in ref.[30]. A similar possibility was also considered in ref. [35] for dissipative systems as modelled by an 1D map.

Our basic idea was to achieve the most complicated renormalization by using a random critical rotation number, a rotation number with a random sequence of the continued fraction entries m_n . As is known from the modern ergodic theory (see, e.g., ref. [20]) this is the case for almost any irrational r . Indeed, we may introduce a sequence of rotation numbers via the Gauss map

$$r = \frac{1}{m + \bar{r}} \ ; \ \bar{r} = \frac{1}{r} \ mod \ 1 \tag{3.26}$$

which is known to be chaotic [20]. Moreover, the basic arithmetical factor in the renormalization (3.22) also obeys the same map

$$\omega = \frac{1}{\bar{\omega}} \ mod \ 1 \ ; \ \omega = \frac{1}{s_q} < 1 \tag{3.27}$$

backwards in renormalization time, and with the 'initial' $\omega_\infty = \tilde{r}$ where $\tilde{r}$ is the irrational with reversed sequence of elements in respect to r . Clearly, the variation of critical structure in this case would be as random and unpredictable as a chaotic trajectory. An example of renormalization chaos is presented in Fig. 6 as described by A and R scaled variables. The irregular character of this renormalization dynamics is clear from Fig. 6, and the proof of its randomness is related to the Gauss map (3.27).

A chaotic trajectory is completely determined, in principle, by the initial conditions via the formal equations of motion. By analogy, we can conjecture that the chaotic variation of critical structure is related to the rotation number r . This would imply that the scaled variables are some universal functions of r . Then, asymptotically, as $n \to \infty$, the renormalization dynamics is described by an infinite dimensional map

$$\overline{A}(r) = A(\bar{r}) \ ; \ \overline{R}(r) = R(\bar{r}) \ etc. \ ; \ \bar{r} = \frac{1}{r} \ mod \ 1 \tag{3.28}$$

Some numerical confirmation of this conjecture was presented in ref. [30]

Thus, particular critical structure essentially depends on r , and in this sense is not universal. Nevertheless, the statistical properties of chaotic renormalization are the same for almost any r . Particularly, the average arithmetical factor (3.22)

$$< s_q > = \ e^{h/2} \approx 3.28 \ ; \ h = \frac{\pi^2}{6 \ln 2} \approx 2.37 \tag{3.29}$$

where h is the KS-entropy of the Gauss map (3.27). This may be compared with a non-generic case of the scale invariance for $r = [m^\infty]$:

$$s_q = \frac{m + \sqrt{4 + m^2}}{2} \to 1.618... \tag{3.30}$$

Numerical value is given for $m = 1$ (golden tail).

A grand example of renormalization chaos is the oscillation of the whole universe near the singularity in the homogeneous but anisotropic cosmological models [36]. So far there is no sign of such oscillations in our early Universe. Yet, the equations of the general relativity allow that type of solution. Remarkably, the very complicated relativistic equations are here approximately reduced to the trivial Gauss map.

5. Higher dimensions

A general picture of overlapping resonances, which destroy KAM tori, holds for arbitrary number of degrees of freedom [7]. This allows us to extend our resonant theory of critical phenomena to higher dimensions. There are two generally different cases of the latter: (i) $N > 2$ degrees of freedom, and (ii) a driving quasiperiodic perturbation of one degree of freedom. In the resonant theory both are similar, the principal parameter being the number of frequencies N [37].

First of all, for arbitrary N the number of all resonances with $\sim q$ harmonics of each basic frequency is $\sim q^N$, hence the detuning $\rho \sim q^{-N}$ (cf. eq. (3.9)), and

$$D = \rho q^N \sim 1 \tag{3.31}$$

Now the main rotation number r is defined with respect to one of the perturbation frequencies. The remaining $N - 2$ rotation numbers enter as driving frequencies $\nu_q \sim q\rho \sim q^{1-N}$ in the critical Hamiltonian (3.12). The resonance width $\Delta\rho \sim \nu_q^{1/2}$, and from the overlap criterion (3.10) the criticality condition is $\nu_q \sim q^{-2N}$, or (cf. eq. (3.14))

$$R \sim V = \nu_q q^{2N} \sim 1 \tag{3.32}$$

Hence, the critical perturbation smoothness $\beta_c = 2N - 1$ increases with N (cf. ref. [37]).

Longitudinal amplitudes $a_q \sim q\nu_q/\nu_q^2 \sim q^{-1}$ of the critical motion $x_c(t)$ do not depend on N, and

$$A = qa_q \sim 1 \tag{3.33}$$

as before. The transverse amplitudes $b_q \sim a_q\nu_q \sim q^{-N} \sim \Delta\rho$ decrease with N but remain of the order of the width of the resonances. Finally, the perturbation scaling does not, approximately, depend on N:

$$P = q \, \Delta \, K_q \sim 1 \tag{3.34}$$

However, in the many-dimensional case $(N > 2)$ there is no simple procedure to single out the principal resonances like for $N = 2$.

The renormalization group in higher dimensions was generally discussed already for dissipative systems (see, e.g., ref. [35]). Yet, I am not aware of any particular results concerning the scaling properties in such systems.

To the best of my knowledge the only numerical data for $N = 3$ (standard map with a time-periodic parameter $K(t)$) were presented recently in ref. [23]. They seem to confirm the scalings related to R, P and A.

On the other hand, the authors did not find the scale invariance in this model, and it seems that it does not exist at all. What is even more important, they discovered a breakdown of the renormalization universality in the sense that irregular oscillations of the critical structure depend, generally, not only on the two rotation numbers but also on the parameters of the model. Thus, many-dimensional renormalization dynamics appears to be even more complicated (chaotic?) as compared to the simplest case $N = 2$.

4 Critical Statistics

The most difficult, and as yet unsolved, problem is the impact of the critical structure at the chaos border on the statistical properties of motion.

1. Smooth perturbation: $\beta < \beta_c$. To begin with, let us consider a simpler problem of a smooth perturbation (1.3) with $\beta < \beta_c = 3$. First, we can calculate β_c directly from the resonance overlap criterion as applied to the original perturbation (1.3). The simplest estimate is as follows. The total width of all primary resonances $r_{pq} = p/q$ on the unit r interval is

$$\sim \sum_q q v_q^{1/2} \sim K^{1/2} \sum_q q^{\frac{1-\beta}{2}} \sim 1 \tag{4.1}$$

This sum diverges for $\beta \le 3$, hence $\beta_c = 3$ in agreement with the previous estimate in section 3.2. The latter estimate in eq. (4.1) determines those resonances which provide the overlapping for $\beta < 3$. The critical $q_c \sim K^{(\beta-3)^{-1}}$. The corresponding resonance width $\Delta \rho_c \sim K^{1/2} q_c^{-(\beta+1)/2}$, and the frequency (cf. eq. (3.17)) $\omega_c \sim q_c \Delta \rho_c$. Hence, the diffusion rate in r (or in y) is

$$D \sim \omega_c (\Delta \rho_c)^2 \sim K^{3/2} q_c^{-\frac{1+3\beta}{2}} \sim K^{\frac{5}{3-\beta}} \tag{4.2}$$

The border case $\beta = 3$ requires more accurate estimates.

Estimate (4.2) agrees with numerical results in ref. [38] for $\beta = 1$ (discontinuous $f(x)$).

2. Critical perturbation [31]. One peculiarity of the standard map (1.6) is the periodicity not only in x but also in y with the same period 2π. As a result there is exact critical perturbation $K = K_G \approx 1$ [33], such that for $K > K_G$ the motion is unbounded in y and diffusive for some initial conditions (section 2.3). The problem I am going to discuss now is to explain the scaling (2.3) for the diffusion rate as $K \to K_G$.

For $K > K_G$ the last (most robust) KAM curve is destroyed and transformed into a chaotic layer comprising all critical scales $q_n \ge q_\epsilon$ where $q_\epsilon \sim \epsilon^{-1}$, and $\epsilon = K - K_G \to 0$ (see eq. (3.19) and around). This chaotic layer is just the critical 'bottleneck' which controls the transition time between integer resonances $r = m$, and, hence, the global diffusion. The time scale in the layer is $\sim q_\epsilon$, and so is the exit time (t) from the layer. However, penetration into this thin layer ($\Delta r_\epsilon \sim q_\epsilon^{-2}$, eq. (3.16)) from a big region ($\Delta r \sim 1$) takes much longer time:

$$t_+ \sim t_- \frac{\Delta r}{\Delta r_\epsilon} \sim q_\epsilon^3 \sim \epsilon^{-3} \sim D^{-1} \tag{4.3}$$

This determines the transition time inversely proportional to the diffusion rate D, in agreement with recent numerical results (see eq. (2.3) and ref. [18]). Notice that the first value for the exponent ≈ 2.6 [7] was not very accurate. The above estimate $\Delta r_\epsilon / t_- \sim \Delta r / t_+$ (4.3) is simply the flow balance in statistical equilibrium.

It is interesting to note that the renormalization group theory [39] gives the value $\ln s_c / \ln s_K = 3.011...$ This is another example of the surprising accuracy of the apparently primitive resonant theory.

3. The chaos border [30]. The impact of the critical structure at the chaos border in phase space on the statistical properties of the chaotic motion is the most difficult, and as yet unsolved, problem. The straightforward approach would be as follows. The transition time τ_n between adjacent scales is proportional to the time scale $t_n \sim q_n$

which, in turn, scales like $\mu_n^{-1/2} \sim C_y^{-1/2}$ where $\mu_n \sim \rho_n \sim q_n^{-2}$ is the sticking measure, and where C_y is the correlation (section 3.1 and 3.2). Hence, $C_y \sim \tau^{-2}$, and $p_c = 2$; $p = 3$. In a more sophisticated way the same result was obtained in ref. [40]. Unfortunately, this is in sheer contradiction with numerical data: $p \approx 1.5 < 2$.

The only way of avoiding this contradiction that I see is to conjecture that at the exact criticality all transition times $\tau_n = \infty$, i.e., that all scales are dynamically disconnected. Why does then a connected chaotic component near the chaos border exists? The natural answer is in that the exact criticality is achieved on the border only, while inside a chaotic region the motion is supercritical. Consider, for example, model (3.1). Locally it is described by the standard map with $K \approx \lambda/y$. In a small vicinity of the border $y = y_b \approx \lambda$, the perturbation K indeed increases like $\triangle K \sim \triangle y \sim \rho \sim q^{-2}$. However, this is not enough to destroy the corresponding scale q_n, as $q_n \triangle K_n \sim q_n^{-1} \ll 1$ (see eq. (3.19)). Only resonances with $q \geq Q_n \sim q_n^2$ would be destroyed and they form a very narrow ($\sim Q_n^{-2} \sim q_n^{-4}$) chaotic layer which could play a role of the bottleneck controlling the transition time τ_n. Similarly to derivation of eq. (4.3) we obtain

$$\tau_n \sim Q_n \frac{Q_n^2}{q_n^2} \sim q_n^4 \sim \mu_n^{-2} \sim C_y^{-2} \tag{4.4}$$

Hence

$$C_y(\tau) \sim \tau^{-1/2} \; ; \; P(\tau) \sim \tau^{-3/2} \tag{4.5}$$

now in agreement with numerical data.

The same result can be obtained in a different, more formal, way. Namely, we can rescale the dependence (4.3) for the transition between integer resonances ($q_n = 1$) to arbitrary scale q_n. To this end we rewrite eq. (4.3) in scaled variables

$$\frac{\tau_n}{t_n} \sim (q_n \triangle K)^{-3} \tag{4.6}$$

With $t_n \sim q_n$ and $\triangle K \sim q_n^{-2}$ we arrive at eq. (4.4). Notice that a different relation $\tau_n(q_n)$ in ref. [22] was due to a mistake in scaling.

A weak point of the latter approach (4.6) is that the scaling (4.3) is asymptotic ($q_n \to \infty$), while the integer resonances ($q_n = 1$) are not. In any event, further studies into the mechanism of critical statistics are certainly required.

In higher dimensions (section 3.5) the supercriticality $\triangle K \sim \rho \sim q^{-N}$; the bottleneck harmonic is $Q \sim (\triangle K)^{-1} \sim q^N$, and the transition time (cf. eq. (4.4)) is given by

$$\tau \sim Q^{N-1} \frac{Q^N}{q^N} \sim Q^{2N-2} \sim \mu^{2-2N} \tag{4.7}$$

Hence

$$C_y \sim \mu \sim \tau^{-\frac{1}{2N-2}} \; ; \; p_c = \frac{1}{2N-2} \tag{4.8}$$

As $N \to \infty$, $p_c \to 0$, and correlations do not decay at all. I am going to come back to this interesting case in section 4.5 below.

Another difficult problem is the arithmetic of rotation numbers r_b of the critical border curves. In ref. [22] it was conjectured that the set of r_b consists of all combinations of only two elements $m = 1$ and 2 in the continued fraction representation. This is sufficient for r_b to be random, and hence to explain the irregular oscillation of the local exponent in the distribution of Poincaré recurrences (section 3.1). This conjecture was partially confirmed numerically in ref. [45]. Our recent refined conjecture is that r_b are the so-called Markov numbers [30].

4. Internal borders. Typically, the central part of principal critical resonances is not destroyed (see, e.g., Fig.5). Hence, in any neighborhood of the main chaos border there is an infinite set of internal chaos borders, each one with its own critical structure. Assuming universality of critical phenomena at any chaos border we arrive at the following estimate in scaled variables

$$(\mu_c q^2) \sim (\frac{\tau}{q})^{-p_c} \tag{4.9}$$

for a principal resonance q where μ_q is the sticking measure at the internal border.

The main difficulty here is that the internal borders exists not only inside the principal resonances but also in many others, near the critical border curve, which are not destroyed by the local supercritical perturbation. To estimate the total number of such resonances we can make use of eq. (3.19) which determines the stability zone $\Delta K_s \sim \rho_s \sim q^{-1}$ for any q (as a very crude approximation, of course). Then, for a given q only $M_q/q \sim 1$ resonances fall into this zone, where $M_q \sim q$ is the total number of resonances p/q for a fixed q. Again, as a crude approximation we can extend the estimates, particularly eq. (4.19), on all undestroyed resonances. As a result, the total internal border contribution to the correlation is

$$\tilde{C}_y \sim \sum_q \mu_q \sim \tau^{-p_c} \sum_q^{\tau} q^{p_c-2} \tag{4.10}$$

where the sum is taken over all q up to τ. This contribution is essential if $p_c \geq 1$. But for $p_c > 1$ the above estimate is not self-consistent as $\tilde{C}_y \sim \tau^{-1}$, contrary to assumed universality. However, the latter holds for $p_c = 1$, to logarithmic accuracy. This was the preliminary conclusion in ref. [41] which was also confirmed in ref. [42].

This would be a nice solution in the spirit of universality of the critical phenomena. Yet, first, the value $p_c = 1$ seems still to be incompatible with numerical data (section 3.1), and second, there is another possibility missed in ref. [41], namely, $p_c < 1$, as is suggested by numerical data. Then, the effect of internal borders is not decisive, at least, for the exponent p_c whose value is determined by another mechanism, for example, the one described in the previous section.

In higher dimensions we have instead of eq. (4.9) (see section 3.5)

$$(\mu_q q^N) \sim (\frac{\tau}{q^{N-1}})^{-p_c} \tag{4.11}$$

In calculating the total contribution of all internal borders we need to take into account that now there are as many as $\sim q^{N-2}$ undestroyed resonances, for a given q, within the stability zone. Hence, the total 'internal' correlation is

$$\tilde{C}_y \sim \sum_q \mu_q q^{N-2} \sim \tau^{-p_c} \sum_q^{\tau^{1/(N-1)}} q^{p_c(N-1)-2} \tag{4.12}$$

The critical value p_c^* of the critical exponent is $p_c^* = (N-1)^{-1}$. Only this value preserves universality based entirely on the internal borders. And, again, there is another possibility that $p_c < p_c^*$ so that internal borders are irrelevant. This is just the case if the above estimate (4.8) is true: $p_c = p_c^*/2$.

Preliminary numerical results obtained in collaboration with V.V. Vecheslavov ($p_c \approx 0.26$ and 0.19 for N=3 and 4, respectively) seem to confirm (or, at least, do not contradict) prediction (4.8).

5. Superfast diffusion [22]. Slow correlation decay with $p_c < 1$ (4.5) may result in a superfast diffusion. Indeed, if this correlation determines the diffusion, the rate

$$D_Z \sim \int C_y(\tau)d\tau$$

formally diverges. Here the diffusion proceeds along a new variable Z, and $\dot{Z} = Y$. The divergence means that the dispersion (the second moment of the distribution function)

$$\sigma^2 \sim \int D d\tau \sim t^{2-p_c} \tag{4.13}$$

grows faster than the time t, hence we describe this as the 'superfast diffusion'. This phenomenon was studied from different points of view in many papers (see, e.g., refs. [43,44]).

The simplest example is again the standard map for special values of the parameter $K \approx 2\pi m$, with any integer $m \neq 0$. At these K the so-called accelerator modes exist [7], i.e., relatively small areas of regular motion with linearly increasing momentum: $y \sim \pm t$, while the phase x is fixed. A chaotic trajectory cannot penetrate into these domains but it does stick to their borders. As a result, a superfast diffusion in y occurs, which was first observed numerically in ref. [46]. Notice that in the above notation $z = y$, while the role of y is now played by a new coordinate normal to the chaos border surrounding the regular regions. According to eq. (4.13)

$$\sigma^2 \approx \alpha \mu_s \frac{K^2}{2} t^{3/2} \sim t^{3/2} \tag{4.14}$$

where $\alpha \approx 0.5$ from numerical data [46] for $K \approx 2\pi$, and relative stable area $\mu_s \approx 0.02$. As $\mu_s \sim K^{-2}$ [7], the rate of this anomalous diffusion ($\sigma^2/t^{3/2}$) does not depend either on $K \to \infty$ or on $\mu_s \to 0$.

In ref. [44,47] more complicated accelerator modes were shown to produce a super-fast diffusion, corresponding to $p_c \approx 2/3$, in reasonable agreement with our numerical data. A simple expression for the growth of all moments of the distribution function was also given in ref. [44], namely:

$$\sigma^k \sim t^{k-p_c} \tag{4.15}$$

for k even. In higher dimensions when $N \to \infty$ and $p_c \to 0$ this relation becomes especially simple but somewhat puzzling. It appears to describe an almost free motion, but in both directions of the Z variable! The limiting case $p_c = 0$ corresponds to the fastest homogeneous diffusion possible.

Further insight into the nature of superfast diffusion can be obtained from the dynamical power spectrum, i.e., the Fourier transform of the correlation [30]. For $\omega \to 0$ we have from eq. (4.8)

$$S_y(\omega) \sim \omega^{p_c-1} \sim \omega^{-\frac{2N-3}{2N-2}} \tag{4.16}$$

As $N \to \infty$, it approaches the famous $1/\omega$ spectrum which, thus, produces the fastest diffusion. If $\dot{Z} = y$, the spectrum of the Z-motion is

$$S_Z = \frac{S_y}{\omega^2} \sim \omega^{p_c-3} \tag{4.17}$$

130

From normalization (Parseval's theorem)

$$\overline{Z^2} = \int S_Z(\omega)d\omega \sim \omega^{p_c-2} \sim t^{2-p_c} \qquad (4.18)$$

If $p_c < 1$ the integral diverges as $\omega \to 0$. For a finite time interval the minimal $\omega \sim t^{-1}$, and the diffusion law (4.13) is recovered, including the limiting $p_c = 0$. However, in the latter case the velocity dispersion $\overline{y^2} \sim \ln \omega$ diverges (see eq. (4.16)). In our models with a chaos border this is impossible, so $p_c > 0$ always.

The theory of superfast diffusion can be applied to a broad variety of different problems. A nice example is the tangle of a long polymeric molecule in a certain environment. Such molecule can be considered approximately as a trajectory of the self-avoiding random walk. The constraint imposes a long-term correlation which can be estimated as follows. Suppose that the molecule length l and the tangle size σ are related by

$$\sigma^2 \sim l^{2\nu} \qquad (4.19)$$

with some so far unknown parameter ν. The correlation due to avoided crossings of the molecule line is then roughly proportional to the probability of the self-crossing:

$$C \sim \frac{l}{\sigma^d} \sim l^{1-\nu d} \qquad (4.20)$$

where the integer d is the space dimension. Hence, we have a power-law correlation with the exponent $p_c = \nu d - 1$. Using eq. (4.13) with $t = l$ and eq. (4.19) we arrive at the relation

$$2\nu = 3 - \nu d \; ; \quad \nu = \frac{3}{2+d} \qquad (4.21)$$

which is known as Flory's formula (see, e.g., ref. [48]). It was derived in a completely different way (from the thermodynamics of a polymeric molecule), and holds for $d \leq 4$, otherwise $\nu = 1/2$. In our dynamical approach the latter limitation follows from the condition $p_c < 1$ for anomalous diffusion. In the border case $\sigma^2 \sim l \ln l$ (see, e.g., T. Geisel et al. in ref. [43]), which slightly differs from Flory's formula.

6. Fractal properties [49]. The critical structure in Hamiltonian systems is also called 'random fractals' (R.Voss, ref. [43]) because of the renormalization chaos (section 3.4), or 'fat fractals' [49] for carrying a finite measure in contrast to dissipative systems. Some fractal properties were studied numerically in ref. [49].

Here I am going to explain one property - the fractal dimension d_L of the set of all chaos borders (mainly internal ones, of course). It is inferred from the dependence of the measure μ_{ch} of a chaotic component on the spatial linear resolution $\epsilon \to 0$:

$$\mu_{ch}(\epsilon) = \mu_{ch}(0) + \alpha \, \epsilon^\beta \qquad (4.22)$$

Here $\mu_{ch}(0) > 0$ is the measure of the whole chaotic component. Hence, its topological dimension is $d_s = 2$. The second term represents the borders whose total length and dimension are

$$L(\epsilon) \sim \epsilon^{\beta-1} \; ; \quad d_L = 2 - \beta \qquad (4.23)$$

The simplest evaluation of this scaling can be done as follows (see section 4.4). Each surviving resonance has internal borders of total length $l_q \sim 1$. This estimate follows from the fact that an individual border is a curve of topological dimension $d_l = 1$. This is so because the ratio of transverse to longitudinal scaling factors of the critical structure $s_b/s_a = q_n \to \infty$ as $n \to \infty$ (see sections 3.2 and 3.3). Thus, the border curve $y_b(x)$ is very smooth (see Fig.5). The number of surviving resonances is of the order of maximal $q = q_{max}$, determined by the resolution $\epsilon \sim q_{max}^{-2}$. Hence

$$L \sim \epsilon^{-1/2} \ ; \ \beta = \frac{1}{2} \ ; \ d_L = \frac{3}{2} \tag{4.24}$$

in reasonable agreement with the numerical result $\beta = 0.4 - 0.7$ [49].

Notice, also, that the total number of resolved borders scales, in this approximation, as

$$N_b \sim q_{max}^2 \sim \epsilon^{-1} \tag{4.25}$$

In higher dimensions with N frequencies we need to consider an N-dimensional map with non-fractal border surfaces of $N-1$ dimensions. Now there are $\sim q^{N-1}$ surviving resonances up to $q_{max} \sim \epsilon^{-1/N}$. The border surface in each such resonance $S_1 \sim 1$, and the total border surface and its dimension are

$$S_b \sim \epsilon^{-(1-\frac{1}{N})} \ ; \ d_S = N - \frac{1}{N} \tag{4.26}$$

The total number of resolved border surfaces, or of the domains with regular motion

$$N_b \sim \epsilon^{-1} \tag{4.27}$$

does not depend on N, and in this sense it is universal.

Acknowledgements

I would like to express my sincere gratitude to D.L.Shepelyansky, my main collaborator in the studies of critical phenomena, to F.Vivaldi who attracted our attention to this interesting problem, and to G.Casati, D.Escande, R.MacKay, I.Percival and Ya.Sinai for stimulating discussions.

References

[1] V.M. Alekseev and M.V. Yakobson, Phys. Reports **75** (1981) 287.

[2] G. Chaitin, Information, Randomness and Incompleteness (World Scientific, 1990).

[3] B.V. Chirikov, F.M. Izrailev and D.L. Shepelyansky, Physica D **33** (1988) 77.

[4] B.V. Chirikov, Time-Dependent Quantum Systems, Proc. Les Houches Summer School on Chaos and Quantum Physics (Elsevier, 1990).

[5] A. Lichtenberg, M. Lieberman, Regular and Stochastic Motion (Springer, 1983).

[6] G.M. Zaslavsky, Chaos in Dynamic Systems (Harwood, 1985).

[7] B.V. Chirikov, Phys. Reports **52** (1979) 263.

[8] B.V. Chirikov and V.V. Vecheslavov, Astron. Astroph. **221** (1989) 146.

[9] G. Casati et al., Phys. rev. A **36** (1987) 3501.

[10] A.A. Chernikov, R.Z. Sagdeev and G.M. Zaslavsky, Physica D **33** (1988) 65.

[11] B.V. Chirikov, Foundations of Physics **16** (1986) 39.

[12] M. Eisenman et al., Lecture Notes in Physics **38** (1975) 112; J.von Hemmen, ibid., **93** (1979) 232.

[13] B.G. Konopelchenko, Nonlinear Integrable Equations, Lecture Notes in Physics **270** (1987).

[14] B.V. Chirikov and V.V. Vecheslavov, KAM integrability, in: Analysis etc. (Academic Press, 1990) p.219.

[15] V.I. Arnold and A. Avez, Ergodic Problems in Classical Mechanics (Benjamin, 1968).

[16] B.V. Chirikov, Proc. Roy. Soc. Lond. A **413** (1987) 145.

[17] A. Rechester et al. Phys. Rev. A **23** (1981) 2664.

[18] B.V. Chirikov, D.L. Shepelyansky, Radiofizika **29** (1986) 1041.

[19] F. Vivaldi, private communication.

[20] I. Kornfeld, S. Fromin and Ya. Sinai, Ergodic Theory (Springer, 1982).

[21] R. MacKay, Physica d **7** (1983) 283.

[22] B.V. Chirikov and D.L. Shepelyansky, Physica D **13** (1984) 395.

[23] R. Artuso, G. Casati and D.L. Shepelyansky, 1990 (to appear).

[24] B.V. Chirikov, D.L. Shepelyansky, Proc. 9th Int. Conf. on Nonlinear Oscillations, Kiev, 1981, Vol.2, p.421. (Kiev, Naukova Dumka, 1983) English translation available as preprint PPL-TRANS-133, Plasma Physics Lab., Princeton Univ., 1983.

[25] S. Channon and J. Lebowitz, Ann. N.Y. Acad. Sci **357** (1980) 108.

[26] G. Paladin and A. Vulpiani, Phys. Reports **156** (1987) 147.

[27] C. Karney. Physica D **8** (1983) 360.

[28] P. Grassberger and I. Procaccia, Physica D **13** (1984) 34.

[29] J. Bene, P. Szépfalusy and A. Fülöp, A generic dynamical phase transition in chaotic Hamiltonian systems, Phys. Rev. Lett. (to appear).

[30] B.V. Chirikov and D.L. Shepelyansky, Chaos Border and Statistical Anomalies, in: Renormalization Group, D.V. Shirkov, D.I. Kazakov and A.A. Vladimirov (eds.), p. 221. (World Scientific, Singapore, 1988).

[31] B.V. Chirikov, Intrinsic Stochasticity, Proc. Int. Conf. on Plasma Physics, Lausanne, 1984, Vol. II, p.761.

[32] G. Schmidt and J. Bialek, Physica D **5** (1982) 397.

[33] J. Greene, J.Math.Phys. **9** (1968) 760; **20** (1979) 1183.

[34] M. Feigenbaum, J.Stat.Phys. **19** (1978) 25; **21** (1979) 669.

[35] S. Ostlund et al.,Physica D **8** (1983) 303.

[36] E.M. Lifshits et al., Zh.Eksp.Teor.Fiz. **59** (1970) 322; ibid (Pisma) **38** (1983) 79; J. Barrow, Phys. Reports **85** (1982) 1.

[37] B.V. Chirikov, The Nature and Properties of the Dynamic Chaos, Proc. 2d Int.Seminar "Group Theory Methods in Physics" (Zveingorod, 1982), Vol.1, p.553. (Harwood, 1985).

[38] I. Dana et al., Phys.Rev.Lett. **62** (1989) 233.

[39] R. MacKay et al., Physica D **13** (1984) 55.

[40] J. Hanson et al. J.Stat.Phys. **39** (1985) 327.

[41] B.V. Chirikov, Lecture Notes in Physics **179** (1983) 29.

[42] J. Meiss and E. Ott, Phys.Rev.Lett. **55** (1985) 2741; Physica D **20** (1986) 387.

[43] P. Lévy, Théorie de l'addition des variables eléatoires. (Gauthier-Villiers, Paris, 1937); T. Geisel et al. Phys.Rev.Lett. **54** (1985) 616; R. Pasmanter, Fluid Dynamic Research **3** (1988) 320; R. Voss, Physica D **38** (1989) 362; G.M. Zaslavsky et al., Zh.Exper.Teor.Fiz. **96** (1989) 1563.

[44] H. Mori et al., Prog.Theor.Phys. Suppl., 1989, No.99, p.1.

[45] J. Greene et al. Physica D **21** (1986) 267.

[46] C. Karney et al., ibid **4** (1982) 425.

[47] Y. Ichikawa et al., ibid **29** (1987) 247.

[48] P. de Gennes, Scaling Concepts in Polymer Physics. (Cornell Univ. Press., 1979).

[49] D. Umberger and D. Farmer, Phys.Rev.Lett. **55** (1985) 661; G. Grebogi et al., Phys.Lett. **A 110** (1985) 1.

SPATIO-TEMPORAL CHAOS

Jean-Pierre Eckmann and Itamar Procaccia

Département de Physique Théorique, University of Geneva, Switzerland
Department of Chemical Physics, The Weizmann Institute of Science, Israel

1. Introduction

Continuous physical systems, such as electromagnetic fields or fluids, are described dynamically by partial differential equations, or field theories. Thus they have infinitely many degrees of freedom; accordingly, it came as a surprise at the end of the seventies that such systems can exhibit low dimensional chaos which is characteristic of systems with few degrees of freedom. In the meanwhile this miracle has been fully understood. By keeping a fluid in a box which is not too large compared to some typical macroscopic scale (like the size of a convection roll) one can maintain spatial coherence but the system will become temporally chaotic. Only a few spatial modes get appreciably excited, and their amplitudes define a low-dimensional "phase-space" in which the chaotic dynamics takes place.

On the other hand, in extended physical systems which are much larger than the typical macroscopic scale, very different phenomena appear. The coherence in space cannot be maintained when complicated phenomena occur. Since extended systems can support many different types of motions locally, these can interact to destroy spatial coherence before or concurrently with the onset of temporal chaos. A new language needs to be developed in order to capture relevant and controllable aspects of this new phenomenology.

A fundamental difficulty which stands in the way of any theory is that systems of this type are described by partial differential equations (PDE's) such as the equations of fluid dynamics. For *ordinary* differential equations, i.e., dynamical systems in *finite dimensional phase space* there exists a rich body of theory of mathematical rigor. For partial differential equations, no general theory is known and only sporadic results for special cases are available. In these lectures, we describe a scheme which partially avoids these difficulties by restricting the questions to solvable problems. At the same time, we feel that the scheme is sufficiently general to capture some essential aspects of spatio-temporal chaos.

There are two essential ingredients in our approach to the problem of spatio-temporal chaos [EP1, EP2]. The first is a judicious choice of the *geometry* that confines the system under consideration. The second is an identification of *spatially chaotic* but *time stationary* states as organizers of the space-time evolution.

The *geometry* we consider will have one direction of *infinite extent*, and zero, one, or two directions of finite extent. This means that we focus attention to physical systems which are confined to infinite lines, strips, or slabs. The infinity of one direction has the same positive effects on the theoretical tractability of the problem as the thermodynamic limit has in Statistical Mechanics. We therefore view finite size experiments as perturbations of these idealized infinite systems. We do not treat finite size effects in these lectures. It should be possible to take these into account systematically, as has been shown recently in an interesting paper [N]. The choice of an infinite system will allow us to offer a rigorous analysis of the spatial structure of time stationary solutions of the PDE's by using dynamical systems theory.

The second ingredient of our method is to use the time stationary solutions which can be shown to be spatially chaotic as *organizers* of the full space time motion. This is similar to the use of hyperbolic fixed points of dynamical systems to understand aspects of chaotic dynamics. Locally, our spatially chaotic, time stationary states have stable and unstable directions for the time-dependent problem. Initial conditions falling close to the stable manifold result in time-evolutions which come close to the spatially chaotic time stationary states. In a later phase of temporal evolution, the orbit leaves along the unstable direction. This scenario pinpoints and captures some relevant aspects of spatio-temporal chaos.

One interesting consequence of this approach is that it clarifies the relation between spatial disorder and the creation of defects. In fact, topological defects now appear in the second stage of following the unstable manifold of a time stationary state which is already chaotic in space. In particular, it will be seen that topological defects are possible by the availability of space directions which are normal to the infinite one, whereas spatial chaos appears already in the 1-dimensional problem. It will turn out that there is *no conceptual difference* between the long wavelength instabilities such as the Eckhaus instability and the skew-varicose instability with its associated formation of topological defects.

The ubiquity of defect-mediated turbulence in hydrodynamic systems with large aspect ratio indicates that the phenomenon calls for an essentially model-free interpretation. Spatio-temporal chaos with non-trivial dynamics of defects has been observed in fluid thermal convection [AB,PCL], in nematics under electrohydrodynamic convection [RJ,LG,RRS,GPRS], in surface waves [K], in numerical simulations of certain partial differential equations in 2+1 space-time dimensions [CGL,BPL] etc. In these lectures we show that indeed the phenomenon is expected to appear when rather general conditions are met. These are:

i) The partial differential equations do not have a Liapunov function, i.e., they are not derivable from a potential.

ii) The system exhibits a fundamental instability towards a one-dimensional cellular pattern, e.g., convection rolls. These patterns exist for a range of parameters and we shall assume that in this range no bounded, biperiodic solutions can exist.

iii) The system has a large aspect ratio in one dimension (to be taken as an infinite x coordinate) and a medium extent in another dimension (the y coordinate). The third dimension (the z direction) is small, and it determines the scale of the cellular pattern. The extent of the system in the y direction will be a central parameter in our theory.

iv) The underlying hydrodynamic equations possess continuous symmetries (such as translation and rotation) which give rise to secondary long wavelength instabilities of the cellular structure, e.g., skew varicose [ZS, GC]. We shall assume that the secondary instability involves transverse and longitudinal modes; this assumption will lead naturally to the creation of defects.

v) The highest-order spatial derivative in the hydrodynamic equations is of order m, and is linear in the fields. This assumption will allow us a direct link to dynamical systems theory.

The systems which we consider have crossed one of the instabilities, described in ii), e.g., Eckhaus or skew-varicose. We shall show that although the fundamental cellular pattern is linearly unstable, a time-stationary, spatially quasiperiodic solution with two independent periods can form. Within perturbation theory, one shows that the nonlinearities will allow this solution to exist at some finite amplitude of the unstable mode. However, perturbation theory fails to reveal that close to these solutions there exist stationary solutions which are chaotic in space. Below, we construct these solutions and discuss their relevance for the space-time dynamics.

2. Models

One model which possesses properties (i)–(v) is the generalized Swift-Hohenberg model in 2 space and 1 time dimensions:

$$\partial_t u + (\vec{U} \cdot \vec{\nabla})u = \big(\alpha - (1 + \partial_x^2 + \partial_y^2)\big)^2 u - u^3 , \tag{2.1}$$

where

$$U = (\partial_y \zeta, -\partial_x \zeta) ,$$

and

$$\nabla^2 \zeta = g(\vec{\nabla}(\nabla^2 u) \times \vec{\nabla}u) \cdot \hat{z} . \tag{2.2}$$

Here, g is a coupling constant, $\alpha \geq 0$ is the driving force and $\hat{z}$ is a unit vector in the z direction. The phase diagrams of the long wavelength instabilities of this and related models have been analyzed in detail in [GC]. All the analysis which we expound in these lectures can be done for the model (2.1)–(2.2). However, since our central ideas are focused on the *local* time properties, the questions of existence of a Liapunov functional, which determines the *global* behavior are not a part of our study. We therefore consider simpler models which have the same local properties as the model (2.1)–(2.2), but which can be extracted much more easily. We draw the reader's attention to the fact that the more general context of [GC] is intended, but what really is being used are properties (ii)–(v) and not any special feature of the PDE.

The simpler model is (2.1) with $g = 0$. This reads, [SH],

$$\partial_t u(x, y, t) = \big(\alpha - (1 + \partial_x^2 + \partial_y^2)^2\big)u(x, y, t) - u^3(x, y, t) . \tag{2.3}$$

We view this equation as some sort of "normal form" for the class of problems covered by our method. It will turn out that the understanding of the 2+1 dimensional problem relies heavily on a proper reformulation of the 1+1 dimensional problem. Therefore, we begin our study with the 1+1 dimensional version of (2.3):

$$\partial_t u(x, t) = \big(\alpha - (1 + \partial_x^2)^2\big)u(x, t) - u^3(x, t) . \tag{2.4}$$

This equation will be called the (1+1 dimensional) Swift-Hohenberg equation (SH). A detailed account of the mathematics for this equation can be found in [CE].

Let us show in which sense (2.4) is a normal form. For example, one can reduce

$$\mu \partial_t v(x, t) = \big(\alpha - (1 + \gamma \partial_x^2)^2\big)v(x, t) - \beta v^3(x, t) , \tag{2.5}$$

to the form (2.4) by a change of variables of the form $v(x, t) = au(bx, ct)$.

What matters for our study is that a frequency ω becomes unstable and gives rise to the pattern formation. In Eq.(2.4) this frequency is normalized to 1. The precise choice of the nonlinearity is irrelevant for what follows, and we really only use the fact that

i) The zero solution is unstable, leading to pattern formation.

ii) The nonlinearity saturates the growth of the unstable solution.

Consider now Eq.(2.4). It is normalized in such a way that $u(x,t) \equiv 0$ is a stationary, i.e., time-independent, solution. The standard procedure starting from any stationary solution is to study its *linear stability*, i.e., the stability under small perturbations, really infinitesimally small perturbations. One of the basic difficulties of the subject is to find an adequate space of perturbations which one wants to consider. In finite dimensional problems, there is in general an obvious choice: $\mathbf{R}^n$ with an arbitrary topology. In infinite dimensional problems, such as the one we are considering here, this is much less obvious. For example, in quantum mechanics, the natural choice is the space of L^2 functions, because of the relation of this space to probability measures. In hydrodynamic problems, where u represents something like a "field" of temperatures or velocities, some class of *bounded* functions, *not decaying to zero at infinity* seems more adequate. We shall take L^∞. This space is "bigger" than L^2 at infinity and hence the boundary conditions at infinity are weaker than in quantum mechanical problems. Thus we expect a richer set of possibilities. On the other hand, equations such as (2.4) are locally very regular and hence we can expect, and will find, smooth solutions.

In L^∞, the linearization of Eq.(2.4) at $u = 0$ is given by

$$\partial_t v(x,t) = \left(\alpha - (1 + \partial_x^2)^2\right)v(x,t) \ . \tag{2.6}$$

Defining the r.h.s. of (2.6) as an operator L, we see that if $v_\sigma(x)$ is an eigenvector with eigenvalue σ of L, then, on the linear level,

$$v(x,t) = e^{t\sigma}v_\sigma(x) \ .$$

Thus, v will grow in time if the real part of σ is positive. It is now easy to see that $v(x) = \exp(i\omega x)$ is an eigenvector of L with eigenvalue

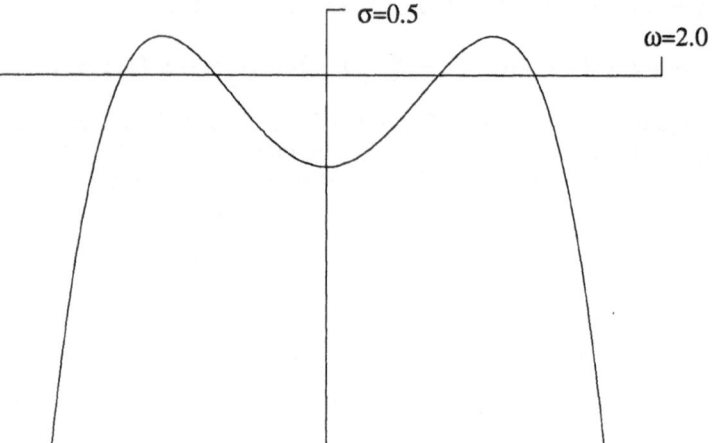

Fig. 1. The spectrum of the operator $\alpha - (1 + \partial_x^2)^2$ in L^∞ for $\alpha > 0$. The horizontal axis is the frequency ω, the vertical axis is the eigenvalue σ.

$$\alpha - (1 - \omega^2)^2 \ ,$$

and we see from Fig. 1 or by direct calculation that *if $\alpha > 0$ then the Eq.(2.6) has unstable modes.* The unstable modes exist for a band of frequencies, namely the set of ω satisfying $\alpha > (1 - \omega^2)^2$.

3. Stationary, Cellular Solutions of the SH-Equations

We next study the existence of stationary solutions (other than 0) in perturbation theory. We *choose* one of the unstable frequencies ω. Furthermore, we insist on finding real solutions to Eq.(2.4). So we study

$$\left(\alpha - (1 + \partial_x^2)^2\right)u(x) - u^3(x) = 0 \ , \tag{3.1}$$

and we look for a cellular, i.e., space-periodic solution of the form

$$u(x) \approx 2\epsilon \cos(\omega x) \ .$$

The task is to determine ϵ in perturbation theory. Note that the nonlinear term couples the positive and negative frequency parts of v and we have

$$u^3 = \epsilon^3 \left(e^{3i\omega x} + 3e^{i\omega x} + 3e^{-i\omega x} + e^{-3i\omega x}\right) \ .$$

Comparing the coefficients of $e^{i\omega x}$ in (3.1), we get the fundamental relation (valid in lowest order) for (some) stationary solutions of (3.1):

$$\alpha = 3\epsilon^2 + (1 - \omega^2)^2 \ .$$

As an exercise, let us look at what happens in the next order in ϵ. We will explain below why ϵ—not α as is often asserted in the literature—is the natural expansion parameter. The nonlinearity has created a new frequency, $\exp(3i\omega x)$ and the linear term on this frequency is $\alpha - (1 - 9\omega^2)^2$. Therefore, if we consider the unknown coefficient γ of $\exp(3i\omega x)$ in the stationary solution, we find the relation

$$0 = \left(\alpha - (1 - 9\omega^2)^2\right)\gamma - 3\gamma^3 - 3\epsilon^2\gamma - \epsilon^3 \ .$$

We now make a basic **assumption**: ω is close to 1. (For our purpose, it is in fact enough to suppose e.g., $|\omega - 1| < 0.3$, but the argument is easier when $|\omega - 1| < \mathcal{O}(\epsilon)$.) Then we find

$$\gamma \approx \frac{\epsilon^3}{64} \ . \tag{3.2}$$

Remarks. Although all this is somewhat trivial, it is important to realize at this point what kind of physical and implicit assumptions are being made. It is really here that the class of allowed solutions is being selected by the method of studying the Eq.(3.1).

i) We have chosen to work in the infinite x-domain. This has the same advantages and disadvantages as the study of the thermodynamic limit in Statistical Mechanics. Finite size boundary conditions should play in this subject the same subordinate role as do finite size effects in Statistical Mechanics.

ii) The price to pay for considering infinitely extended domains is the presence of *continuous spectrum* for the operator L. This makes rigorous treatments of the time-dependent problem (2.4) very hard to control.

iii) The way perturbation theory was started above *discretizes* the continuous spectrum by the selection of *one of the possible* frequencies ω. This frequency will couple to a *discrete* set of other frequencies, namely $n\omega$, $n \in \mathbf{Z}$. For all these other frequencies the denominator corresponding to 64 in Eq.(3.2) will be $\alpha - (1 - n^2\omega^2)^2$. This quantity is *bounded away from* 0. Therefore, perturbation theory exists order by order in ϵ. We call this condition of non-vanishing of the denominator a non-resonance condition. Note that many experiments also "artificially" select one among several possible spatial frequencies.

iv) Once we have fixed the frequency and its harmonics and if we require the solution to be an even function of x, breaking thus the translation invariance of the problem, the spectrum of L becomes *discrete* and at $\alpha = 0$ there is one simple eigenvalue which is exactly zero. All other eigenvalues are negative. As α crosses zero, this eigenvalue also crosses zero and we are in the presence of bifurcation from a simple eigenvalue, cf. [CE]. This is the reason why ϵ, the *amplitude* of the solution, is the natural expansion parameter.

Guided by perturbation theory, we search stationary solutions of the form

$$u(x) = \sum_{n \in \mathbf{Z}} e^{i\omega n x} u_n , \qquad (3.3)$$

with the condition $u_n = \bar{u}_{-n}$ making the solution real. The unknowns are thus the $u_n, n = 0, 1, \ldots$. The trick is now to fix ϵ and ω and to consider α as the free parameter of the problem. At the end of the analysis, one can use the implicit function theorem to change back to the free parameters α and ω which are common in experiments. The unknown coefficients should be viewed as elements of a Banach space $\mathbf{S}$, and an adequate norm for $u = \{u_n\}_{n \in \mathbf{Z}}$ is

$$\|u\|_\rho = \sum_{n \in \mathbf{Z}} \rho^{|n|} |u_n| , \qquad (3.4)$$

for some $\rho > 1$.

Exercise. Show that if $v(x) = u^1(x)u^2(x)$, then $\|v\|_\rho \leq \|u^1\|_\rho \cdot \|u^2\|_\rho$.

The existence proof for a solution to Eq.(3.1) is obtained as follows. Write $u(x) = 2\epsilon \cos(\omega x) + \epsilon v(x) \equiv \epsilon(v^0(x) + v(x))$ with $v(x) = \sum_{n \in \mathbf{Z}} e^{i\omega n x} v_n$ and $v_{\pm 1} = 0$. Then the stationary Eq.(3.1) takes the form—upon reparametrizing $\alpha = \lambda + (1 - \omega^2)^2$:

$$\left(\lambda + (1 - \omega^2)^2 - (1 + \partial_x^2)^2\right) u - u^3 = 0 , \qquad (3.5)$$

in the unknowns λ and v. Define the operators $G_n : \mathbf{S} \to \mathbf{C}$ by

$$G_n(v) = \left((v^0 + v)^3\right)_n , \qquad (3.6)$$

where $(\cdot)_n$ denotes the coefficient of $\exp(i\omega n x)$. Comparing the coefficients of the various Fourier modes, the equation (3.5) is equivalent to the system

$$\lambda + (1 - \omega^2)^2 - (1 - \omega^2)^2 = \lambda = \epsilon^2 G_1(v) , \quad \text{for } |n| = 1 , \qquad (3.7)$$
$$(\lambda + (1 - \omega^2)^2 - (1 - n^2\omega^2)^2)v_n = \epsilon^2 G_n(v) , \quad \text{for } |n| \neq 1 . \qquad (3.8)$$

The first equality in (3.7) follows from the parametrization of α, and it determines λ. Since

$$v^0(x) = 1 \cdot e^{i\omega x} + 1 \cdot e^{-i\omega x} ,$$

we have

$$\|v^0\|_\rho = 2\rho .$$

Therefore, we get

$$\|G(v)\|_\rho \leq (2\rho + \|v\|_\rho)^3 , \tag{3.9}$$

$$\|\partial_v G(v)\delta v\|_\rho \leq 3(2\rho + \|v\|_\rho)^2 \|\delta v\|_\rho , \tag{3.10}$$

$$\|G(0)\|_\rho = \|v_0^3\|_\rho \leq (2\rho)^3 . \tag{3.11}$$

We have already discussed the inverse of $\lambda + (1 - \omega^2)^2 - (1 + \partial_x^2)^2$ when doing perturbation theory, and it suffices to retain the result

$$\inf_{|n|\neq 1} |(1 - \omega^2)^2 - (1 - n^2\omega^2)^2| \geq D > 0.$$

Consider now the map $\mathcal{L} : \mathbf{R} \times \mathbf{S} \to \mathbf{R} \times \mathbf{S}$, defined by

$$(\lambda, v) \mapsto \mathcal{L}(\lambda, v)$$

$$= \left(\epsilon^2 G_1(v), \left\{\frac{\epsilon^2}{\lambda + (1 - \omega^2)^2 - (1 - n^2\omega^2)^2} G_n(v)\right\}_{|n|\neq 1}\right) .$$

In the equality, we use the Fourier components to define $\mathcal{L}$. Our problem is the fixed point problem

$$(\lambda, v) = \mathcal{L}(\lambda, v) . \tag{3.12}$$

Using the results of perturbation theory, it is easy to show (cf. [CE]) that $\mathcal{L}$ contracts a small ball in $\mathbf{R} \times \mathbf{S}$, centered at the origin, into itself. Therefore, $\mathcal{L}$ has a unique fixed point in this ball which is the solution of the stationary Swift-Hohenberg equation for a given amplitude ϵ and a given frequency ω.

Theorem 3.1. *Let $a < \sqrt{3}$ and let $|\omega^2 - 1| < a\tau$. Let $\alpha = 3\tau^2$. Then, if τ is sufficiently small, the Eq.(2.4) has a stationary solution with frequency ω of the form (3.3), with $u_1 = u_{-1} = \epsilon$ and*

$$\epsilon = \sqrt{\frac{\alpha - (1 - \omega^2)^2}{3}} + \mathcal{O}(\tau^2) .$$

Remark. The line $\alpha = (1 - \omega^2)^2$ is called the *neutral line*.

Remark. The proof above shows that the solution is a real analytic function of x and that the domain of analyticity grows as $\epsilon \to 0$. In fact, $|u_n| = \mathcal{O}(\epsilon^{|n|})$.

4. Secondary Instabilities

In this section, we consider instabilities of the stationary cellular solutions found earlier. We shall find that, in a certain parameter range, long wavelength perturbations become unstable and grow in time. Eventually, this temporal instability will lead to stationary states of more complicated patterns. These will be analyzed starting from Section 4.4.

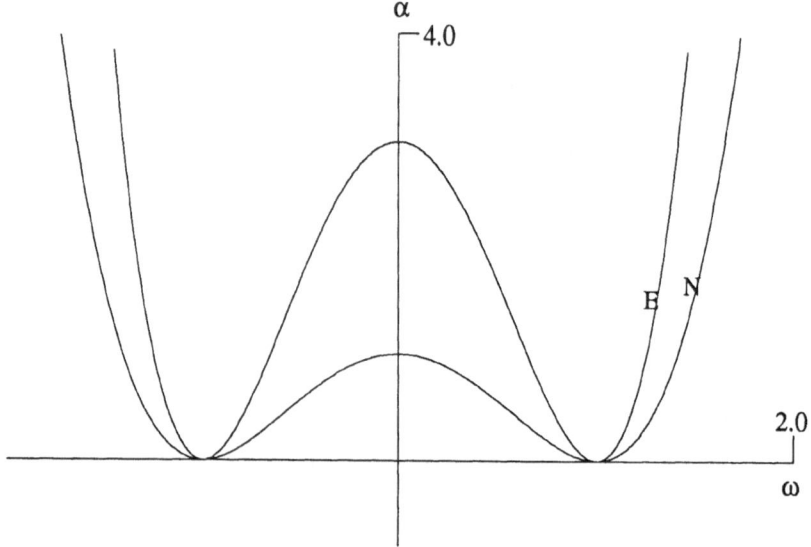

Fig. 2. The phase space of solutions for the SH equation. The picture is only rigorous for small α. Periodic solutions exist above the neutral curve N, and they are stable above the Eckhaus line E. They are unstable to long-wavelength perturbations between E and N.

4.1. The Eckhaus Instability

Consider a stationary solution u, make the ansatz $u + f$, substitute into (2.4) and expand to first order in f. Then one gets the equation

$$\partial_t f(x,t) = \big(\alpha - (1 + \partial_x^2)^2\big)f(x,t) - 3u^2(x)f(x,t) \ .$$

Thus one is led to study the operator

$$f \mapsto \mathcal{L}_u f(x) \equiv \big(\alpha - (1 + \partial_x^2)^2\big)f(x) - 3u^2(x)f(x) \ . \tag{4.1}$$

If $\mathcal{L}_u$ has an eigenvector $f(x)$ with eigenvalue σ, then the initial condition $u(x) + \eta f(x)$ will evolve as $u(x) + \eta e^{\sigma t}f(x)$ in time and will grow if σ has positive real part. This is then called an instability.

The translation invariance of the SH equation and its relatives is at the origin of a set of secondary instabilities, called the Eckhaus instability. By translation invariance, if $u(x)$ is a solution of (3.1), then so is $u(x + \xi)$ for any fixed ξ. Differentiating the Eq.(3.1) with respect to x (or ξ), we see that u' satisfies the equation

$$\big(\alpha - (1 + \partial_x^2)^2\big)u'(x) - 3u^2(x)u'(x) = 0 \ .$$

This means that u' *is an eigenvector with eigenvalue 0 of* $\mathcal{L}_u$. In other words, $\eta u'$ is a small perturbation of u which does neither grow nor decay in time. A secondary instability will occur if we can find close to u' another eigenvector w whose eigenvalue σ has positive real part.

Remark. There is another symmetry of the problem, related to a change in wavelength of the stationary solution. To lowest order in perturbation theory, we have, for fixed α,

$$u_\omega(x) = 2\sqrt{\frac{\alpha - (1-\omega^2)^2}{3}}\, \cos(\omega x)\,,$$

and differentiating w.r.t. ω we get an *unbounded* eigenfunction with eigenvalue 0.

Coming back to the translation mode, we seek now eigenvectors of the operator $\mathcal{L}_u$, where u is one of the stationary solutions. In view of the above discussion, we make an ansatz of the form

$$v(x) = \epsilon\omega e^{i\omega x}\left(a(\varkappa)e^{ix} + b(\varkappa)e^{-ix}\right) + \text{c.c.}\,.$$

Here, $a(0) = i, b(0) = 0$, since we have already seen that $u'(x)$ is an eigenvector with eigenvalue 0. Note that we immediately include *both* positive and negative $\varkappa$-frequencies, since the nonlinearity and the reality requirement will generate anyway a coupling between them. More precisely, we look first for (non-real) eigenfunctions of the form $ae^{i(\omega+\varkappa)x} + be^{i(-\omega+\varkappa)x}$ and will then make them real by adding the complex conjugate. In the basis whose components are $e^{i(\omega+\varkappa)x}$ and $e^{i(-\omega+\varkappa)x}$, the operator $\mathcal{L}_u$ is approximately equal to the 2×2-matrix

$$\begin{pmatrix} \alpha - (1-(\omega+\varkappa)^2)^2 - 6\epsilon^2 & -3\epsilon^2 \\ -3\epsilon^2 & \alpha - (1-(-\omega+\varkappa)^2)^2 - 6\epsilon^2 \end{pmatrix}\,.$$

For example, the term $-6\epsilon^2$ is the coefficient of $e^{i(\omega+\varkappa)x}$ in the term $-3u^2 e^{i(-\omega+\varkappa)x}$.

Exercise. Verify the above statement.

At this point, it is useful to change the scale of the variables ω, and $\varkappa$, which is in fact dictated by the relation $\alpha \approx 3\epsilon^2 + (1-\omega^2)^2$. We take

$$\omega = \sqrt{1 + \sqrt{3}\epsilon W}\,, \qquad \varkappa = \sqrt{3}\epsilon K/2\,.$$

Straightforward algebra leads to the more appealing form for the matrix:

$$3\epsilon^2 \begin{pmatrix} -1 - K^2 - 2KW & -1 \\ -1 & -1 - K^2 + 2KW \end{pmatrix}\,.$$

Note that when $K = 0$ then $(1,-1)$ is an eigenvector with eigenvalue 0, and $(1,1)$ has eigenvalue $-6\epsilon^2$. The eigenvector $(1,-1)$ corresponds to the function u'. In the general case, we get the eigenvalues

$$\sigma_\pm = 3\epsilon^2\left(-K^2 - 1 \pm \sqrt{4W^2K^2 + 1}\right)\,. \tag{4.2}$$

In particular,

$$\sigma_+ = 3\epsilon^2(2W^2 - 1)K^2 + \mathcal{O}(K^4)\,.$$

These equations describe the Eckhaus instability. If $W^2 > 1/2$ the spectrum is unstable, and if $W^2 < 1/2$ the spectrum is stable, cf. Fig. 3.

143

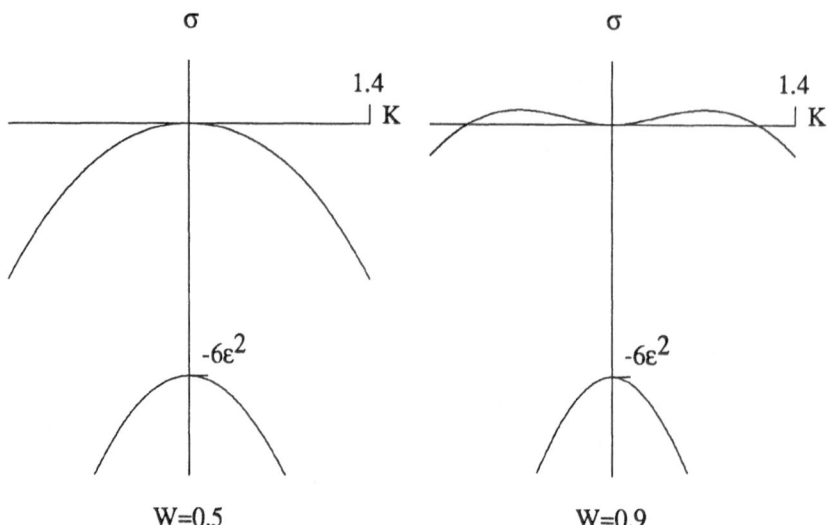

Fig. 3. The shape of the spectrum of excitations with a frequency difference close to zero from the basic frequency. Left: Eckhaus stable domain. Right: Eckhaus unstable domain.

Remarks

i) The condition $W^2 > 1/2$ can be interpreted as saying $3\epsilon^2 < 2(1 - \omega^2)^2$, which says that the *amplitude is small*. We thus find the interpretation of the Eckhaus instability as saying that at small amplitudes the wavelength is unstable, i.e., it has not yet been completely fixed due to the soft mode in the system.

ii) The time scale on which this change of wavelength takes place is given by the maximal value of σ_+, and it is of order ϵ^2.

iii) The above argument is *not* rigorous since it is done in perturbation theory. However, it can be made rigorous in the space of functions **S** considered above. The problem consists in taking care of the higher order terms in the function u and in the expansion of the eigenvector, see [CE] for details.

Theorem 4.1. *Let $a < \sqrt{3/2}$ and let $|\omega^2 - 1| < a\epsilon$. Let u_ϵ be one of the periodic stationary solutions described before. Then the operator $\mathcal{L}_{u_\epsilon}$ is stable for sufficiently small ϵ. If $a > \sqrt{3/2}$ the operator $\mathcal{L}_{u_\epsilon}$ is unstable. The qualitative shape of the spectrum is as in Fig. 3.*

Remark. The line $\alpha/3 = (1 - \omega^2)^2$ is called the *Eckhaus line*. The periodic solutions are stable inside the Eckhaus line, are unstable between the Eckhaus line and the neutral line and cease to exist outside the neutral line.

4.2. Exponentially Decaying Solutions

The analysis of the preceding section was based mainly on perturbation theory. We can exploit this to construct other perturbations of the periodic solution when ω is in the Eckhaus stable domain, which *decay exponentially towards them* as $x \to \pm\infty$. We will see that the exponentially decaying solutions correspond exactly to those imaginary values of K for which $K = \sqrt{4W^2 - 2}$. This is the (purely imaginary) point where the curve $\sigma_+(K)$ hits zero, when W is in the Eckhaus stable domain.

We now explain how this is connected to the method of Bloch waves, which can be found in every quantum mechanics textbook. In Section 4.4, we use this result in a dynamical systems picture.

The search for eigenfunctions of the operator $\mathcal{L}_u$ can be considered in the language of periodic potentials. It is then useful to consider the evolution over one period, i.e., $2\pi/\omega$ in our case. (In fact, because of the very special form of the nonlinearity the period is really π/ω.) In the theory of Bloch waves, one looks for eigenfunctions of the form $v(x) = w(x)e^{-kx}$, or $v(x) = w(x)e^{+kx}$ or $v(x) = w(x)e^{\pm ikx}$, with w a $2\pi/\omega$-periodic function. In the preceding section, we considered the last of these cases, now we concentrate on the first, with $k > 0$. Note that if $v(x)$ is to be an eigenfunction of $\mathcal{L}_u$ with eigenvalue 0, then we must find that, starting from an initial condition at $x = 0$, this initial condition must reproduce after a period—up to a constant factor. This means that w is an eigenvector with eigenvalue zero of the operator $\mathcal{L}_{u,k}$ defined by

$$\mathcal{L}_{u,k}f(x) = e^{kx}\big(\mathcal{L}_u(g \cdot f)\big)(x) \, ,$$

where $g(x) = e^{-kx}$.

Exercise. Verify the above statement.

From the calculations of the last section, we already know that $\mathcal{L}_{u,k}$ has an eigenvalue zero when σ_+ is zero, i.e., when K satisfies $K = \sqrt{4W^2 - 2}$, or $K = 0$. Since we want k to be real, we require K to be purely imaginary (and not zero), and this happens exactly for $W^2 < 1/2$ and then we find $K = \pm i\sqrt{2 - 4W^2}$. In unscaled variables, we find that the SH equation has solutions of the form

$$u(x) \approx 2\epsilon\big(\cos(\omega x) + \eta e^{-kx} w(x)\big) \, , \qquad (4.3)$$

with $k = \epsilon\sqrt{3/2 - 3W^2}$, $w(x) = 2\epsilon \cos(x + \theta)$, and $\theta = \frac{1}{2}\arccos(1 - 4W^2)$ [EZ]. To verify that (4.3) is a solution of Eq.(3.1), we note that upon substituting (4.3) in (3.1), we get

$$(\alpha - (1 + \partial_x^2)^2)u - u^3 = 2\epsilon\eta e^{-kx}\mathcal{L}_{2\epsilon\cos(\omega x),k}w(x) + \mathcal{O}(\epsilon\eta + \epsilon^2)$$
$$= 0 + \mathcal{O}(\epsilon\eta + \epsilon^2) \, .$$

Remark. These solutions are, so far, only valid in perturbation theory. It will follow from the section on invariant manifolds why there are, for small ϵ, solutions close to the ones we found in perturbation theory.

4.3. Biperiodic Solutions

We have already seen in Eq.(4.2) that the SH-equation possesses unstable eigenvalues

$$\sigma_+ = 3\epsilon^2\big(-K^2 - 1 + \sqrt{4W^2K^2 + 1}\big) \, ,$$

in the Eckhaus unstable domain $W^2 > 1/2$. One can find the corresponding eigenvector: It is of the form

$$v(x) = e^{ix(\omega +)} - c(\omega, \varkappa)e^{ix(-\omega +)} \, , \qquad (4.4)$$

and its complex conjugate. The function c is given by

$$c(\omega, \varkappa) = 2WK + \sqrt{1 + 4K^2 W^2} \,,$$

If we substitute the real version of this eigenvector into (3.1), (i.e., $v + \bar{v}$) we find that *in perturbation theory*, a stationary solution exists which is of the form

$$2\epsilon \cos(\omega x) + \rho\big(e^{iz(\omega+)} - c(\omega, \varkappa)e^{iz(-\omega+)} + \text{c.c.}\big) \,, \qquad (4.5)$$

with ρ given by

$$\rho^2 \approx \frac{1}{9}\sigma(W, K) \,, \qquad (4.6)$$

and correction terms which are small with K and $W^2 - 1/2$. Such solutions have been found earlier, on the level of the amplitude equation in [KSZ]. We see the following important feature emerge: Although the periodic solution is *unstable* for perturbations in the direction of v, in the time-dependent picture (2.4), there is a solution with *saturated* amplitude ρ. This solution is *stable* as far as the ρ-variations are concerned, but *unstable* for more general perturbations to be discussed below. Thus, it is a saddle point for the time-dependent problem (2.4). By choosing adequate initial conditions, it should be possible to follow the stable manifold of these fixed points on orbits which come very close to them. Thus, they are experimentally accessible as transient states. This is very similar to the situation encountered in nucleation phenomena, cf. [L].

4.4. Interpretation of Time Stationary Solutions in Dynamical Systems Language

The existence of biperiodic solutions has been shown in perturbation theory in the amplitudes. But in fact, beyond perturbation theory, more interesting solutions appear. In order to reach this more complete theory we reinterpret the problem of stationary solutions in terms of a dynamical systems representation of the x-coordinate. We warn the reader not to be confused between this dynamical system which we shall call the S **problem** for the Eq.(3.1) and the true time dependence of the PDE Eq.(2.4) which we call the T **problem**. For example, stability and instability in the S problem have no direct relation with the same notions for the T problem.

We furthermore point out that information about the original problem (3.1) is lost if one passes to an amplitude equation such as

$$\partial_t u(x, t) = \partial_x^2 u(x, t) + u(x, t) - u^3(x, t) \,, \qquad (4.7)$$

as is sometimes done in the literature.

Reconsider now the results of the preceding section in a dynamical systems picture. This picture easily generalizes to more complicated equations, and remains valid with obvious modifications.

The method is really quite easy in the present case and is well known in Hamiltonian dynamics. It is clear that one can reduce the problem (3.1) to a 4-dimensional phase space $\mathcal{B}$ with coordinates X, where X is defined by

$$X(x) = \{X_p(x)\}_{p=0,1,2,3} = \{\partial_x^p u(x)\}_{p=0,1,2,3} \,. \qquad (4.8)$$

Thus, the coordinates are just the first 3 derivatives of the amplitude u. One now views x as a kind of "time."

Remark. This is the place where we use that the problem is one-dimensional and that we are considering the time-independent equation (3.1) rather than (2.4). The novice might think that direct information about the partial differential equation (2.4) can be gained from this approach, but this is not the case. Rather, one should think of this approach as the analogue of studying first the fixed points of a dynamical system and using them, and their local hyperbolicity properties as a guide for the description of the global flow [ACEGP]. We will see this approach in action in the qualitative discussion of the scenario for weak turbulence.

Eq.(3.1) can be rewritten as

$$\partial_z X = MX + N(X) , \qquad (4.9)$$

where

$$M = \begin{pmatrix} 0 & 1 & 0 & 0 \\ 0 & 0 & 1 & 0 \\ 0 & 0 & 0 & 1 \\ \alpha - 1 & 0 & -2 & 0 \end{pmatrix} , \qquad (4.10)$$

and

$$N(X) = \begin{pmatrix} 0 \\ 0 \\ 0 \\ -X_0^3 \end{pmatrix} .$$

The matrix M and the Eq.(4.9) describe in their first three rows the relation between the derivatives and the last row expresses the equation (3.1).

Remark. The dynamical system (4.9) is reversible, i.e., it is the same if x is replaced by $-x$, because of the space reflection symmetry of Eq.(2.4). This implies that the eigenvalues of M come in pairs $\sigma, -\sigma$.

The picture which emerges from the results of the last section is as follows: Each periodic solution corresponds to a periodic orbit in the space $\mathcal{B}$.

Thus, according to Theorem 3.1, there is a one-parameter family of such periodic orbits. Going along the orbit means translating in x-space. The exponentially decaying small solutions are stable and unstable manifolds of these periodic orbits. They can be best viewed in the Poincaré section of the circles. Each time we go around the circle, the amplitude of this perturbation increases or decreases by a factor $e^{2\pi\sigma/\omega}$. This is illustrated in Fig. 5.

We have seen that the above picture holds in the Eckhaus stable domain $W^2 < 1/2$. What do we expect in the unstable case? As W^2 crosses $1/2$, the value $\varkappa$ will go to zero, become imaginary, and lead to a rotation (rather than the hyperbolic motion found so far). This will be discussed in the next section.

4.5. Quasiperiodic and Chaotic Stationary Solutions

It is useful to note that we have a *two-parameter* family of biperiodic solutions of (3.1), with ω and K (resp. $\varkappa$) being the two parameters. We can view ω and ρ as the two independent parameters and then we see that from Eq.(4.5) the picture of Fig. 6 emerges: There is a line of periodic orbits, essentially $2\epsilon \cos(\omega x)$ and a secondary oscillation of amplitude ρ with a *frequency which depends on ω and $\varkappa$.*

Remark. Keep in mind that ϵ is not an independent variable above, since it is fixed by a relation of the form $\alpha \approx 3\epsilon^2 + (1 - \omega^2)^2$.

Up to now, we have only done perturbation theory. How will the higher order terms influence the picture of Fig. 6? To analyze this question, we note that the

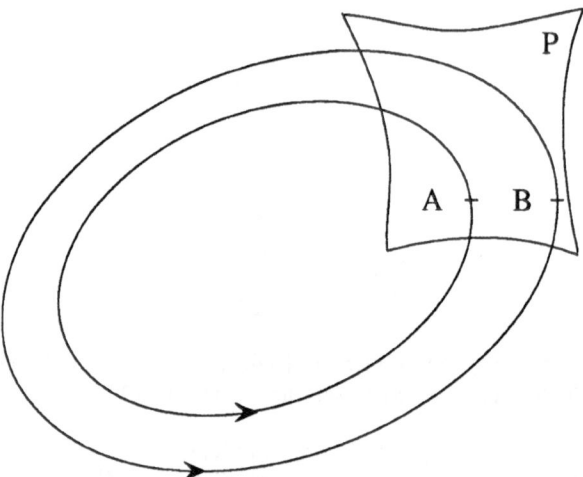

Fig. 4. The flow, with x as the "time" coordinate in the space $\mathcal{B}$. The periodic solutions of (3.1) correspond to periodic orbits. There is a band of such orbits, with different frequencies. We perform a Poincaré section through the three-dimensional hyperplane $\mathbf{P}$.

frequency $\varkappa$ is a non-trivial function of the amplitude ρ. This means that there is some *twist* in the map in the Poincaré section $\mathbf{P}$. Having a twist implies that resonances will occur as a function of the commensurability or incommensurability of the frequencies ω and $\varkappa$ and we expect a KAM picture to emerge in each such plane, due to the existence of these non-linear twists [Mo]. Circles with irrational winding numbers will coexist with chaotic regions formed by the breakup of circles with rational winding numbers. Translated back to the spatial setting, this means that we have:

Proposition 4.2. *The Swift-Hohenberg equation (3.1) has, in the Eckhaus unstable domain, quasiperiodic and chaotic stationary solutions.*

Remark. To our knowledge, there are two different situations where a proof is available. Near the neutral line, the methods of Moser [Mo] can be applied since both

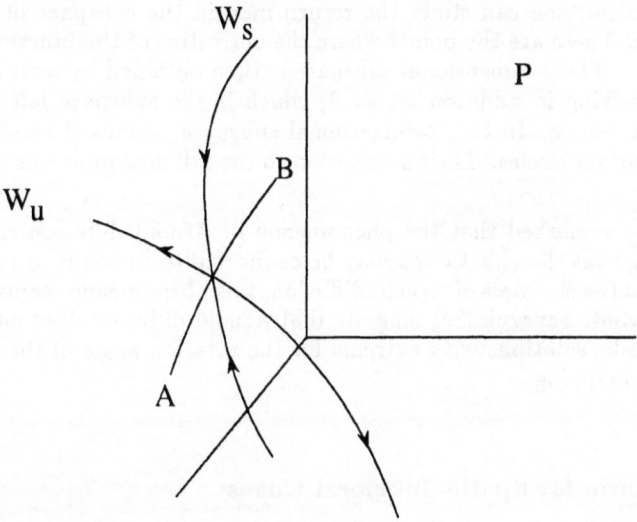

Fig. 5. The motion in the Poincaré section **P**, when $W^2 < 1/2$, i.e., in the Eckhaus stable domain. Each periodic solution of (3.1) corresponds to a hyperbolic fixed point of the Poincaré map.

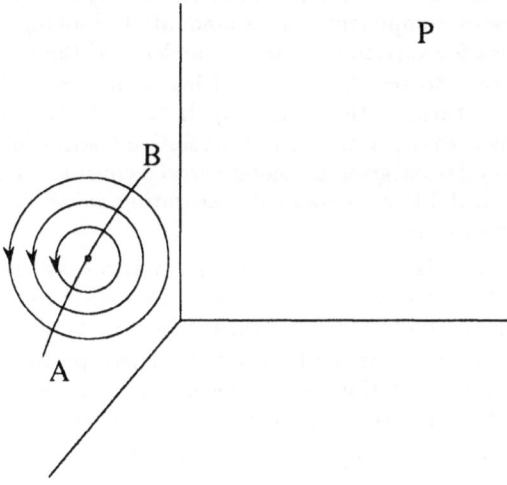

Fig. 6. The motion in the Poincaré section **P**, when $W^2 > 1/2$, i.e., in the Eckhaus unstable domain. Each periodic solution of (3.1) corresponds to an elliptic fixed point of the Poincaré map.

amplitudes, the one of the basic frequency and the modulation amplitude, are small. One then expects frequencies close to $\sqrt{1 \pm \sqrt{\alpha}}$. On the other hand, near the Eckhaus line, one is perturbing around a periodic orbit. This situation was studied by Sevryuk [Se, p.170], who used the reversibility of the original problem to reduce the three dimensional Poincaré section to two dimensions. More precisely, for the Swift-Hohenberg problem, one can study the return map in the subspace of $\mathcal{B}$ satisfying $X_1 = 0$, $X_0 > 0$. These are the points where the derivative of the function u vanishes and u is positive. The 2-dimensional subspace is then obtained by restricting further to those X satisfying in addition $X_3 = 0$, which is the subspace left invariant by the involution $x \to -x$. In this 2-dimensional subspace, standard results imply the existence of invariant circles. Their suspension to the full flow produces quasiperiodic solutions.

It should be remarked that the phenomenon of Arnold diffusion cannot be excluded, meaning that the chaotic sea may be connected to unbounded parts of phase space. In view of the slowness of Arnold diffusion, this phenomenon seems to us experimentally irrelevant. Sevryuk [Se] suggests that Arnold diffusion does not occur near those quasiperiodic solutions with extrema for the rotation angle of the linearization of the Poincaré mapping.

4.6. The Scenario for Spatio-Temporal Chaos

At this point we want to emphasize the role of these stationary solutions in terms of the T problem, Eq.(2.4). We shall argue below that these solutions have stable and unstable manifolds in function space. Accordingly, they are important in organizing the dynamics, like hyperbolic fixed points of low dimensional strange attractors. The time evolution will approach one of the stationary states on the stable manifold, will remain close to it for some time and leave it along an unstable direction.

To see the existence of a stable manifold for the T problem is easy. One examines the stability of the solution (4.5) (in perturbation theory) to perturbations of the amplitudes of the various components. This amounts to looking at the dynamics in the truncated equations for variations of the coefficients of the six components $e^{\pm i \omega x}$ and $e^{\pm i(\omega \pm)x}$. It is easy to see that the resulting dynamics is linearly stable, and the perturbed solution returns to the form (4.5). In fact, the translation invariance of the equations gives one zero eigenvalue, and the relative translation between the main frequency and the modulation gives a second zero eigenvalue. The third eigenvalue is associated with the stability of ρ about its saturation value, cf. (4.5). Three more eigenvalues are clearly negative.

To see that there are also unstable directions is also easy. In fact, the solution (4.5) is, at least for W^2 close to the Eckhaus value, very close to the periodic solution (with $\rho = 0$). Therefore, it has still an Eckhaus instability for (very small) $\eta > 0$ which consists of perturbations of the form $e^{\pm i \eta x}$. These perturbations cannot lead to quasiperiodic stationary solutions since the phase space has only 4 dimensions (which are "used up" by the first two frequencies ω, and $\varkappa$).

The above discussion can be summarized in the following way: Cellular solutions with wavelengths belonging to the Eckhaus unstable domain are destabilized by long wavelength perturbations. Close to these periodic solutions there exist other stationary solutions which are either quasiperiodic or chaotic in space. Since these stationary have a stable direction for the T problem, it is reasonable to expect that the dynamics will allow a crossover from the cellular state to the quasiperiodic or spatially chaotic states. In a second step, this solution is going to evolve along its unstable direction.

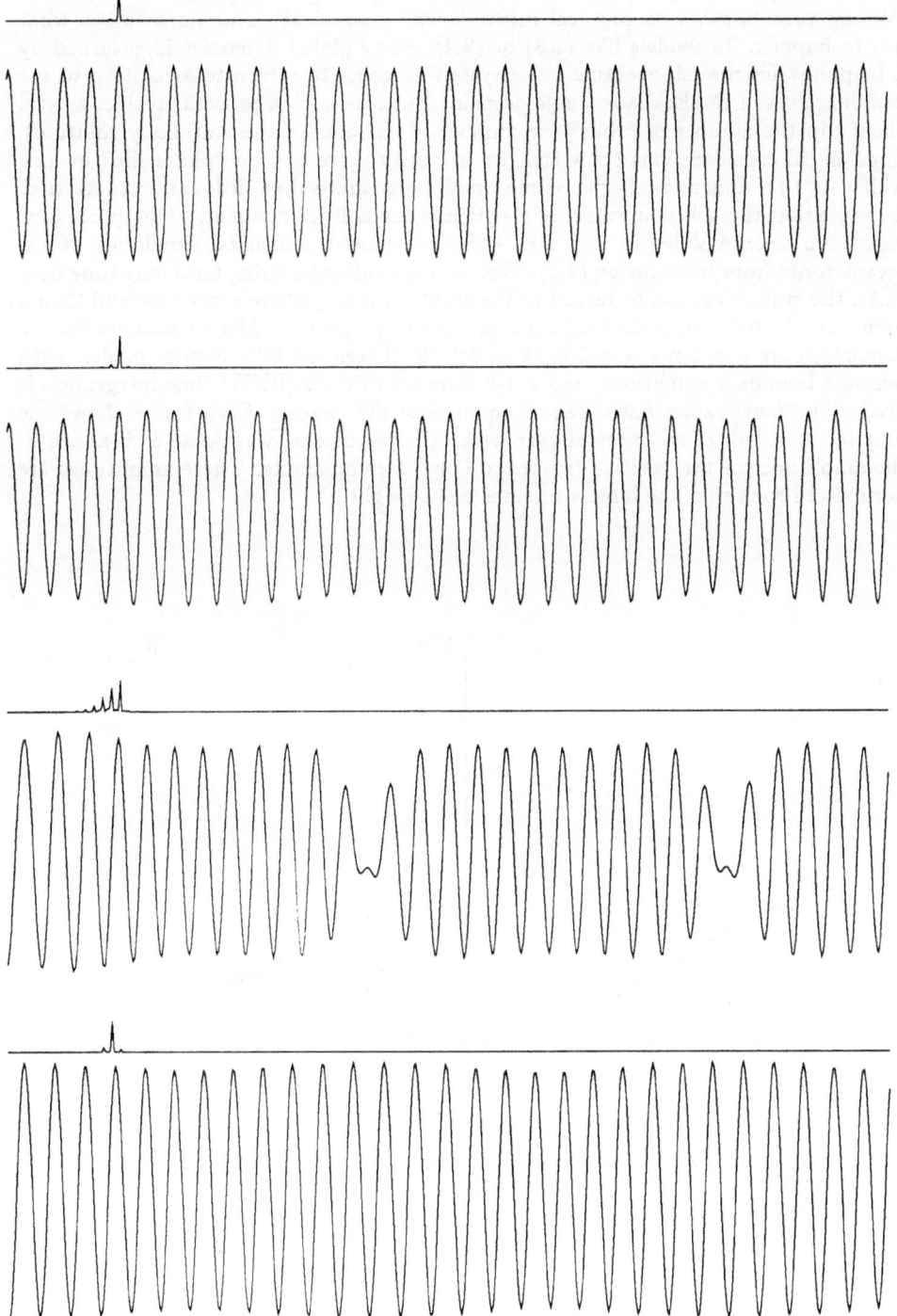

Fig. 7. Four frames of the time-evolution for the SH equation, see text.

The final state after the escape in the unstable directions cannot be calculated from these local considerations. It is determined by the global dynamics of the PDE. We can turn however to physical intuition and numerical simulations to see what has to happen. In models like (2.3) or (2.4) whose global dynamics is governed by a Liapunov functional, eventually the system is going to return to a solution whose wavelength is in the Eckhaus stable domain. Even so, such a process involves a total change in the number of "rolls" (or number of extrema of the stationary solution). Topological considerations show that by necessity there will be "phase slip" events, which in 1+1 dimensional space-time amount to space-time dislocations. At such a dislocation the solution vanishes identically for a brief moment. This process of phase slip is exemplified in Fig. 7 in which we show a numerical simulation of the escape route from the solution (4.5). One sees that after hovering for a long time near (4.5), the system evolves to return to the stable domain, where every now and then a period in the solution is shed off via a process of phase slip. The parameters for the simulation are as follows: $\epsilon = 0.2$, $W = 0.9789$. There are 1024 Fourier modes, with periodic boundary conditions, and a 4-5 Dormand-Prince [HNW] time integration is used. The four frames are taken at equal time differences. Each frame shows the function u in half of the interval over which the calculation was done, and, above it, the amplitudes of the positive frequency Fourier components. These amplitudes are normalized such that the largest one has fixed height.

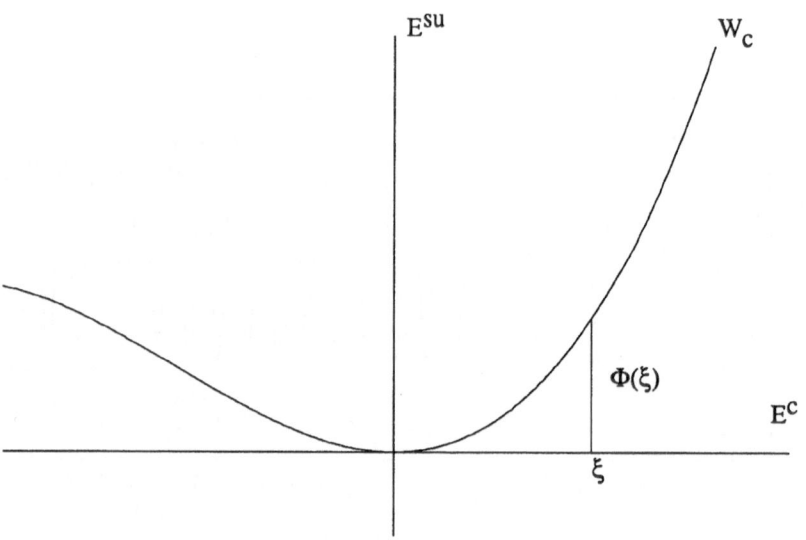

Fig. 8. The center manifold W_c is tangent to the center subspace $\mathcal{E}^c$. It is the graph of the function $\Phi : \mathcal{E}^c \to \mathcal{E}^{su}$.

Thus, we have completed the picture sketched in the Introduction for the 1+1 dimensional example. Next, we shall show that this understanding explains also the phenomenology of systems in 2+1 dimensions which are confined to strips. To do this, we need to introduce some further background material. The reader who is familiar with the Center Manifold Theorem, or is satisfied with the "adiabatic approximation" version of it, can go directly to Section 7.

5. The Center Manifold Theorem

Before we go on to more complicated examples, we want to lay their mathematical foundations. This is well known material, and we reproduce it here in the form given in [EW]. The idea behind all "invariant manifold theorems" is that the flow of a nonlinear dynamical system near a fixed point is very similar to the flow of the system obtained by looking only at the linearized part. For a linear system, the linear subspaces (eigenspaces) of the tangent operator are invariant subspaces. For a nonlinear system, the situation is similar, but the invariant subspaces are deformed to "invariant manifolds." The mathematical content of the invariant manifold theorems is the *existence* of these manifolds. This existence, and in particular the smoothness of such manifolds is not evident, since resonances between the eigenvalues can mess up things. Another technical problem which is always present in applications is the *size* of the region in which the invariant manifold can be shown to exist. This is the radius of the region in which the intuitive picture of Fig. 8 can be shown to apply. This radius depends on the nature of the nonlinearities, in a way which we want to make precise below.

We consider a differential equation on the Banach space $\mathcal{E}$, and we assume that $\mathcal{E} = \mathcal{E}^c \oplus \mathcal{E}^s \oplus \mathcal{E}^u$, which we refer to respectively as the center, the stable, and the unstable subspaces. Let (z^c, z^s, z^u) be coordinates for these subspaces. We write the differential equation in the form:

$$
\begin{aligned}
\frac{dz^c}{dt} &= A^c z^c + f^c(z^c, z^s, z^u) \,, \\
\frac{dz^s}{dt} &= A^s z^s + f^s(z^c, z^s, z^u) \,, \\
\frac{dz^u}{dt} &= A^u z^u + f^u(z^c, z^s, z^u) \,.
\end{aligned}
\tag{5.1}
$$

Any combination of the indices c, s, u, will refer to the direct sum of the corresponding subspaces, and the corresponding norm will be the maximum. For example, $\mathcal{E}^{cs} = \mathcal{E}^c \oplus \mathcal{E}^s$ with the norm $\|z^c \oplus z^s\|_{cs} = \max(\|z^c\|_c, \|z^s\|_s)$. Define B_r^c to be the ball of radius r, centered at the origin in $\mathcal{E}^c$, with B_r^u and B_r^s analogously defined. We also need $B_r^{cs} = B_r^c \oplus B_r^s$ and similar combinations. Denote by P_c, P_s and P_u the projections onto the components of $\mathcal{E}$.

We make the following hypotheses.

H1: The linear operators A^s and A^u define continuous semigroups on $\mathcal{E}^s$ and $\mathcal{E}^u$, for $t \geq 0$, and $t \leq 0$, respectively. Furthermore, we assume that there exist positive constants λ, and D, such that $\sup_{t \geq 0} \max(e^{\lambda t}\|e^{A^s t}\|_s, e^{\lambda t}\|e^{-A^u t}\|_u) \leq D$.

H2: The linear operator A^c defines a flow on $\mathcal{E}^c$, with $\|e^{A^c t} z^c\|_c \leq a(1 + |t|^k)\|z^c\|_c$ for all $t \in \mathbf{R}$.

H3: The functions f^c, and f^s, and f^u are Lipschitz functions from $\mathcal{E}$ to $\mathcal{E}^c$, $\mathcal{E}^s$, and $\mathcal{E}^u$, respectively. The Lipschitz constants of these three functions on the ball of radius r about the origin in $\mathcal{E}$ will be denoted by $\ell_c(r)$, $\ell_s(r)$, and $\ell_u(r)$. By this we mean, e.g.,

$$
\|f^s(z) - f^s(\tilde{z})\|_s \leq \ell_s(r)\|z - \tilde{z}\|_{csu} \,,
$$

for all z, $\tilde{z} \in B_r^{csu}$. We require that all three functions (ℓ_s etc.) vanish as $r \to 0$.

Theorem 5.1. *Assume that there exists a constant $\sigma > 1$, and positive constants β and r such that the inequalities C1–C5 below hold. Then there exists a Lipschitz function h defined on some neighborhood, U^c, of the origin in $\mathcal{E}^c$ and mapping U^c to $\mathcal{E}^s \oplus \mathcal{E}^u$. Furthermore, $h(0) = 0$ and the graph of h is left invariant by (5.1). If the* non-linearity (f^c, f^s, f^u) in (5.1) is C^{m+1}, $1 \le m < \infty$, if $\mathcal{E}$ has the C^{m+1} extension property, then the function h is C^m, and the mth derivative is Lipschitz, possibly on a smaller neighborhood $\tilde{U}^c$.

Remark. A Banach space $\mathcal{E}$ has the C^m extension property if there exists a function $\chi \in C^m(\mathcal{E}, \mathbf{R})$ such that $\chi(z) = 1$ if $\|z\| < 1/2$, and $\chi(z) = 0$ if $\|z\| > 1$. If $\mathcal{E}$ is finite dimensional or a Hilbert space then it has the C^m extension property for all $m = 1, 2, \ldots, \infty$.

Remark. One could assume that the vector field $f^{c,s,u}$ in Theorem 5.1 was only C^m, with Lipschitz m^{th} derivative.

Sketch of Proof of Theorem 5.1. We begin by describing the construction of the center-stable manifold. The construction is very similar to that of [CE], modified only to take account of the presence of the "center" directions (cf. [M]). We look for an invariant manifold, W^{cs}, which is the graph of a function $h : \mathcal{E}^{cs} \to \mathcal{E}^u$. To simplify notation, we denote points in $\mathcal{E}^{cs}$ by ξ, and let their projections onto $\mathcal{E}^c$, and $\mathcal{E}^s$ be denoted by ξ^c and ξ^s respectively. Let Ψ_t^h be the projection onto $\mathcal{E}^{cs}$ of the flow on the graph of h.

Then formally, Ψ_t^h satisfies

$$\Psi_t^h = e^{A^{cs}t} + \int_0^t d\tau \, e^{A^{cs}(t-\tau)} f_h^{cs} \circ \Psi_\tau^h . \qquad (5.2)$$

Here, $A^{cs} = A^c \oplus A^s$, and

$$f_h^{cs}(\xi) = \begin{pmatrix} f^c(\xi, h(\xi)) \\ f^s(\xi, h(\xi)) \end{pmatrix} .$$

In order for W^{cs} to remain invariant under the flow ϕ_t we must have

$$\phi_t(\xi, h(\xi)) = \begin{pmatrix} \Psi_t^h(\xi) \\ h(\Psi_t^h(\xi)) \end{pmatrix} ,$$

from which it follows that

$$h = -\int_0^\infty d\tau \, e^{-A^u \tau} f_h^u \circ \Psi_\tau^h . \qquad (5.3)$$

We note that the calculations leading to (5.2) and (5.3) are formal ones—for instance, the flow ϕ_t will in general not be defined for all initial conditions in a neighborhood of the origin. One can prove that the calculations make sense by first showing that given h, (5.2) has a solution which one then substitutes into the right hand side of (5.3). One then considers the right hand side of (5.3) as defining a transformation of the function h, and one shows that it has a fixed point. That fixed point defines our center-stable manifold.

We begin by defining the spaces in which h and Ψ^h will be shown to lie. The space for h is given by

$$H_\sigma = \{ h : \mathcal{E}^{cs} \to \mathcal{E}^u \mid h(0) = 0 ,$$
$$\|h(\xi) - h(\tilde{\xi})\|_u \le \sigma \|\xi - \tilde{\xi}\|_{cs}, \ \forall \xi, \ \tilde{\xi} \in \mathcal{E}^{cs} \} .$$

If h and $\tilde{h}$ are elements of H_σ, we define the Lipschitz metric

$$d_H(h, \tilde{h}) = \sup_{\substack{\xi \in \mathcal{E}^{cs} \\ \xi \neq 0}} \frac{\|h(\xi) - \tilde{h}(\xi)\|_u}{\|\xi\|_{cs}} .$$

We next define the space in which the map Ψ^h is going to live. Due to the bad control over the center direction, we will allow for a small exponential divergence, (cf. Mielke[M].) Fix $\beta > 0$ (small). We require that $\beta < \lambda$. Define

$$K_{\sigma,\beta} = \{\psi : \mathbf{R}^+ \times \mathcal{E}^{cs} \to \mathcal{E}^{cs} \mid \psi_0(\xi) = \xi , \text{ for all } \xi \in \mathcal{E}^{cs} ,$$

$$\psi_t(0) = 0 , \text{ for all } t \geq 0 ,$$

$$\|\psi_t(\xi) - \psi_t(\tilde{\xi})\|_{cs} \leq \sigma e^{\beta t} \|\xi - \tilde{\xi}\|_{cs} , \text{ for all } \xi, \tilde{\xi} \in \mathcal{E}^{cs}\} .$$

We also define the Lipschitz norm

$$\|\psi\|_K = \sup_{t \geq 0} \sup_{\substack{\xi \in \mathcal{E}^{cs} \\ \xi \neq 0}} \frac{\|\psi_t(\xi)\|_{cs} e^{-\beta t}}{\|\xi\|_{cs}} ,$$

and the induced Lipschitz metric on $K_{\sigma,\beta}$,

$$d_K(\psi, \tilde{\psi}) = \|\psi - \tilde{\psi}\|_K .$$

Remark. H_σ and $K_{\sigma,\beta}$ are complete metric spaces with respect to d_H and d_K.

The construction of center manifolds makes the use of a cutoff necessary. We do not assume that f itself has a restricted domain, but work instead with an explicit cutoff function χ. This has the advantage of making the bounds on Lipschitz constants more explicit. We choose a smooth function $\chi_0 : \mathbf{R} \to \mathbf{R}$, satisfying $0 \leq \chi_0 \leq 1$, $\chi_0(x) = 1$ when $|x| \leq 1/2$, $\chi_0(x) = 0$ when $|x| > 1$. We assume for definiteness that the derivative satisfies $|\chi_0'(z)| \leq 3$. Let $z = (z^c, z^s, z^u)$. We then define $\chi(z) = \chi_0(\|z\|)$.

For r a positive real number, we define the functions

$$g^s(z) = f^s(z) \cdot \chi(z/r) ,$$
$$g^c(z) = f^c(z) \cdot \chi(z/r) ,$$
$$g^u(z) = f^u(z) \cdot \chi(z/r) .$$

Then g^s, g^c, and g^u are all Lipschitz functions, and they have Lipschitz constants on all of $\mathcal{E}$ which are bounded by

$$\ell_g = 4 \max(\ell_s(r), \ell_c(r), \ell_u(r)) .$$

Note that the functions $g^{s,c,u}$ defined above are not necessarily smooth. In fact, there exist Banach spaces on which there exists no smooth function $\chi(z)$, such that $\chi(z) = 0$ if $\|z\| > 1$, and $\chi(z) = 1$ if $\|z\| < 1/2$. If we wish to prove the existence of smooth center manifolds, we must assume that $\mathcal{E}$ has the C^m extension property, which guarantees that there exists a C^m function $\chi : \mathcal{E} \to \mathbf{R}$, which has these properties. We then redefine the functions $g^{s,c,u} = f^{s,c,u}(z)\chi(z/r)$, and set $\ell_g = 2c_\chi \max(\ell_s(r), \ell_c(r), \ell_u(r))$, where c_χ is the Lipschitz constant of χ.

Define

$$\mathcal{G}_h(\psi)_t = e^{A^{cs}t} + \int_0^t d\tau\, e^{A^{cs}(t-\tau)} g_h^{cs} \circ \psi_\tau$$

$$= \begin{pmatrix} e^{A^c t} + \int_0^t d\tau\, e^{A^c(t-\tau)} g_h^c \circ \psi_\tau \\ e^{A^s t} + \int_0^t d\tau\, e^{A^s(t-\tau)} g_h^s \circ \psi_\tau \end{pmatrix}, \qquad (5.4)$$

where $g_h(\xi) = g(\xi, h(\xi))$. To prove the existence of the center-stable manifold, one shows that $\mathcal{G}_h$ has a unique fixed point in $K_{\sigma,\beta}$, see [EW]. This fixed point gives us a center-stable manifold. We now restrict the differential equation to this manifold and consider the evolution of the reduced form of (5.1) for $t \leq 0$, or equivalently we replace t by $-t$. With this change the equations take the form

$$\frac{dz^c}{dt} = -A^c z^c - f^c(z^c, z^s, h(z^c, z^s)),$$

$$\frac{dz^s}{dt} = -A^s z^s - f^s(z^c, z^s, h(z^c, z^s)). \qquad (5.5)$$

To emphasize the similarity with the previous situation, define $\tilde{A}^u = -A^s$, $\tilde{f}^u(z^c, z^s) = -f^s(z^c, z^s, h(z^c, z^s))$ and $\tilde{f}^c(z^c, z^s) = -f^c(z^c, z^s, h(z^c, z^s))$ so that (5.5) becomes

$$\frac{dz^c}{dt} = -A^c z^c + \tilde{f}^c(z^c, z^s),$$

$$\frac{dz^s}{dt} = \tilde{A}^u z^s + \tilde{f}^u(z^c, z^s). \qquad (5.6)$$

Thus, restricted to the center manifold, (and with t replaced by $-t$) the equation has a "center" part and an "unstable" part, but no "stable" part. One proves again the existence of a fixed point for an operator similar to $\mathcal{G}_h$. We omit the proof, but we give the detailed assumptions which are needed to make the theorem work. They express really bounds on the nonlinearities and on how close the spectrum comes to zero.

The conditions for the existence of the center manifold are:

C1: $\qquad \dfrac{a(k!+1)}{\beta^k}\left(1 + \dfrac{1}{\beta}\sigma^2 \ell_g(1+\sigma)\right) < \sigma$,

C2: $\qquad D + \dfrac{D}{(\lambda+\beta)}\sigma^2 \ell_g(1+\sigma) < \sigma$,

C3: $\qquad \lambda > \beta > 0$,

C4: $\qquad \dfrac{D}{\lambda - \beta}\sigma \ell_g(1+\sigma) < 1/2$,

C5: $\qquad \left(\dfrac{a(k!+1)}{\beta^{k+1}} + \dfrac{D}{\lambda+\beta}\right)\sigma \ell_g(1+\sigma) < 1/2$.

Remark. In order to start perturbation theory, one just replaces the center manifold W_c by the center subspace $\mathcal{E}^c$. We shall do this below for an example.

Remark. There are also theorems for stable and unstable manifolds, which are really just simpler versions of Theorem 5.1, see [CE].

6. A Typical Application of the Center Manifold Theorem: The Existence of Oscillating Fronts

In this section, we exemplify the use of the center manifold theorem for a problem that is very tractable. The example we are presenting here is one of the very few where a rigorous analysis is possible. In particular, we show how to derive the effective dynamical system inside the center manifold (Eq.(6.12) below). The material of this section is otherwise not needed for the understanding of the remainder of these lectures.

The material is taken from [CE, EW]. We look at the Eq.(2.4) *with* time-dependence, in two simultaneous frames, the laboratory frame and a frame advancing with constant speed c to the right. We define *front solutions* for Eq.(2.4) as solutions of the form

$$u(x,t) = W(x, x - ct) \tag{6.1}$$

with the boundary conditions at infinity

$$\lim_{y \to -\infty} W(x,y) = S(x) , \qquad \lim_{y \to +\infty} W(x,y) = 0 ,$$

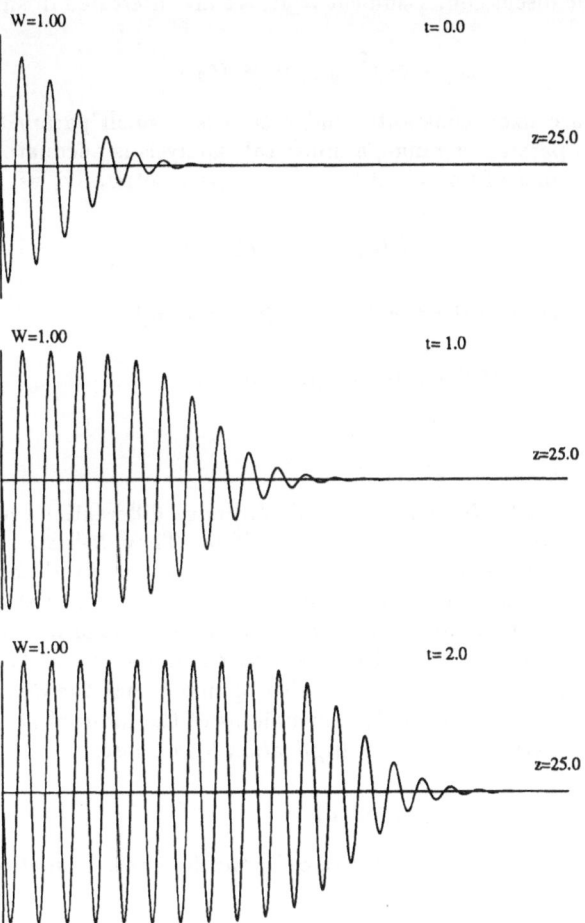

Fig. 9. The front for the SH equation in the laboratory frame. Note that in this frame, the oscillations stand still, but the envelope advances. The time step is a multiple of the period.

where S is one of the stationary solutions whose existence we proved in Theorem 3.1. The solutions W look like a fixed envelope advancing in the laboratory frame and leaving a periodic pattern (looking like the stationary solution) behind. These look like: In order to describe informally the nature of the center manifold and the relevant dynamics in it, we use a Fourier decomposition. We write

$$S(x) = \sum_{n \in \mathbf{Z}} S_n e^{-in\omega z} ,$$

and a corresponding decomposition of $W(x,y)$,

$$W(x,y) = \sum_{n \in \mathbf{Z}} W_n(y) e^{-in\omega z} .$$

Then the boundary conditions for the W_n are

$$\lim_{y \to -\infty} W_n(y) = S_n , \quad \lim_{y \to \infty} W_n(y) = 0 .$$

To simplify the discussion, assume $\omega = 1$. We are interested in small, positive α, and we assume

$$\alpha = \epsilon^2 \alpha_0 , \quad c = \epsilon c_0 .$$

Here, α_0 and c_0 are fixed constants, and $\epsilon > 0$ is a small parameter. It is well-known that in this parameter range, a multiscale analysis is adequate. The principal contribution to W comes from $n = \pm 1$, and we reparametrize W_1 as

$$U(\xi) = \epsilon W_1(\epsilon \xi) .$$

Then U approximately satisfies the "amplitude equation,"

$$4U'' + c_0 U' + \alpha_0 U - 3U|U|^2 = 0 . \tag{6.2}$$

Exercise. Verify the above statement.

The two components, U and U', of this ordinary differential equation form essentially the two coordinates of the center manifold. They define a two-dimensional (complex) dynamical system in the center manifold, whose *fixed points* correspond to the stationary solutions of (2.4) and whose *saddle connections* between these fixed points and the zero solution will correspond to the front solutions. It should be noted that the dynamical system obtained in the center manifold depends on the speed c which is imposed on the front. It should also be noted that the original Eq.(2.4) is translation invariant. This will imply a symmetry of the induced flow on the center manifold, namely *covariance under multiplication by a phase*.

6.1. Reduction to a Center Manifold

For fixed ω, and in a frame moving with speed c the differential equation for W_n takes the form

$$(\alpha + c\partial_\xi - (1 + (-i\omega n + \partial_\xi)^2)^2)W_n(\xi) = \sum_{p+q+r=n} W_p(\xi)W_q(\xi)W_r(\xi) . \tag{6.3}$$

We can write this fourth order equation as a system of four first order equations, and we then view ξ as the "dynamical variable." We shall label the variables as follows

$$X_{nj} \longleftrightarrow \partial_\xi^j W_n \ , \ j = 0, \ldots, 3 \ , \ n \in \mathbf{Z} \ .$$

The system of equations (6.3) takes then the form

$$\partial_\xi X_n \ = \ M_n X_n + F_n(X) \ ,$$

where the matrix M_n is of the form

$$M_n \ = \ \begin{pmatrix} 0 & 1 & 0 & 0 \\ 0 & 0 & 1 & 0 \\ 0 & 0 & 0 & 1 \\ A & B & C & D \end{pmatrix} \ , \tag{6.4}$$

with

$$
\begin{aligned}
A &= -(1 - \mu^2)^2 + \alpha \ , \\
B &= 4i\mu(1 - \mu^2) + c \ , \\
C &= 6\mu^2 - 2 \ , \\
D &= 4i\mu \ ,
\end{aligned}
$$

and $\mu = \omega n$. The non-linear part is given by the vector

$$F_n(X) \ = \ \begin{pmatrix} 0 \\ 0 \\ 0 \\ E_n \end{pmatrix} \ , \tag{6.5}$$

where

$$E_n \ = \ - \sum_{p+q+r=n} W_p W_q W_r \ = \ - \sum_{p+q+r=n} X_{p0} X_{q0} X_{r0} \ .$$

6.2. The Linear Operator

The calculations involving M_n can be done easily by observing that the characteristic polynomial of M_n is

$$p_n(\lambda) \ = \ \lambda^4 - D\lambda^3 - C\lambda^2 - B\lambda - A \ . \tag{6.6}$$

We begin by studying the spectrum of M_n for small values of α and c by performing perturbation theory in α and c. When α and c are zero, the characteristic polynomial factors as $p_n(\lambda) = (\lambda - i(\mu + 1))^2(\lambda - i(\mu - 1))^2$ and the spectrum of M_n consists of the double eigenvalues $i(\mu \pm 1)$, see [EW] for details. Thus, for $\alpha = 0$, $c = 0$, the linear part of the problem has purely imaginary spectrum.

We analyze how this spectrum evolves as the parameters are varied. We shall see that for $n = \pm 1$ the spectrum will leave the imaginary axis by an amount which is an order of magnitude smaller than for the other values of n.

We consider the case of small α and c. In order to make perturbative statements we set

$$\alpha = \epsilon^2 \alpha_0 \, ,$$

$$c = \epsilon c_0 \, .$$

For simplicity, we set $\omega = 1$; see below for a generalization to $\omega \neq 1$. It is useful to define

$$\Delta = \sqrt{c_0^2 - 16\alpha_0} \, .$$

The characteristic polynomial may be written as

$$p_n(\lambda) = (\lambda - i(\mu + 1))^2 (\lambda - i(\mu - 1))^2 - c_0 \epsilon \lambda - \alpha_0 \epsilon^2, \qquad (6.7)$$

where $\mu = n\omega = n$. If $\epsilon\mu$ is small, we may write $\lambda = i(\mu \pm 1) + \delta$, and solve approximately for δ, and one finds the asymptotic formulae

$$\lambda_{\mu,\pm}^+ = \begin{cases} \epsilon(-c_0 \pm \Delta)/8 + \mathcal{O}(\epsilon^2) & \text{when } n = -1 \\ i(\mu + 1) \pm \epsilon^{1/2} i^{3/2} \sqrt{c_0(\mu + 1)}/2 + \mathcal{O}(\epsilon) & \text{when } n \neq -1 \end{cases}$$

$$\lambda_{\mu,\pm}^- = \begin{cases} \epsilon(-c_0 \pm \Delta)/8 + \mathcal{O}(\epsilon^2) & \text{when } n = 1 \\ i(\mu - 1) \pm \epsilon^{1/2} i^{3/2} \sqrt{c_0(\mu - 1)}/2 + \mathcal{O}(\epsilon) & \text{when } n \neq 1 \, , \end{cases}$$

for the eigenvalues. If, on the other hand, $\epsilon\mu > \tilde{c}$ for some small constant $\tilde{c}$, then one shows easily that there exists some small positive constant r_1 such $|\operatorname{Re} \lambda_{\mu,\pm}^{\pm}| > r_1$. We assume now $\epsilon > 0$, $\alpha_0 > 0$ and $c_0^2 > 16\alpha_0$. Then Δ is real. (The other case for c_0 corresponds to fronts which are slower than the minimal speed for which the amplitude equation has positive solutions.) We see that for $\omega = 1$,

$$\operatorname{Re} \lambda_{-1,\pm}^+ = \epsilon(-c_0 \pm \Delta)/8 + \mathcal{O}(\epsilon^2) \, ,$$

$$\operatorname{Re} \lambda_{+1,\pm}^- = \epsilon(-c_0 \pm \Delta)/8 + \mathcal{O}(\epsilon^2) \, .$$

For all other choices of $n \in \mathbf{Z}$ and of the sign $s \in \{+, -\}$, we have

$$\operatorname{Re} \lambda_{n,s}^{\pm} = \mathcal{O}(\epsilon^{1/2}) \, .$$

This means that $\lambda_{-1,\pm}^+$ and $\lambda_{+1,\pm}^-$ are "more central" than all other eigenvalues. Note that if $\omega^2 - 1 = \omega_0 \epsilon$, the preceding observations remain valid.

Theorem 6.1. *Given $c_0^2 > 16\alpha_0 > 0$, there is an $\epsilon_0 > 0$ such that for all ϵ in $(0, \epsilon_0)$ the Eq.(2.4) has stationary solutions and front solutions of frequency $\omega = 1$, and of the form (6.1). The amplitudes of these solutions are close to saddle connections for Eq.(6.10).*

Note that all our considerations above still apply if we replace $\omega = 1$, with any ω for which $|\omega^2 - 1| < \epsilon^\nu$, for some $\nu > 1$. Thus we have the

Corollary 6.2. *Given $c_0^2 > 16\alpha_0 > 0$, there is an $\epsilon_0 > 0$ such that for all ϵ in $(0, \epsilon_0)$ the Eq.(2.4) has stationary solutions and front solutions of frequency ω, for all $|\omega^2 - 1| < \epsilon^\nu$, where $\nu > 1$. The amplitudes of these solutions are close to saddle connections for Eq.(6.10).*

6.3. The Flow on the Center Manifold

We begin the reduction of our system of equations to the center-manifold by studying in more detail the eigenvectors and eigenvalues of the linearized piece which correspond to the "center" directions of the equations. By the center directions, we mean the subspace of $\mathcal{E}$ corresponding to the eigenvalues whose real part are of order $\mathcal{O}(\epsilon)$.

We first note that the submatrices M_n of the linearized operator leave a four-dimensional subspace of $\mathcal{E}$ invariant. We refer to this subspace as the "n^{th} sector." We note further that the center directions are confined to the first sector.

Given the matrix M_1, we will use the fact that if ϕ is the eigenvector corresponding to eigenvalue λ, and if $M_1^\dagger$, is the adjoint of M_1, with ψ the adjoint eigenvector corresponding to the eigenvalue $\bar{\lambda}$, then $\langle \psi, \phi \rangle = p_1'(\lambda)$, the derivative of the characteristic function of M_1.

Because of the definition of the subspace $\mathcal{E}$ on which we work, there are only two center directions, corresponding to the eigenvalues $\lambda_{+1,\pm}^-$. We will henceforth refer to these eigenvalues as $\lambda_\pm$ when there is no possibility of confusion, and to their corresponding eigenvectors and adjoint eigenvectors as $\phi_\pm$ and $\psi_\pm$.

In deriving the reduced equations on the center manifold we follow the method of Kirchgässner[K] and Mielke[M]. We introduce the projection operators

$$\mathbf{P} = c_+ |\phi_+\rangle \langle \psi_+| + c_- |\phi_-\rangle \langle \psi_-| \quad ; \quad \mathbf{P}^\perp = 1 - \mathbf{P} .$$

In this definition $c_\pm$ are normalization constants chosen so that $c_\pm \langle \psi_\pm, \phi_\pm \rangle = 1$. Note further that $\langle \psi_\mp, \phi_\pm \rangle = 0$, since $\lambda_+ \neq \lambda_-$ when $\alpha_0 \neq 0$ and ϵ is sufficiently small.

We now rewrite the variable X in our differential equation as $X = w + w^\perp$, where $w = \mathbf{P}X$, and $w^\perp = \mathbf{P}^\perp X$. Note that w is always contained in the first sector. The new equations are:

$$\partial_\xi w = Mw + \mathbf{P}F(w + w^\perp) , \tag{6.8}$$

$$\partial_\xi w^\perp = Mw^\perp + \mathbf{P}^\perp F(w + w^\perp) . \tag{6.9}$$

The center manifold theorem implies that there exists an invariant manifold of the form $(w, h(w))$, with h a C^1 map from $\mathbf{P}\mathcal{E}_0$ to $\mathbf{P}^\perp \mathcal{E}_0$ defined in a neighborhood of the origin of size $\mathcal{O}(\epsilon^{\frac{3}{4}+\gamma})$, for any $\gamma > 0$, provided ϵ is sufficiently small. Since we are interested in solutions of size $\mathcal{O}(\epsilon)$ this is a sufficiently large neighborhood for our purpose. If we substitute $h(w)$ for $w^\perp$ in (6.9), we obtain, from the invariance under the flow, the equation

$$h'(w) \cdot \partial_\xi w = Mh(w) + \mathbf{P}^\perp F(w + h(w)) .$$

We substitute for $\partial_\xi w$ from (6.8) and obtain

$$h'(w)Mw - Mh(w) = -h'(w)\mathbf{P}F(w + h(w)) + \mathbf{P}^\perp F(w + h(w)) .$$

If we now assume that $h(w) \approx \mathcal{O}(w^m)$ near the origin, then the left hand side of this equation is $\mathcal{O}(w^m)$, while the right hand is $\mathcal{O}(w^{m+2}) + \mathcal{O}(w^3)$. (This uses the fact that $F(w) \approx \mathcal{O}(w^3)$.) Hence, we may choose $h(w) \approx \mathcal{O}(w^3)$ near the origin. Thus we find the following

Rule: "Adiabatic Approximation". *If we ignore terms of order $\mathcal{O}(w^4)$ and higher we obtain approximate equations of motion in the center manifold, just by setting $w^\perp = 0$ in (6.8).*

We introduce coordinates $x_\pm$ on the center manifold. We write $w = x_+\phi_+ + x_-\phi_-$ and recall that $X = w + w^\perp$, so that when $w^\perp = 0$, $x_+ + x_- = X_{10}$. In order to calculate the form of the nonlinear term in the x variables, we note that

$$\mathbf{P}F(w) = c_+|\phi_+\rangle\,\langle\psi_+, F(w)\rangle + c_-|\phi_-\rangle\,\langle\psi_-, F(w)\rangle$$
$$= c_+|\phi_+\rangle E_1(w) + c_-|\phi_-\rangle E_1(w) \;.$$

The last of these equalities came from the explicit formulae for $\psi_\pm$ and $F(w)$. Rewriting (6.8) in terms of the x's, using this information we obtain

$$\begin{aligned} \dot{x}_+ &= \lambda_+ x_+ - 3c_+(x_+ + x_-)|x_+ + x_-|^2 \;, \\ \dot{x}_- &= \lambda_- x_- - 3c_-(x_+ + x_-)|x_+ + x_-|^2 \;. \end{aligned} \tag{6.10}$$

6.4. Identification with the Amplitude Equation

These quantities can be worked out in powers of ϵ. Note that we are assuming throughout $c_0^2 > 16\alpha_0$, so that Δ is real. For our purposes it is only necessary to use expansions up to order $\mathcal{O}(\epsilon)$. In Section 6.2, we already computed $\lambda_\pm$:

$$\lambda_\pm = \epsilon(-c_0 \pm \Delta)/8 + \mathcal{O}(\epsilon^2) \;.$$

Using the remark made earlier we have $c_\pm = 1/\langle\psi_\pm, \phi_\pm\rangle = 1/p_1'(\lambda_\pm)$. Using the expression (6.6) for $p_1(\lambda)$, and the asymptotic formula for $\lambda_\pm$ from above, we find that

$$c_\pm = \frac{\mp 1}{\epsilon\Delta}(1 + \mathcal{O}(\epsilon)) \;. \tag{6.11}$$

We now change coordinates in (6.10). We choose

$$\chi = (x_+ + x_-), \quad \eta = (x_+ - x_-) \;.$$

Neglecting terms of higher order, this leads to

$$\begin{aligned} \dot{\chi} &= \dot{x}_+ + \dot{x}_- = -\epsilon\frac{c_0}{8}\chi + \epsilon\frac{\Delta}{8}\eta \;, \\ \dot{\eta} &= \dot{x}_+ - \dot{x}_- = \epsilon\frac{\Delta}{8}\chi - \epsilon\frac{c_0}{8}\eta + \frac{6}{\epsilon\Delta}\chi|\chi^2| \;. \end{aligned}$$

We next rescale the "time:" Define

$$\chi(t) = \epsilon u(\epsilon t), \quad \eta(t) = \epsilon v(\epsilon t) \;.$$

Then

$$\dot{\chi}(t) = \epsilon^2 \dot{u}(\epsilon t), \quad \chi^3(t) = \epsilon^3 u^3(\epsilon t) \;,$$

and the equations take the form

$$\begin{aligned} \dot{u} &= -\frac{c_0}{8}u + \frac{\Delta}{8}v \;, \\ \dot{v} &= \frac{\Delta}{8}u - \frac{c_0}{8}v + \frac{6}{\Delta}u|u^2| \;. \end{aligned}$$

Our next transformation is

$$u = \frac{q}{2}, \quad v = \frac{c_0}{2\Delta}q + \frac{4}{\Delta}p.$$

Then the equations take the form

$$\dot{q} = p,$$
$$\dot{p} = \frac{1}{4}\left(-\alpha_0 q - c_0 p + 3q|q|^2\right). \tag{6.12}$$

Clearly, the system (6.12) is equivalent to (6.2).

6.5. Stationary Solutions and Fronts

The vector field described by (6.12) is very well studied in the literature, see for example [AW, CE]. We need here a discussion of a slightly perturbed system in the complex domain. Although the equation (6.12) could be viewed as a real equation, the correction terms of higher order will force the solution to acquire imaginary components even if the initial data at $x = -\infty$ are real. This phenomenon describes in fact the *corrections of the positions of the nodes* of the solution when the amplitudes change. The Eq.(6.12) has a fixed point at $p = q = 0$, and a circle of fixed points at $p = 0$, $|q| = (\alpha_0/3)^{1/2}$. (We assume always $\alpha_0 > 0$.) We need to know what happens to these fixed points when we perturb slightly the Eq.(6.12). It will be seen that the fixed point at 0 is hyperbolic and that the circle of fixed points is normally hyperbolic. This will imply that under a small perturbation, the fixed point at zero persists, and the circle *remains an invariant circle*. However, no abstract argument guarantees that it remains a circle of fixed points. We have shown in Theorem 3.1 that the invariant circle is made up of fixed points. We stress again that the other alternative is physically not uninteresting and has been observed in other, realistic examples, cf. [CE]. It corresponds to convective instabilities.

6.6. Stationary or Convective Solutions

We now study (6.12) as an equation in $\mathbf{C}^2$. Writing out the equations for the real and imaginary parts of p and q, we see that the linearization of the vector field at the origin has two double eigenvalues, $\frac{-c_0 \pm \sqrt{c_0^2 - 16\alpha_0}}{8}$, (which both have real part less than zero), and the linearization at $q = \sqrt{\alpha_0/3}$ has eigenvalues 0, $-c_0/4$, and $\frac{-c_0 \pm \sqrt{c_0^2 + 32\alpha_0}}{8}$. Note that here we have one unstable direction, one neutral direction, corresponding to motion along the circle of fixed points, and two stable directions. Thus, the circle is normally hyperbolic cf. [HPS].

We next study what happens to the fixed point and the circle as we perturb the vector field. Let $\mathbf{X}_0$ denote the vector field in (6.12), and let $\mathbf{X}_\epsilon$ be the $\mathcal{C}^1$ small perturbation of $\mathbf{X}_0$ obtained by changing the parameters to $\epsilon > 0$. Then we have:

Lemma 6.3.

X1: $\mathbf{X}_\epsilon$ *is covariant under the transformation* $q \to e^{i\phi}q$ *and* $p \to e^{i\phi}p$, *and*

X2: $\mathbf{X}_\epsilon$ *has a fixed point which approaches* $q = \sqrt{\alpha_0/3}$, $p = 0$ *as* ϵ *approaches zero.*

6.7. Front Solutions

The circle of fixed points with $|q| > 0$ corresponds to stationary solutions, and any two of these solutions differ only by a phase (which corresponds to translation in x for the original equation (6.2)). The point $p = q = 0$ corresponds to the zero solution of Eq.(6.2). For the unperturbed system, (6.12), a phase space analysis (see [AW]) shows that for every q on the circle $|q|^2 = \alpha_0/3$, there is a saddle connection, tangent to the unstable direction at that point, which connects it to the point $p = q = 0$. *These solutions are front solutions for the amplitude equation.* The connections for different "initial points" q are again related by a phase, and if $q^2 = \alpha_0/3$, they are real. The next result shows that these front solutions persist when we perturb the Eq.(6.12).

Proposition 6.4. *Under the conclusions of Lemma 6.3,* $\mathbf{X}_\epsilon$ *has a hyperbolic fixed point at $p = 0$, $q = 0$, and a circle of fixed points near $|q|^2 = \alpha_0/3$. Furthermore, $\mathbf{X}_\epsilon$ has a family of front solutions (related to one another via $q \to e^{i\phi}q$ and $p \to e^{i\phi}p$) which are saddle connections between the circle of fixed points and the origin.*

For a proof, see [EW].

7. Systems in a Strip

We now apply the methods of center manifolds to systems in a strip. In this case, we choose a basis of functions which is somewhat different from that for fronts. There, the basis had 1 space and one time direction, now we are dealing with 0 time and two space directions.

7.1. Preliminary Remarks

Although chaotic spatial solutions may appear already in systems with 1+1 dimensions (through the Eckhaus instability), we are really interested in 2+1 dimensions, in which weak turbulence is intimately related to defects. The hydrodynamic systems we have in mind have long-wavelength instabilities, such as the skew-varicose instability. By numerical simulations, it has been shown [ZS] that the Boussinesq equations of fluid convection lead to a skew-varicose instability due to the coupling of the dynamics of the convection pattern to the z-component of vorticity. In addition, it has been shown that Swift-Hohenberg like models in 2+1 dimensions show a skew-varicose instability upon adding terms that model the coupling to vertical vorticity. For example, in [GC] the following model has been analyzed:

$$\partial_t u + (\vec{U} \cdot \vec{\nabla})u = \left(\alpha - (1 + \partial_x^2 + \partial_y^2)\right)^2 u - u^3 , \qquad (2.1)$$

where

$$U = (\partial_y \zeta, -\partial_x \zeta) ,$$

and

$$\nabla^2 \zeta = g\left(\vec{\nabla}(\nabla^2 u) \times \vec{\nabla} u\right) \cdot \hat{z} . \qquad (2.2)$$

Here, g is a coupling constant.

This model shows, for suitable parameter values, a skew-varicose instability which sets in *before* the Eckhaus instability, because wiggly perturbations of the convective pattern are destabilized by the additional term. By changing the nonlinear term u^3

to $3\nabla^2 u(\vec{\nabla} u)^2$, the analysis leads to a phase diagram that resembles very closely the fluid stability diagram, again with a skew-varicose instability occurring *before* the Eckhaus instability.

These findings are relevant from our point of view, since we shall show that it is the y-component of the long-wavelength instability which leads naturally to defects. However, to expose this idea more easily, we shall analyze the Swift-Hohenberg model in 2+1 dimensions, *without* coupling to a second field, although the order of the appearence of instabilities is then reversed. Keeping this in mind, it will be much easier to show how to reduce the model to a dynamical system and to explain why the y-components are so crucial. We have checked that the main ideas apply to the more complicated model as well, but the details of this calculation contain no new ideas and will not be given here. (It might seem at first sight that the inverse Laplacian which is implicitly contained in Eq.(2.2) destroys the reducibility to a dynamical system since it is in principle a non-local operator. However, the natural basis in a strip makes the Laplacian diagonal and allows for an explicit inversion in the basis of Fourier modes given below.)

7.2. Reduction to a Dynamical System

As noted before, in 2+1-dimensions even the equations for stationary solutions are PDE's. Therefore they are not directly amenable to techniques of dynamical systems theory. In this subsection, we show that in the case of infinitely long strips of *finite* width in the y-direction, a reduction to finitely many variables is still possible through the use of a center manifold. By expanding the fields in a Fourier *series* in the y-direction (not an integral!) we shall be able to perform this reduction to a dynamical system, where the x-direction stands for time. It will turn out that the dimension of the corresponding dynamical system is infinite, with a few degrees of freedom corresponding to each y-Fourier mode. But slightly beyond a secondary hydrodynamic instability, there exists a center manifold of finite dimension on which the relevant physics takes place. We stress again that *this mechanism of reduction is made possible by the two steps that lead to compactification of the problem: The finite strip, and the periodic basic instability in the x-direction* [K, M, EW].

Coming back to our problem, we will see that the dimension of the center manifold increases with the width of the strip. To make these notions explicit, we again consider the Swift-Hohenberg equation, because of its simplicity, but now in 2+1 dimensions. The main ideas are immediately translated to any model with only one infinite direction. The equation for stationary solutions is now

$$\left(\alpha - (1 + \partial_x^2 + \partial_y^2)^2\right)u(x,y) - u^3(x,y) = 0 \, . \tag{7.1}$$

Consider this in a strip and, for further simplification, consider periodic boundary conditions in the y-direction. Expanding in Fourier modes, we write

$$u(x,y) = \sum_{n=-\infty}^{\infty} e^{iny/h} u_n(x) \, , \tag{7.2}$$

where the width of the strip is $2\pi h$.

Extending the method of Section 4.4, we introduce now an infinite dimensional phase space $\mathcal{B}$, with coordinates X. Of course, one should make $\mathcal{B}$ into a Banach or Hilbert space by the choice of an adequate topology, but this depends on the details of the problem, see (3.4) or [EW] for a typical choice. Recalling that the "time" of the dynamical system is going to be x, we will have at every instant a value of the coordinate X. Thus, X is defined as

$$X(x) = \{X_{n,p}(x)\}_{n \in \mathbf{Z},\ p=0,1,2,3} = \{\partial_x^p u_n(x)\}_{n \in \mathbf{Z},\ p=0,1,2,3}\ . \tag{7.3}$$

Thus, the coordinates are just the first 3 derivatives of the amplitudes u_n. The "dynamics" in this phase space is then

$$\dot{X} = \mathbf{M}X + \mathbf{N}(X)\ , \tag{7.4}$$

where $\mathbf{M}$ is the linear part, and $\mathbf{N}$ the nonlinear part. The infinite dimensional operator $\mathbf{M}$ is block diagonal in the natural basis defined in (7.3) with the nth block being M_n,

$$M_n = \begin{pmatrix} 0 & 1 & 0 & 0 \\ 0 & 0 & 1 & 0 \\ 0 & 0 & 0 & 1 \\ \alpha - (1 - \frac{n^2}{h^2})^2 & 0 & -2(1 - \frac{n^2}{h^2}) & 0 \end{pmatrix}\ . \tag{7.5}$$

Note that $\mathbf{M}_{n=0}$ coincides with the operator $\mathbf{M}$ in (4.10). The nonlinear part $\mathbf{N}$ is just the third power written in components, viz.,

$$\mathbf{N}(X)_{n,p} = 0\ , \quad \text{if } p \neq 3\ ,$$

$$\mathbf{N}(X)_{n,3} = - \sum_{n_1 + n_2 + n_3 = n} X_{n_1,0} X_{n_2,0} X_{n_3,0}\ .$$

Note that $\mathbf{N}$ is *not* diagonal in n. The characteristic polynomial P_n of M_n is

$$P_n(\lambda) = -\alpha + \left(1 - \left(\frac{n}{h}\right)^2 + \lambda^2\right)^2\ . \tag{7.6}$$

Note that when $n \neq 0$ and h is very small, the roots of P_n are on the real axis and of very large modulus. Since the corresponding eigenfunctions are $e^{\lambda x}$, they will grow very fast as either $x \to \infty$ or $x \to -\infty$. On the other hand, when h is small but positive, we find imaginary solutions for λ if and only if

$$1 \pm \sqrt{\alpha} > \frac{n^2}{h^2}\ , \tag{7.7}$$

(and $\alpha > 0$), and then these roots are

$$\lambda = \pm i \sqrt{1 \pm \sqrt{\alpha} - \frac{n^2}{h^2}}\ .$$

Thus, we expect a number of modes of the order of h to exist which correspond to bounded solutions. *These modes will make up the center manifold.*

Assume now α and h are such that the inequality (7.7) is only satisfied for $n = 0, \pm 1$. Since all other modes are linearly unstable when $x \to \infty$ or $x \to -\infty$, it is reasonable to restrict attention to the center manifold. As before, let $\mathcal{E}^c$ denote the linear space spanned by the modes with imaginary eigenvalues. By assumption, these are necessarily arising from blocks with $|n| \leq 1$. Note that for wider strips there would be more such modes. We denote by $\mathcal{E}^{su}$ a supplementary subspace in $\mathcal{B}$. By definition, the center manifold, W_c, is an invariant manifold tangent to $\mathcal{E}^c$ at zero. In particular, $\dim W_c = \dim \mathcal{E}^c$. Fig. 8 illustrates the geometry of this setup. We denote the coordinate along $\mathcal{E}^c$ by ξ. The invariant manifold is the graph of a function Φ

which maps $\mathcal{E}^c$ to $\mathcal{E}^{su}$. A point X in W_c can be written as

$$X = \xi + \Phi(\xi) ,$$

so that the equations of motion (7.4) and the condition of invariance of W_c lead to the system of equations

$$\dot{\xi} = \mathbf{M}\xi + \mathbf{P}_c \mathbf{N}(\xi + \Phi(\xi)) , \qquad (7.8)$$

$$D\Phi(\xi)\dot{\xi} = \mathbf{M}\Phi(\xi) + (1 - \mathbf{P}_c)\mathbf{N}(\xi + \Phi(\xi)) . \qquad (7.9)$$

Here, $\mathbf{P}_c$ is the projection onto $\mathcal{E}^c$ along $\mathcal{E}^{su}$. Some remarks on (7.8)–(7.9) are in order.

i) Since $\mathcal{E}^c$ is spanned by eigenfunctions of $\mathbf{M}$, we have $\mathbf{M}\xi = \mathbf{P}_c \mathbf{M}\xi$, and hence all the terms of (7.8) are in $\mathcal{E}^c$. Similarly, all terms in (7.9) are in $\mathcal{E}^{su}$.

ii) Since W_c is tangent to $\mathcal{E}^c$, we expect $\Phi(\xi) = \mathcal{O}(\xi^2)$. (This will be true if Φ is sufficiently smooth.) Therefore, to lowest order in ξ, one can omit the $\Phi(\xi)$ in Eq. (7.8). Then it reduces to

$$\dot{\xi} = \mathbf{M}\xi + \mathbf{P}_c \mathbf{N}(\xi) , \qquad (7.10)$$

which is just a rewriting of the standard amplitude equations (for stationary solutions) in a dynamical systems guise.

As indicated above, the analysis of this section pertains equally well to the models (2.1) in which a coupling to a second field is introduced, since this coupling is purely non-linear and does not change the linear operator $\mathbf{M}$.

7.3. Quasiperiodic and Chaotic Solutions

Having built the appropriate machinery, we wish to re-apply the considerations of Section 4. Therefore, we first have to check that biperiodic solutions exist in perturbation theory. This is a model dependent question. It is easy to verify that the nonlinearities in Eq.(7.1) saturate the growth of biperiodic solutions exactly as in the developments of Section 4.3. Using a computer algebra package, it is also straightforward, if tedious, to check this saturation for the Eq.(2.1). These calculations again do not prove the existence of biperiodic solutions, but they allow to verify the twist condition, i.e., that the rate of rotation depends on the amplitude. Therefore the KAM picture developed in Section 4.5 applies in the present higher dimensional phase space of coordinates in $\mathcal{E}^c$. The quasiperiodic solutions that survive live on two-dimensional tori and have incommensurate frequencies. Between the tori there is a chaotic sea.* Note that the chaotic sea is now of higher dimension due to the larger dimension of phase space. Therefore, Arnold diffusion is unavoidable, but again we assume that it is experimentally irrelevant. Also, *bounded* chaotic solutions may very well exist in this chaotic sea.

---

* Strictly speaking, we need here a version of the KAM theorem for invariant tori whose dimension is less than half the space dimension. Such a theorem has been proven in the Hamiltonian case in [LW]. We are not aware of an existing proof in the reversible, but non-Hamiltonian case.

To draw conclusions about the nature of the stationary solutions obtained in this way, we need to clarify the relation between the spatial structure of the problem and the variables on the center manifold. Recall that ξ is a system of coordinates in $\mathcal{E}^c$, parametrizing the center manifold W_c. Assume now as a first example that the dynamics of ξ is periodic with period 1, i.e.,

$$\xi(x+1) = \xi(x) .$$

In this case, $\xi(x) + \Phi(\xi(x))$ describes again a periodic orbit of period 1 which lies entirely in W_c. Identify now the coordinates of $\mathcal{B}$ with the y-Fourier components of $u(x,y)$ and their x-derivatives. Clearly, with the notation of (7.3), we have

$$u(x,y) = \sum_{n=-\infty}^{\infty} X_{n,0}(x)e^{iny/h} .$$

Because W_c is tangent to $\mathcal{E}^c$ at zero, all the amplitudes except those corresponding to modes in $\mathcal{E}^c$ will be quadratically small. Therefore, $u(x,y)$ is approximately of the form

$$u(x,y) \approx \sum_{|n| \leq n_0} X_{n,0}(x)e^{iny/h} , \tag{7.11}$$

where n_0 depends on the width of the strip. Because ξ was assumed to be periodic, all the amplitudes $X_{n,0}$ are seen to be periodic of the same period and $u(x,y)$ is periodic in x with period 1.

Next consider quasiperiodic solutions, with two frequencies. We can choose a parametrization of $\mathcal{E}^c$ for which the two coordinates ξ_1 and ξ_2 will be periodic with period $(2\pi)^{-1}$ and $(2\pi\gamma)^{-1}$, respectively. The simplest example of this kind is

$$\xi_1(x) = \cos(x) ,$$
$$\xi_2(x) = \cos(\gamma x) .$$

The same reduction as before leads to the representation (7.11), which for $n_0 = 1$ reads

$$u(x,y) = X_{0,0}(x) + X_{1,0}(x)e^{iy/h} + X_{-1,0}(x)e^{-iy/h} . \tag{7.12}$$

Assuming that the solution has to be real, we have the condition $X_{1,0} = \bar{X}_{-1,0}$. Our assumptions on ξ lead to the general form

$$X_{0,0} = A_0 \cos(x) + B_0 \cos(\gamma x) ,$$
$$X_{1,0} = A_1 \cos(x) + B_1 \cos(\gamma x) ,$$

up to higher order terms in the amplitudes. The values of A_i and B_i depend on the orientation of the coordinates ξ_i in $\mathcal{E}^c$. In the next section we shall argue that solutions of this type give rise to defects in the spatial structure. It will then turn out that the values of the coefficients A_i and B_i determine the distribution of these defects in space.

In the case of chaotic solutions the same small number of coordinates ξ_i are relevant, and they look almost quasiperiodic, but with a slow phase diffusion. Roughly speaking this means that ξ_i is of the form

$$\xi_i(x) = \big(1 + \psi_i(x)\big) \cos\big((\gamma_i + \phi_i(x)) \cdot x\big) ,$$

where $\phi_i(x)$ and $\psi_i(x)$ are slowly varying. The translation of this means that $u(x,y)$ will have Fourier coefficients $X_{n,0}$ of a similar form. This leads to a chaotic spatial structure.

8. Defects

The surprising result of the analysis presented so far is that defects in the spatial structure show up naturally. In fact, this is an unavoidable consequence of our general framework. The phenomenon is most easily seen with a typical function of the form of the solution (7.11). To make the features visible, consider the following example:

$$u(x,y) = \cos(x) + A\cos(y) \cdot \cos(\gamma x) , \qquad (8.1)$$

which clearly is of the type (7.12). In Fig. 10, the cellular structure described by this function is displayed and one sees the extra pairs of "rolls" inserted in the structure.

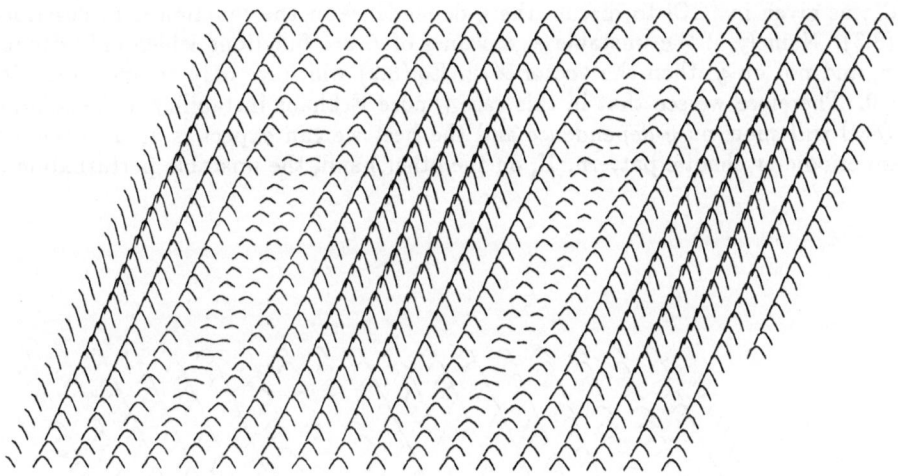

Fig. 10. The positive part of the function $\cos(x) + 1.3\cos(y)\cos(\sqrt{1.26}x)$ for $y \in [-\pi/2, \pi/2]$ and $x \in [0, 120]$.

The spatial points where maxima and minima merge will be distributed quasiperiodically for this example. We refer to these points as **defects**. The positions of these defects along the rolls is special for this case because of the simple form of the ansatz (8.1). In general, skewed arrangements of defects will occur as is manifested in Fig. 11.

It should be stressed that the appearance of defects is necessarily linked to the availability of the y-direction. The question whether pairs of defects are formed and whether they are aligned with the axis of the rolls ("climb" configuration) depends on the nature of the eigenvectors of $\mathbf{M}$.

To make this explicit, we write the center eigenvectors of $\mathbf{M}$ in the form

$$X^{(m)} = \left\{ X_{n,p}^{(m)} \right\}_{n \in \mathbf{Z}, p=0,1,2,3} ,$$

for $m = 1, \ldots, m_0$. Here, m_0 is the number of center eigenvectors, and the $X_{n,p}^{(m)}$ are the (complex) components of $X^{(m)}$. The coordinates parametrizing the center manifold are denoted ξ_m, $m = 1, \ldots, m_0$ so that we recognize m_0 as the dimension of $\mathcal{E}^c$. We can write the general element in $\mathcal{E}^c$ as

$$X = \sum_{m=1}^{m_0} \xi_m X^{(m)} . \qquad (8.2)$$

169

Assume now that $\xi = \{\xi_1, ..., \xi_{m_0}\}$ describes a periodic or quasiperiodic orbit in $\mathcal{E}^c$, and denote this orbit by $\xi(x)$. Then the corresponding stationary solution of the Swift-Hohenberg equation takes the approximate form

$$u(x,y) = \sum_{m=1}^{m_0} \xi_m(x) \sum_{n \in \mathbf{Z}} X_{n,0}^{(m)} e^{iny/h} . \qquad (8.3)$$

This makes the relation (7.11) explicit. Note next that operators, such as $\mathcal{L}_u$ can be studied in the center manifold whose coordinates are naturally given by the eigenvectors of the matrix M. We can use the classification of the eigenvectors and eigenvalues of $\mathcal{L}_u$, as given in [GC] to discuss the y-dependence of the function u as described by (8.3). Namely, if the unstable eigenvalues of $\mathcal{L}$ are functions which only depend on x, and not on y, then the vector X in Eq.(8.2) will have only components with $n = 0$. Therefore we see that in this case ("pure Eckhaus instability") *the solution Eq.(8.3) will show no y-dependence and the best we can expect is an x-dependent quasiperiodic or chaotic pattern.* If, on the other hand, the unstable perturbation of

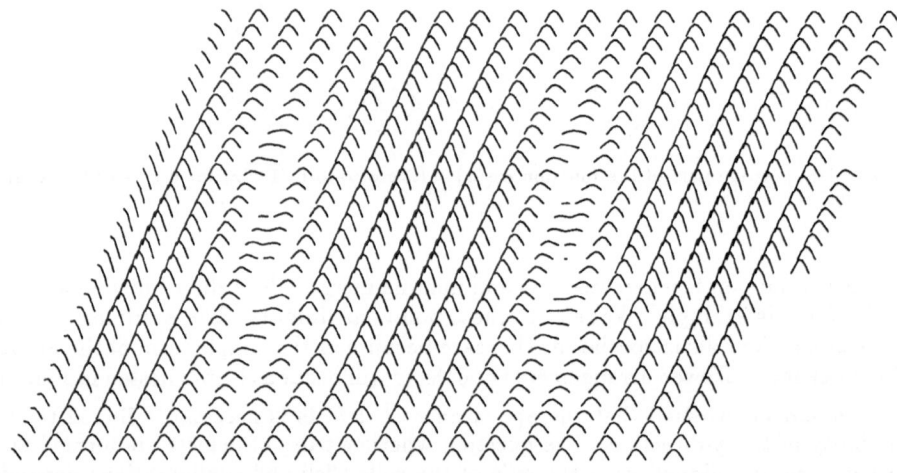

Fig. 11. The positive part of the function $\cos(x) + \cos(y)\cos(1.123x + 1.134y)$ for $y \in [-\pi/2, \pi/2]$ and $x \in [0, 120]$. Note the skewed arrangement of the defects.

the original x dependent pattern is either a function of the form $e^{ik_y y}$ ("zig-zag instability") or of the form $e^{i(k_x x + k_y y)}$ ("skew-varicose instability"), then the stationary solution will have a non-trivial x- and y-component in general. It follows that in these cases, $u(x,y)$ will have a term with some non-zero n-component. This shows that the example of (8.1) is *generic in the case of a skew-varicose instability*, since then, there is a y-component *and* two different x-components because of the presence of $k_x \neq 0$. On the other hand, the zig-zag instability, while being y-dependent, will not generally form two frequencies in the x-direction. Thus, we see that *a non-zero width of the strip is necessary, and generically sufficient for the generation of defects.*

At this point we should connect the discussion of the scenario for the onset of space-time chaos in systems in 1+1 dimension to that of systems in a strip. We have argued in Section 4.6 that in 1+1 dimension the tendency to change the local wavevector results in space-time dislocations whenever a wavelength is shed off or gained. The appearance of these space-time dislocations was not the reason for spatial chaos, but was really generated by the solution to the S problem (stationary state) which was seen to organize the motion of the T problem.

In very much the same way we are thinking about the events that occur in systems in strips. The secondary instability is again responsible for sending the system through the stable direction towards a state that is stationary in time, but chaotic in the one infinite coordinate x. The use of the Center Manifold Theorem allowed us to prove that indeed such states yield to a KAM mechanism of chaotization in space. Again, the temporal orbit is going to evolve along the unstable direction, leading as in 1+1 dimension to changes of wavelengths in the x direction. The difference is that rather than generating space-time dislocations, now the availability of a y-coordinate can result in topologically stable dislocations in the x-y plane.

These can appear in one of the two following ways:

i) The process of phase slip is confined by a non-trivial y-dependence (as in $\cos(x) + 1.3\cos(y)\cos(\sqrt{1.26}x)$, cf. Fig. 10) to a portion of the strip, meaning that the wavelength is changed in a "bubble" but not all along the y coordinate. As a result the number of "rolls" at the boundaries is not the same as in the middle of the strip, and such a situation can only be accommodated by topological defects.

ii) The phase slip process causes the solution to vanish identically on a line rather than at a point as in 1+1 dimension. We expect such a line to be unstable to y-fluctuations which would result in the formation of pairs of points of vanishing solutions, and these are again topologically stable.

It should be stressed that the precise mechanism for the nucleation of topological defects should be studied independently, see [EGP], and the above discussion is far from exhausting the issue. The point is however that we understand that the production of defects is a necessary *consequence* of the escape along the unstable manifold of our already spatially chaotic stationary state.

We thus summarize our message that understanding the 1+1 dimensional system and the nature of its stationary, spatially chaotic solutions brings one a long way in understanding the onset of defect-mediated turbulence in 2+1 dimensions. We have avoided the difficult questions of defect formation (see [LF]), and of a global study of the PDE, and have realized that the randomly appearing defects are a secondary consequence of the existence of chaotic stationary solutions.

Acknowledgments. We would like to thank S. Alexander, G. Goren, and R. Zeitak for very pleasant and helpful discussions on the subject of this paper. We thank C.E. Wayne for useful information on the status of the KAM theory. The questions and comments of the participants of the Como school were helpful for the elaboration of these notes. JPE was partially supported by the Einstein Center of Theoretical Physics at the Weizmann Institute. This work was partially supported by GIF, the German-Israeli Foundation, Grant #I-97-169.7188, and the Fonds National Suisse.

References

[AB] Ahlers G. and R.P.Behringer: Prog. Theor. Phys. Suppl. **64**, 186 (1979).

[ACEGP] Auerbach D., P.Cvitanović, J.-P.Eckmann, G.Gunaratne, and I.Procaccia: Exploring chaotic motions through periodic orbits. Phys. Rev. Lett **58**, 2387–2389 (1987).

[AW] Aronson, D. and H.Weinberger: Multidimensional nonlinear diffusion arising in population genetics. Adv. Math. **30**, 33–76 (1978).

[BPL] Bodenschatz E., W.Pesch and L.Kramer: Physica **32D**, 135 (1988).

[CE] Collet P. and J.-P.Eckmann: *Instabilities and Fronts in Extended Systems*, Princeton University Press (1990).

[CGL] Coullet P., L.Gil and F.Lega: Phys. Rev. Lett. **62**, 1619 (1989).

[EGP] Eckmann J.-P., G.Goren and I.Procaccia: Nonequilibrium nucleation of topological defects as a deterministic phenomenon. Preprint. (Weizmann Institute 1991).

[EP1] Eckmann J.-P. and I.Procaccia: The onset of defect mediated turbulence. Preprint. (Weizmann Institute 1990).

[EP2] Eckmann J.-P. and I.Procaccia: The generation of spatio-temporal chaos in large aspect ratio hydrodynamics. Preprint. (Weizmann Institute 1990).

[EW] Eckmann J.-P. and C.E.Wayne: Propagating fronts and the center manifold theorem. Preprint. (University of Geneva 1990).

[EZ] Eckmann J.-P. and M.Zamora: Stationary solutions for the Swift-Hohenberg equation in nonuniform backgrounds. Preprint. (University of Geneva 1990).

[GPRS] Goren G., I.Procaccia, S.Rasenat, and V.Steinberg: Interactions and dynamics of topological defects: Theory and experiments near the onset of weak turbulence. Phys. Rev. Lett. **63**, 1237–1240 (1989).

[GC] Greenside H.S. and M.C.Cross: Stability analysis of two-dimensional models of three-dimensional convection. Phys. Rev. A **31**, 2492–2501 (1985).

[HNW] E.Hairer, S.P.Nørsett, and G.Wanner: *Solving Ordinary Differential Equations I*, Berlin, Heidelberg, New York, Springer (1987).

[HPS] M.W.Hirsch, C.C.Pugh, and M.Shub: *Invariant Manifolds*, Lecture Notes in Mathematics Vol. 583, Berlin, Heidelberg, New York, Springer (1977).

[K] Kirchgässner K.: Nonlinearly Resonant Surface Waves and Homoclinic Bifurcation. Adv. Appl. Mech. **26**, 135–181 (1988).

[KSZ] Kramer L., H.R.Schober, and W.Zimmermann: Pattern competition and the decay of unstable patterns in quasi-one-dimensional systems. Physica **D31**, 212–226 (1988).

[L] Langer J.S.: Theory of the condensation point. Ann. Physics **41**, 108–157 (1967).

[LF] Langer J.S. and M.E.Fisher: Intrinsic critical velocity of a superfluid. Phys. Rev. Lett. **19**, 560–563 (1967).

[LW] Llave R. and C.E.Wayne: Whiskered and low dimensional tori for nearly integrable hamiltonian systems. (To appear). Nonlinearity.

[LG] Lowe M. and J.P.Gollub: Pattern selection near the onset of convection: The Eckhaus instability. Phys. Rev. Lett. **55**, 2575–2578 (1985).

[M] Mielke A.: Reduction of Quasilinear Elliptic Equations in Cylindrical Domains with Applications. Math. Meth. Appl. Sci. **10**, 51–66 (1988).

[N] Newell A.: (To appear). Phys. Rev. Lett.

[Mo] Moser J.: *Stable and random motions in dynamical systems: with special emphasis on mechanics*, Princeton University Press (1973).

[PCL] Pocheau A., V.Croquette, and O.LeGal: Phys. Rev. Lett. **55**, 1099 (1985).

[RRS] Rehberg I., S.Rasenat, and V.Steinberg: Phys. Rev. Lett. **62**, 756 (1989).

[RJ] Ribotta R., A.Joets: In *Cellular Structures and Instabilities*, (Wesfreid J.E. and S.Zaleski eds.). Springer (1984).

[Se] Sevryuk M.B.: *Reversible Systems*, Lecture Notes in Mathematics, Vol. 1211, Springer (1986).

[SH] Swift J. and P.C.Hohenberg: Phys. Rev. **A15**, 319 (1977).

[ZS] Zippelius A. and E.D.Siggia: Stability of finite-amplitude convection. Phys. Fluids **26**, 2905–2915 (1983).

TOPICS IN PATTERN FORMATION PROBLEMS AND RELATED QUESTIONS*

Yves Pomeau

Laboratoire de Physique Statistique
24, rue Lhomond, 75231 Paris Cedex 05
France

Abstract

Our thinking in pattern formation problems is more guided by the solution of simple models than by direct investigation of the basic equations for a well identified process. The clearest example of this situation is in instabilities of fluid flows: the Navier-Stokes equations are rather hopeless, but one can understand a lot of phenomena occuring in real flows by a reduced amplitude approach.

Those amplitude equations aim at describing the dynamics of weak fluctuations and are defined usually under restrictive assumptions: large aspect ratio (many typical wavelength), supercritical or weakly subcritical instabilities, control parameter near threshold. Dropping anyone of those assumptions leads to major difficulties linked to the summation of poorly controlled perturbation series. It seems better then to try to study simple amplitude models, with the most general structure one can think of, and to compare their predictions with real life experiments.

This is the line of thought followed here. To be a little more concrete, I shall consider first a now classical question, which is the connection between the subcritical character of the bifurcation in parallel flows and the occurence of localised patches of turbulence therein. This can be explained with a simple model of the reaction diffusion (RD) type. However this model is quite special in the sense that it has a variational structure. This entails some nongeneric qualitative properties, that are no more valid for more general system, as one expects the turbulent flows are. I shall consider some of those nongeneric properties, and refer briefly to other works on other ones. I will single out a remarkable theoretical discovery, that is the possibility of stable localised structures

---

*This is a written summary of a talk presented at the conference held at the NATO advanced scientific institute held in Como, Italy on "Chaos, Order and Patterns", June 25-July 6, 1990.

Chaos, Order, and Patterns, Edited by R. Artuso *et al.*
Plenum Press, New York, 1991

in a nonvariational system, although in gradient/variational systems those localised structures are always linearly unstable. Then I shall examine an application of this outside of the realm of fluid mechanics, that is the dynamics of a so-called Bloch wall in a ferromagnet. Those Bloch walls are between two magnetic domains of opposite polarization. Under the action of a convenient external magnetic field, those walls drift in a direction depending on their core structure. The time dependence in the external field breaks the variational structure of the system: in variational systems walls move to replace a state on one side of the wall by the other with a lower energy, and the direction of this motion cannot depend on the core structure of the defect.

Finally I will show how to analyse, again by using general arguments only, the dynamics of a periodic structure, as a set of Rayleigh-Bénard rolls under external stress. This makes appear a rather interesting mathematical structure, linked perhaps to the famous von Karman conjecture on the buckling of plates.

1. Finite amplitude instabilities in open flows: generic features

Fluid mechanics is a fascinating domain of science: the basic equation are known now for more than a century and they remain still many unsolved problems, the least one being not fully developed turbulence. One may get an idea of the difficulty of the field by thinking how much ingenuity and nontrivial mathematics it took to understand (via strange attractors and chaotic dynamics) the scenarios of transition to chaos in "small boxes".

Below I do not intend to consider such a grand problem as fully developed turbulence, but instead report some significant progress made in the understanding of transition to turbulence in parallel flows. This is a very much studied field, but very frustrating too, because the best analytical theory one can do, by using formally the basic Navier-Stokes equations only, is the linear and weakly linear stability analysis, that does say very little or nothing at all on real world experiments. This comes from a very fundamental character of the relevant instabilities, that is that they are subcritical (or hysteretic-the pair of words subcritical/supercritical are sometimes used with a different meaning, referring to the absolute/convective character of instabilities in the linear theory). As already noticed by Osborne Reynolds, circular pipe flows are seemingly stable against small amplitude perturbations but unstable against large amplitude, for the same Reynolds number, if large enough. This seemingly innocuous remark about the subcritical character of the instability (in modern words) can be put at the basis of a more formal theory explaining simply many experimental facts at the price of a moderate analytical investment. This requires an act of faith, because one cannot derive everything from the Navier-Stokes equations all the way down to finite amplitude equations. In some sense, the situation is like in dynamical system theory applied (successfully) to "small boxes": one assumes there is nothing special to the Navier-Stokes equations and then try to get "universal" results, applicable to fluid flows from model systems.

To make more concrete what I have in mind, let me consider a classical phenomenon in Fluid Mechanics, that is the formation of localized patches of turbulence in an otherwise laminar flow. As I argued[1] some time ago this may be interpreted by analogy with the behavior of solutions of reaction-diffusion equations. Let us assume that the local state of the fluid can be represented via an 'order parameter', a scalar quantity that changes continuously from a definite value in the turbulent state to another one in the laminar state. The most simple example of such a quantity would be the amplitude of the turbulent fluctuations. The dynamics of this amplitude may be derived through a systematic small amplitude expansion "a la Landau", although in a more rigorous theory(still to come) this amplitude would be an intrinsic parameter in the statistical ensemble describing the turbulent fluctuations. There is likely a long way to go before we can do that starting from the mere Navier-Stokes equations.

Suppose now that both the turbulent state and the laminar one are stable against small amplitude fluctuations, thus it is natural to represent the dynamics of the order parameter as a gradient flow in a two well potential, each well corresponding to a locally stable state of the global system. Then it is still natural to represent the tendency of the system to become spatially uniform, say because of the molecular diffusion effects, by adding a linear diffusion term to the equation of motion for the order parameter. This yields a reaction-diffusion equation of a kind that has been much studied[2]. In good cases it is even possible to derive this sort of equation from the first principles in a convenient limit [3]. The main property that interests us here is that those reaction-diffusion(like) equations have, as an asymptotic solution to a rather large class of initial datas, a front separating the two possible uniform stable states of the system and moving at a constant speed. But this is not the full story, mainly for two reasons:

i) the reaction-diffusion equations (or the amplitude equations) are local in space, whereas the turbulent fluctuations couple to large scale flow, via the Reynolds stress [9]. This may be included into the picture by writing a coupled system of reaction-equation plus an elliptic equation for the large scale flow with the Reynolds stress as source term. This induces a feedback between the growth of the localised patches of turbulence and their control parameter (the Reynolds number, for instance) and this feedback may even stop the growth of the turbulent spot at a given size. This is what is (partly) responsible of the well definite size of the turbulent spirals in counterarotating Taylor-Couette flows [4].

ii) the reaction-diffusion equation with a single unknown, as well the amplitude equation with real coefficients have an important non-generic property of being variational. This implies in particular very specific qualitative properties. One of those properties is that the localised structures are always linearly unstable against a growth/shrinking mode. On the contrary, it has been observed in computer experiments[5] and explained analytically [6] that *non*variational systems may have linearly stable localised structures. This mathematical observation is relevant to the transition in pipe flows, where they are either continuously expanding "puffs" or localised globally stable structures, called "slugs".

175

As explained in [1] one can understand the dynamics of the localized patches in turbulence through the solutions of reaction diffusion (RD) equations with the general structure:

$$\frac{\partial A}{\partial t} = D_{ij} \frac{\partial^2 A}{\partial r_i \partial r_j} - \frac{dV}{dA} \quad , \tag{1}$$

where D_{ij} is a symmetric real and positive "diffusion tensor" and where $\frac{\partial^2 A}{\partial r_i \partial r_j}$ is the second derivative of the amplitude A with respect to the Cartesian coordinates of index i and j. In equation (1) one has applied the Einstein convention for the index summation and V(A) is the two well potential. The equation (1) has a variational structure in the sense that it may be written formally as:

$$\frac{\partial A}{\partial t} = -\frac{\delta G}{\delta A},$$

where G[.] is the real functional of A(.) defined as:

$$G[A] = \int d\mathbf{r} \; [D_{ij} \frac{\partial A}{\partial r_i} \frac{\partial A}{\partial r_j} + V(A)]$$

The evolution described by equation (1) tends to lower as much as possible the functional G[.], so that the system tends to be as uniform as possible in space, because the gradient term $D_{ij} \frac{\partial A}{\partial r_i} \frac{\partial A}{\partial r_j}$ is positive and it tends too to put A at the lowest possible value of V(.), as far as the boundary conditions allow it.

As already said, this equation (1) has solutions attracting a large set of initial datas and representing moving fronts separating two linearly stable regions. Those fronts are sometimes called the ZFK[7] fronts to help to distinguish them from the KPP[8] fronts separating stable from linearly unstable regions. Fronts moving at speed u are represented by solutions of (1) with a dependence on one space coordinate, say x, and on time through the combination z=(x-ut), and with the condition that A(z) reaches each of the two stable equilibria at z equal plus and minus infinity. There are many important differences between this picture and the real front separating two different flow regimes as for instance observed in pipes. Before to come to this let us emphasize however some striking similarities: first of all, in the RD picture as for real flows those fronts are rather independent on the details of the initial conditions because of the metastability of the system, a consequence itself of the subcritical character of the bifurcation yielding one of the two states from the other one. Then the front velocity is a smooth function of any parameter controlling deformations of the potential V(.). Now there are also some qualitative differences between real fronts as observed in pipes for instance and what follows from the RD picture. Below I shall consider them one after the other.

(i) Indeed one expects that in most parallel flows, but for plane Couette, such fronts are convected at some constant speed by the mean flow. Those advection terms appear through first order derivative in space added to equation (1). They have been already introduced[9] in the amplitude theory for inclined Taylor vortices by Tabeling. This is not enough, as even with those advection terms the trailing and leading edges are deduced from each other through a reflection, although all experimental datas point to the absence of such a symmetry[10]. This symmetry may be broken, for instance by adding third order derivatives in space to the amplitude equations, as allowed from the basic symmetries of parallel flows, since no Galilean frame can get rid of all possible effects of the advection.

(ii) The amplitude equations are actually written for complex amplitudes and, when some of their coefficients are complex they do not have a Lyapunov functional, as equation (1). Complex amplitudes represents linearly unstable dispersive waves with an amplitude dependent dispersion relation. However even in those nonvariational cases the notion of linearly stable state persists and one may reasonnably assume that the moving front solutions still attract a large class of initial conditions with the two different states at + and - infinity. As one cannot compare the energies of the two possible states to determine which one is metastable or stable this is decided now (i.e. for those nonvariational systems without any energy as the functional G) through the sign of the velocity of the front. It was predicted in reference [1] and verified[11] on a model of turbulent phase dynamics that this velocity is a critical quantity (in the sense of critical phenomena in statistical physics) when it almost vanishes and for fronts separating regular from turbulent domains.

(iii) Another problem which has not yet been looked at is the extension of this theory to the development of "turbulent" patches of turbulence in anisotropic 2d (as plane Poiseuille or Couette) or 3d flows (as a Blasius boundary layer). Indeed a theory in the form of equation (1) can model at least in part the anisotropy of such flows through a nonisotropic tensor D. However this is not enough. As said before, in this gradient picture the sign of the front velocity is completely determined by the potential difference between the two equilibria. This implies in particular that this velocity vanishes in all possible directions for the same value of the control parameter. This is presumably a nongeneric situation for nonpotential systems [see ref. 12 for more details on this].

(iv) In the next section 2, I shall consider in more details the theory of stable localised states (called s-waves) in a complexified version of equation (1).

2. Stable localised structures in nonvariational systems[+]

As shown below, variational system described by amplitude equations may present localised and stationary structures, but those structures are always unstable. On the contrary, both in transition flows [10 and references therein] as well as in various

---

+ This reports a joint work with V. Hakim and P. Jakobsen.

theoretical models, one has observed such localised structures that are thus certainly linearly stable. Theoretically this occurence of solitary (s-) patterns or waves (we refrain from using the word "soliton", as it implies usually a non generic property of integrability that we shall not be concerned with here) has been for some time more like a set of isolated facts found when studying model equations, as Benney-Lin for the Kapitza instability[13] or Ginzburg-Landau (G.L. is a generic name for amplitude equations with complex coefficients) for nonlinear waves in shear flows [5]. Below I present a rather detailed analytical discussion of those s-waves in the framework of this G.L. equations. As often, this approach is mostly perturbative, and it deals with two extreme cases: first the neighbourhood of the fully conservative case, where one can show the possibility of stable s-waves, then the vicinity of a gradient flow system (or fully dissipative). Those two limits have to be dealt with rather different mathematical techniques that may have their own interest.

Our purpose below will be to study a class of solutions of the G.L. equation for a complex amplitude q :

$$\frac{\partial q}{\partial t} = i\frac{\partial^2 q}{\partial x^2} + 2i|q|^2 q + \varepsilon \left(-\frac{\partial U}{\partial q^*} + \beta\frac{\partial^2 q}{\partial x^2}\right) \quad . \tag{2}$$

In this equation, t is the time, x the position in a frame of reference moving with the group velocity of the unstable waves, $U(|q|)$ is a real function of $|q|$ and β a real positive number. The equation (2) becomes at $\varepsilon = 0$ the nonlinear Schrödinger (NLS) equation:

$$\frac{\partial q}{\partial t} = i\frac{\partial^2 q}{\partial x^2} + 2i|q|^2 q \quad . \tag{3}$$

At $\varepsilon \longrightarrow +\infty$ (ε is always real), equation (2) becomes the gradient flow equation with scaled time $T = \varepsilon t$:

$$\frac{\partial q}{\partial T} = -\frac{\partial U}{\partial q^*} + \beta\frac{\partial^2 q}{\partial x^2} \quad . \tag{4}$$

As shown by Thual and Fauve [5] an equation similar to (2) has solutions representing linearly stable s-waves of zero speed. These s-waves appear when the "potential" $U(|q|)$ has two sets of stable minima, one at $|q| = 0$, and a continuum at $|q_0|e^{i\phi}$, ϕ real arbitrary and $|q_0|$ real $\neq 0$. Let us consider now some relevant solutions of (2) and (3).

i)Solitons of NLS equation

Although the NLS-equation (3) is known to be integrable, we shall not really use this property here, and even we could as well consider a nonintegrable equation instead of NLS, but with a unitary (and not catastrophic) evolution. This would amount to replace the last term on the r.h.s. of the equation (3) by an arbitrary function $2iF(|q|^2)q$, F real, positive, vanishing at zero argument and increasing not too fast at infinity, to avoid catastrophic self-focusing. NLS has a continuous set of 1-soliton solutions in the general form :

$$q(x,t)=Q_0(x;\Omega) \, e^{i\Omega t}$$

where Q_0, as a function of x, is a solution of

$$(\Omega - \frac{d^2}{dx^2})Q_0 = 2|Q_0|^2 Q_0 \qquad , \qquad (5)$$

together with the boundary condition (b.c.) $Q_0 \longrightarrow 0$, $x \longrightarrow \pm\infty$. This yields

$$Q_0(x;\Omega) = \Omega^{1/2} \text{sech}(\Omega^{1/2}x) \, e^{i\phi} \qquad , \qquad (6)$$

ϕ being an arbitrary phase.

ii) s- waves and moving-front solutions of (4)

In a sense, the long-term dynamics of solutions of (4) is trivial, because this equation has a Lyapunov functional that has to decay. But this does not exclude non trivial unstable and metastable solutions either.

Here we shall be interested in 2 kinds of solutions of (4). First, we introduce the moving front solution. This tends to one minimum of $U(|q|)$, say $U(q_0)$ at $x = -\infty$ and to $q = 0$ at $x = +\infty$. For large times, any initial condition satisfying these b.c. will become a front moving at a constant speed u, the sign of u is such that the optimal "state" -i.e., the one with the lowest $U(.)$ replaces at a constant rate the metastable state. This moving front solution is a solution of (4) of the form $q(x,T) = R(\xi = x - uT)$, R real such that

$$-u\frac{dR}{d\xi} = -\frac{dU}{dR} + \beta \frac{d^2R}{d\xi^2} \qquad , \qquad (7)$$

together with the b.c. $R \rightarrow q_0(/0)$ at $\xi \rightarrow -\infty$ (/ $+\infty$). The velocity u in (7) is then uniquely determined as a nonlinear eigenvalue.

An s-wave solution would be a homoclinic trajectory in x-space of solutions of (7) with $u = 0$ and $R \rightarrow 0$ as $x \rightarrow \pm\infty$. Non trivial (i.e., not identically zero) solutions of (4) satisfying those conditions exist if $U(q_0)<U(0)$ only, as can be seen from the conservation of "energy" in the Newton-like equation

179

$$\beta \frac{d^2H}{dx^2} - \frac{dU}{dH} = 0 \qquad , \qquad (8)$$

when the even function $U(H)$ has three minima, one $H = 0$, and two symmetric ones at $H = \pm q_0$. When it exists the relevant solution of (8) may be found by quadratures. However it can be shown that this s-wave solution is always linearly unstable. Consider a solution of (4) in the form $H(x) + \tilde{q}(x)e^{\sigma t}$, $\tilde{q}$ small. This perturbation is the solution of

$$\sigma \tilde{q} = - \frac{d^2U}{dq^2}\Big|_{q=H} \tilde{q} + \beta \tilde{q}_{xx} \qquad , \qquad (9)$$

with $\tilde{q} \rightarrow 0$ as $x \rightarrow \pm \infty$.

The eigenvalue problem (9) has $\tilde{q} = H_x$, $\sigma = 0$ as solution. From Sturm-Liouville this is not be ground state because H_x has at least one zero, due to the b.c. $H \rightarrow 0$ at $x \rightarrow \pm \infty$. This means that the ground state of (9) (the solution with the largest σ with our signs) has a positive σ, which shows that $H(x)$ is a linearly *unstable* solution of (4).

Now I shall interpolate in between the properties of the limit forms of equation (2), by considering two "generic" parameters: ε and another one, not made explicit and representing the continuous changes of $U(.)$. This last parameter could be, for instance the level difference ΔU of the the two stable equilibria of $U(.)$: $\Delta U = U(q_0) - U(0)$.

By perturbation near $\varepsilon = 0$, I show first that the continuum of NLS solitons collapse into finitely many s-waves . Then, in the opposite limit $(1/\varepsilon \rightarrow 0)$ I show that an unstable and a stable s-wave merge at $\Delta U = U(q_0) - U(0) = 0 = 1/\varepsilon$ at the same point in parameter space as the steady front.

(2.1) s-waves near $\varepsilon = 0$

Below, I show by perturbation that, as ε is turned one from 0, the continuum of NLS solitons collapse on a finite number of s-waves. This can be shown in a number of ways, in particular [14] by relying upon the integrability of NLS. A more elementary approach is sufficient here. The NLS part on the right-hand side of (2) conserves the L^2-norm $\int |q|^2 dx$ in the course of time. Multiplying (2) by $q*$ and adding the complex conjugates, one gets an evolution equation for this norm depending formally on the nonunitary piece :

$$\frac{d}{dt}\int |q|^2\, dx = -\varepsilon \int_{-\infty}^{+\infty} dx\ [\beta\ |\frac{dq}{dx}|^2 + q*\frac{dU}{dq*} + q\frac{dU}{dq}] \qquad . \qquad (10)$$

We are looking now for s-wave solutions of (2). By analogy with the NLS case, those solutions have to have the form

$$q(x,t) = e^{i\Omega't} Q'(x,\Omega')$$

with

$$Q' \to 0, \; x \to \pm\infty$$

If such a solution (2) exists, it has to cancel the right-hand side of (10). This is indeed insufficient to find it in general (that is for ε finite), but if ε is close to zero, one expects $Q'(x, \Omega')$ to be very close to the NLS-soliton solution $Q_0(x, \Omega)$ as given by equation (6). Thus, in this limit, the dominant contribution to the integral on the right-hand side of (10) is obtained by plugging in a NLS-soliton. The condition that the result is equal to zero yields an equation for Ω. To be more definite, let us take a potential in the form (all coefficients are real):

$$U(|q|) = \mu \, |q|^2 + \frac{\alpha}{2} |q|^4 + \frac{\Upsilon}{3} |q|^6 \qquad (11)$$

They are two stable equilibria at $q = 0$ and $q = q_0$ whenever, $\mu < 0, \alpha^2 > 4\Upsilon\mu$, $\Upsilon > 0$ and $\alpha < 0$. Moreover the condition that the r.h.s. of (10) vanishes when computed with a NLS soliton reads :

$$2 \frac{\Omega}{5}^{1/2} \left(20\mu + \Omega(3\beta + 4\alpha) + \frac{16\Upsilon\Omega^2}{21} \right) = 0 \qquad (12)$$

which has two real roots, as sought, if

$$(3\beta + 4\,\alpha)^2 > \frac{1280}{21} \Upsilon\mu$$

A more refined calculation based upon a systematic perturbation expansion in ε, and done following the method of ref. [14] shows that a finite number of solitons only survive the ε-perturbation, and their amplitude is given by equation (12). Dropping the term of degree 6 in U(.), one gets only one (instead of 2) s-waves, and this one is always unstable. On the contrary, when a pair of s-waves exist with the 6th degree term, one is linearly unstable and the other is linearly stable. This branch of linearly stable s-wave is the one found by Thual and Fauve [5].

(2.2) s-waves and fronts at ε large

We know already that, if ΔU is negative, there is an unstable s-wave at 1/ε=0, as given by the relevant solution of equation (8). When ΔU→0., the width of this unstable s-wave diverges and we shall show that , in the parameter space, 1/ε=ΔU=O is at the merging of two branches: one branch of steady front solutions and another one where stable and unstable s-waves become identical and metastable.

Under the heading (2.2.i) below, I shall study first the steady front solutions near $1/\varepsilon=0$ and then in (2.2.ii) the s-wave solutions to validate the above picture.

2.2.i : Front wave solutions near $1/\varepsilon=0$

For $\varepsilon \to \infty$, the equation describing steady fronts in the same as eq. (8) but with a possible nonuniform phase of Q :

$$- \frac{dU}{dQ^*} + \beta \frac{d^2Q}{dx^2} = 0 \qquad , \qquad (13)$$

together with the b. c. $Q \to 0$ at $x \to -\infty$ and $Q \to q_0$ at $x \to +\infty$. Putting $q = r e^{i\phi}$ into (13), r and ϕ real functions of x:

$$- \frac{dU}{dr} + \beta \left(\frac{d^2r}{dx^2} - r \frac{d\phi^2}{dx} \right) = 0 \qquad , \qquad (14.a)$$

$$2 \frac{d\phi}{dx} \frac{dr}{dx} + r \frac{d^2\phi}{dx^2} = 0 \qquad . \qquad (14.b)$$

From (14.b), $r^2 \phi_x$ has to be independent of x and is zero from the b.c. $r \to 0$ at $x \to -\infty$. This is possible if $\phi_x = 0$ for a non trivial (=not r = 0 everywhere) solution. Thus, with those b.c., equations (14) becomes equivalent to (8) with r = H. Indeed this has a steady front solution iff $U(q_0) = U(0)$ only, that is if the potential $U(.)$ has two minima of equal depth at q = 0 and q = q_0. The extension of (14) to a non zero ε reads :

$$- \frac{dU}{dr} + \beta \left(\frac{d^2r}{dx^2} - r \frac{d\phi^2}{dx} \right) = \frac{1}{\varepsilon} \left(2 \frac{dr}{dx} \frac{d\phi}{dx} + r \frac{d\phi^2}{dx} \right) \qquad (15.a)$$

$$\frac{\beta}{\varepsilon} r^2 \frac{d\phi}{dx} = \int_{-\infty}^{x} dx' \, r \, (\Omega r - 2r^3 - r_{xx} + r\phi_x^2) \qquad , \qquad (15.b)$$

with the b.c. $r \to 0$ at $x \to -\infty$ (subscripts are for derivatives: $\phi_x = \frac{d\phi}{dx}$).

We shall analyse now the steady front solution of equation (15) by expansion near $1/\varepsilon = 0$. Let r(x) be expanded in Taylor series in $1/\varepsilon$ as :

$$r(x) = r^{(0)} \, (x) + \frac{1}{\varepsilon} \, r^{(1)} \, (x) + ...,$$

Similar expressions hold for ϕ_x as well as for other quantities. We have shown before that $\frac{d\phi^{(0)}}{dx} = 0$ and that $r^{(0)}$ is the solution of

$$- U_r^{(0)} + \beta \, r_{xx}^{(0)} = 0 \qquad (16)$$

182

with the b.c. $r^{(0)} \rightarrow 0$ ($/q_0$) at $x \rightarrow +$ ($/-$)∞. Thus we are looking for a solution of (15) such that $r(x) \rightarrow 0$($/Q_0$) at $x \rightarrow +$($/-$)∞, where Q_0 is the asymptotic value of r, not necessarily q_0 if $\varepsilon \neq \infty$ and that $\phi_x \rightarrow$ constant at $x \rightarrow +\infty$. With those b.c. and to make the integral on the r.h.s. of (15.b) convergent at large x positive, Ω has to be equal to

$$\Omega = 2Q_0^2 + (\frac{d\phi}{dx}\big|_{as})^2 \qquad , \qquad (17)$$

where $\frac{d\phi}{dx}\big|_{as}$ is the limit value of $\frac{d\phi}{dx}$ as x tends to infinity.

Keeping now the dominant terms on the r.h.s. of (15.b) one gets:

$$\frac{d\phi^{(1)}}{dx}\big|_{as} = \frac{1}{\beta \, q_0^2} \int\limits_{-\infty}^{+\infty} dx' \{2(r^{(0)})^2 \, [q_0^2 - (r^{(0)})^2] + [\frac{dr^{(0)}}{dx}]^2\} \quad . \qquad (18)$$

From (15.a), the "equilibrium" at $x = +\infty$ has to be shifted from $q = q_0$ to $Q_0 = q_0 + q^{(1)} + q^{(2)} + ...$where $q^{(1)} = 0$, although $q_{(2)}$ is given by

$$- U_q \big|_{Q_0} q^{(2)} = \beta \, q_0 \, [\frac{d\phi^{(1)}}{dx}\big|_{as}]^2 \qquad . \qquad (19)$$

But this "equilibrium" condition is not the only one to be imposed to $r(x)$ at $x \rightarrow +\infty$. One can deduce from (15.a) a relation extending to non zero $\frac{1}{\varepsilon}$ the previous condition $U(0) = U(q_0)$. By multiplying (15.a) by r_x, one gets :

$$U(0) - U(Q_0) = \int\limits_{-\infty}^{+\infty} [\frac{1}{\varepsilon} \frac{dr}{dx} (2\frac{d\phi}{dx}\frac{dr}{dx} + r \frac{d^2\phi}{dx^2}) + \beta r \frac{dr}{dx} (\frac{d\phi}{dx})^2] \quad . \qquad (20)$$

This condition is independent on (19), as can be checked on particular examples by expansion in $1/\varepsilon$ and by noticing that the dominant order for the r.h.s. of (20) can be expressed in terms of the zeroth order solution. In other terms, the potential U(.) has to satisfy a non trivial condition to show front solutions, and this condition becomes $U(0) = U(q_0)$ in the gradient limit ($\varepsilon = \infty$). We shall assume that this still holds at finite values of ε, so that a codimension 1 manifold in the ($\varepsilon, \Delta U$) space defines the locus where steady fronts exist, this manifold ending at $U(q_0) = U(0)$ at $\varepsilon = \infty$. In this framework, the branch of stable s-wave (of codimension zero !) meets this manifold at a finite ε, say ε_c as ε changes at constant U.

2.2.ii: s-waves near $1/\varepsilon = 0$

Near $\Delta U = 1/\varepsilon = 0$, s-waves are made of widely separated shelfs bounding a plateau where Q is very close to q_0 . The analysis of this situation proceeds in two steps: first one computes ϕ_x inside the s-wave by solving (15.b) in the appropriate limit and then plugs the result into (15.a) to take into account the perturbation brought to the modulus r(x) by this nonconstant phase.

The phase equation (15.b) defines first the boundary value of ϕ_x at the two ends of the plateau. At the dominant order in $1/\epsilon$ the variation of ϕ_x across the shelf is:

$$[\phi_x] = \frac{\epsilon}{\beta \ q_0{}^2} \int_{-\infty}^{+\infty} dx \ [\Omega (r^{(0)})^2 - r^{(0)} r^{(0)}{}_{xx} - (r^{(0)})^3] \qquad , \qquad (21)$$

where $r^{(0)}$ (x) is the modulus of the front solution of (4) at $\Delta U = 0$. Note that the contribution proportional to Ω on the r.h.s. of (21) will turn to be always subdominant with the estimates found below for ϵ large. As $\phi_x = 0$ at $x = +$ or $- \infty$, (21) defines the b.c. for ϕ_x at the two ends of the plateau, as announced. On this plateau, $r \approx q_0$ and, from (15.b):

$$\epsilon \beta \phi_{xx} \approx \Omega + \phi_x{}^2 \qquad . \qquad (22)$$

From the previous remarks the b.c. for (22) are $\pm [\phi_x]$ at the two ends of the plateau, that is at $x = \pm L/2$, L being the width of the s-wave, unknown for the moment. By solving (22) with the appropriate b.c. one relates Ω and L. From (21) ϕ_x is of order ϵ^{-1} and we shall assume that Ω is the dominant term on the r.h.s. of (22)., that is $\Omega >> \epsilon^{-2}$, which implies $L << \epsilon^2$, as we shall assume. This is consistent with the final estimate $L \sim \ln(\epsilon)$ at ϵ large. With those assumptions the solution of (22) to be retained is $\phi_x = \frac{\Omega}{\epsilon \beta}(x + x_0)$ (23).

The b.c. impose $x_0 = 0$ and $\Omega L = 2 \epsilon \beta \ [\phi_x]$. Now we shall put into the equation for r(x) the above expression for ϕ_x. At large ϵ the dominant perturbation with respect to the pure gradient flow comes from the $-\beta r \phi_x{}^2$ term on the l.h.s. of (15.a), and so we shall try to find a s-wave solution of:

$$-U_{rr} + \beta r_{xx} = \beta r \phi_x{}^2 \qquad , \qquad (24)$$

where ϕ_x is given by (23) in the plateau region. There r(x) is close to q_0 and by putting $r \approx q_0 + \delta(x)$, we may linearize (24) near q_0 and assume that $\delta(.)$ is of the same order of magnitude as the r.h.s.. This yields:

$$-\alpha \delta + \Delta + \beta \delta_{xx} = \beta q_0 \phi_x{}^2 \qquad . \qquad (25)$$

In this equation α is for $2U_{rr}$ at $r = q_0$ although the constant Δ is for the possibility that U(.) is a priori not exactly minimum at q_0, as we are exploring the vicinity of $\Delta U = 0$ in the parameter space. The integration of the linear equation (25) is straightforward and the free parameter left (Ω or L) is found by matching the solution with the asymptotics of the front solution limiting the plateau.

A possible form for this solution, consistent with the parity $\delta(x)=\delta(-x)$, x=0 being the center of the s-wave, is:

$$\delta(x)=\frac{\Delta}{\alpha}+a(e^{kx}+e^{-kx})+\beta q_0 e^{kx}\int_0^x dx'\ e^{-2kx'}\int_0^{x'} dx''e^{kx''}\ \phi_{x''} \qquad , \qquad (26)$$

where $k=(\beta/\alpha)^{1/2}$. The parameter α is determined by the condition that this solution has to fit the exponential approach toward q_0 of the front solution near $x=\pm L/2$. This imposes $a=a'e^{kL/2}$, where a' is negative and of order 1 when measured with the quantities entering into U and with β. This coefficient a' can be computed by considering the extension of the equations for U in the complex x-plane. The same extension to the complex x-plane shows that any other contribution to $\delta(x)$ has to cancel in this matching region. They are three such contributions : the constant Δ/α, a term of order $a'e^{-kL}$ coming from the second term on the r.h.s. of (26) and another one, with a positive sign coming from the asymptotics of the double integral and of order $\frac{\beta a''}{\epsilon^2}$, a" being a positive constant of order 1 (that is independent of L and ϵ). Those three contributions cancel if:

$$\Delta+\frac{\beta a''}{\epsilon^2}e^{-kL/2}-a'e^{kL}=0 \qquad . \qquad (27)$$

A more precise theory would also make appear a change of Δ of order $1/\epsilon$, that we shall not detail here. As a' and a" in (27) are both positive, this equation has two roots for L when Δ is negative, and small.

Coming back to the meaning of this last result, two s-waves exist and one is stable (the wider one) and one unstable. They merge as announced for Δ of order $1/\epsilon^2$. Furthermore when Δ tends to zero one of the s-wave becomes infinitely wide and thus disappear for the value of the parameters where a steady front exists.

3. An example of nonvariational effect in Magnetism **

This third section is about a problem belonging to a domain of physics quite different from the class considered in the first and second section (i.e. hydrodynamics in a broad sense), since I shall examine the dynamics of a Bloch wall under the effect of an external time dependent magnetic field. As we shall see it, the interest of this model is that the time dependence of the external field brings a typical nonvariational effect, i.e. that the

---

** This work has been done in collaboration with P. Coullet and J. Lega. An extended version is submitted for publication elsewhere.

direction of the motion of a domain wall depends on its core structure, although it should not in a variational system.

Consider a ferromagnet with a direction of easy magnetization lying in a plane of easy magnetization too. This means that in a ferromagnetic state, the spontaneous magnetization points to anyone of the two opposite directions along the easy magnetization axis, and that pertubations of this will send the magnetization to another orientation, but still in the easy plane. If the magnetization is not uniform and points to either of the two opposite preferred directions, so called "domain walls" draw the border between regions of opposite orientation. Inside those walls, the magnetization interpolate continuously between those two orientations(but for neglected lattice effects). By imposing the local magnetization to stay in the easy (magnetization) plane, one still finds two possible wall structures: in the Néel wall, the magnetization stays in the preferred direction, but decays to zero before to grow in the opposite direction. On the contrary, in the Bloch wall, the magnetization keeps a more or less a fixed modulus, but rotates continuously from up to down in the easy magnetization plane. Néel wall is preferred at large in-plane anisotropies, whereas the Bloch wall is preferrred at low anisotropies.

The way in which nonequilibrium comes into the story is not completely obvious. When a wall (Bloch or Néel) drifts in a ferromagnet, this is usually interpreted by saying that the magnet lower its free energy by replacing (at constant rate if the wall move at constant speed) domains with the bad orientation by the good one. This is what happens in an external magnetic field: good/optimal domains parallel to the external field expand at the expense of bad domains antiparallel to the magnetic field. This optimization of the free energy is of course rather independent of the helicity of the Bloch wall, if such a wall separates the two domains. Below I show a possible situation where the direction of motion of a Bloch wall depend on its helicity, something that could not be found in a variational system(=close to equilibrium), where this direction is dictated by an optimization principle.

To write the equation for this problem, I start from the same equations as Ginzburg[15]. Let us represent the local magnetization in the easy plane as a complex number χ, that may be thus a function of time and/or space. The modulus is the strength of the magnetization and the angle is for the orientation. Below the Curie temperature, this magnetization takes at equilibrium two possible opposite values, ±1 by a convenient choice of unit. Forgetting for the moment the anisotropy in the easy plane, a natural equation for the dynamics of magnetization is of the relaxation (or variational)type and reads:

$$\frac{d\chi}{dt} = \chi - \chi^*\chi^2 \, ,$$

where χ^* is the complex conjugate of χ. This predicts that any initial condition will keep a constant angle (because it is invariant under a change of phase) and relax to the modulus 1, after a time of order 1, again by a proper choice of time scale. The anisotropy in the easy plane is introduced by adding a non-phase invariant term. The most simple such term is $\gamma\chi^*$, γ coefficient measuring the in-plane anisotropy:

$$\frac{d\chi}{dt} = \chi - \chi^*\chi^2 + \gamma\chi^*.$$

Depending on the initial value of the angle, solutions to this equations tend at large times to either of the symmetric stable equilibria $\chi = \pm i (1+\gamma)^{1/2}$. There are also two unstable equilibria $\chi = \pm(1-\gamma)^{1/2}$ and the unstable "paramagnetic" state $\chi = 0$. Notice that this requires $\gamma^2 < 1$, as we shall assume.

In this theory the anisotropy is measured by γ. A small γ means a small anisotropy, and a large γ a large one. To describe walls in this framework, one has to consider a nonuniform χ. The physics says that the magnetization tends to be as uniform as possible. This is represented by adding a diffusion term to the dynamical equation: diffusion tends to make even the distribution of magnetization in space. Whence the new equation (a partial differential equation now):

$$\frac{\partial\chi}{\partial t} = \chi - \chi^*\chi^2 + \gamma\chi^* + \frac{\partial^2\chi}{\partial x^2} + h. \tag{28}$$

This is written in 1D (coordinate x) and the length unit has been chosen to set to 1 the diffusion coefficient. Furthermore the effect of a possible external magnetic field was introduced through the last term on the right hand side, h, proportional to the intensity of this field. This equation has a variational structure(as well as its predecessors) in the sense that it can be written in the form:

$$\frac{\partial\chi}{\partial t} = -\frac{\delta H}{\delta\chi^*},$$

where $\dfrac{\delta H}{\delta\chi^*}$ is the Fréchet derivative of the functional

$$H[\chi] = \int dx \left[\frac{1}{2}|\chi|^4 - |\chi|^2 - \frac{\gamma}{2}(\chi^2+\chi^{*2}) + \left|\frac{\partial\chi}{\partial x}\right|^2 - h(\chi+\chi^*) \right].$$

The (free) energy H tends to values as low as possible in the relaxation dynamics of the system toward its equilibrium state. The stable uniform states are indeed minima of the integrand of H, for uniform solutions.

Domain walls are stationary solutions of (28) with h=0, joining two domains where the magnetization takes different stable values, that are, for instance $\chi = (1+\gamma)^{1/2} e^{i\phi}$, with, say $\phi=0$ at $x=-\infty$ and $\phi=\pi$ at $x=+\infty$. Depending on the value of γ, this problem may have [15] either one solution, the stable Néel wall at large γ or three solutions at small γ: the now unstable Néel wall and the stable left- and rigth handed Bloch walls. The transition is continuous and occurs at $\gamma=\frac{1}{3}$. The Bloch wall solution is [15]:

$$\chi_0 = X_0 + iY_0 \qquad , \qquad (29)$$

with $X_0 = (1+\gamma)^{1/2}$ th(ax), $Y_0 = \pm(1-3\gamma)^{1/2}$sech (ax) and $a^2 = 2\gamma$.

The left-right hand (or mirror) symmetry is the complex conjugation in this formalism, and it is also equivalent to a change of sign of the quantity a. Thus the Bloch wall have the same kind of symmetry as a corkscrew and are images of each other in a mirror. This leads quite naturally to the following idea. Consider an uniform magnetic field turning at constant speed while staying in the easy magnetization plane. The effect of this field on a Bloch wall will be a bit like a constant rotation on a corkscrew, that is a displacement at constant speed in a direction depending on the handedness of the screw. From this we expect a constant drift for Bloch wall solutions of equation (28) with a time dependent h representing a rotating field. With our representation this is realised by putting for h in (28) $h_0 e^{i\nu t}$.

The equation so obtained, that reads:

$$\frac{\partial \chi}{\partial t} = \chi - \chi^* \chi^2 + \gamma \chi^* + \frac{\partial^2 \chi}{\partial x^2} + h_0 e^{i\nu t}, \qquad (30)$$

is time dependent and does not seem to have any simple solution, even uniform in space. So I shall limit myself to a perturbative approach, in the small h_0 limit. To show my point, I shall take as unperturbed solution anyone of the two Bloch wall solutions (that are written as χ_0) given in (29), and then expand a solution of (30) in powers of h_0.

At the first oder in h_0, one finds small oscillations with the frequency ν around the Bloch wall solution (29). Those oscillations are described by the solution of (30) that reads at zeroth and first order as:

$$\chi = \chi_0 + h_0[\alpha_s \sin(\nu t) + \alpha_c \cos(\nu t) + i \beta_s \sin(\nu t) + i\beta_c \cos(\nu t)],$$

where the α's and β's are real functions of x defined by the solution of a set of coupled inhomogeneous linear equations, deduced by a straightforward algebra from (30):

$$\Lambda_+ \alpha_s - 2X_0 Y_0 \beta_s = -\nu \alpha_c \qquad , \qquad (31.a)$$
$$\Lambda_+ \alpha_c - 2X_0 Y_0 \beta_c = \nu \alpha_s - 1 \qquad , \qquad (31.b)$$
$$\Lambda_- \beta_s - 2X_0 Y_0 \alpha_s = -\nu \beta_c - 1 \qquad , \qquad (31.c)$$
$$\Lambda_- \beta_c - 2X_0 Y_0 \alpha_c = \nu \beta_s \qquad , \qquad (31.d)$$

where the linear operators Λ_+ and Λ_- are defined by their action on a test function Γ as:

$$\Lambda_+ \Gamma(x) = (1+\gamma-3X_0^2-Y_0^2)\Gamma + \frac{d^2\Gamma}{dx^2}$$

and

$$\Lambda_- \Gamma(x) = (1-\gamma-X_0^2-3Y_0^2)\Gamma + \frac{d^2\Gamma}{dx^2} \quad .$$

Counting the number of free parameters in (31), they have the right number of constraints when one imposes to their solution to decay far away from the Bloch wall, that is at $x \rightarrow \pm\infty$. This does not prove however that (31) has an unique solution. If two (or more) solutions of (31) existed, the difference between them would represent small oscillations of the Bloch wall, in the absence of external forcing. This is clearly impossible, because of the relaxational character of the original equation. Furthermore, it is possible to prove that, at least for small frequencies this equation has certainly a solution, although this not an easy result.

By continuing the expansion of the solution of (30) to higher orders in h_0, one meets, as often in this sort of expansion, a solvability condition. This is because, in the formal expansion of the solution of (30), one solves at the n-th order a linear problem in the form:

$\Omega\chi_n = $ (right hand side),

where the r.h.s. is a known finite algebraic combination of the lowest order contributions $\chi_{n-1}, \chi_{n-2}, \chi_{n-3},...$where the index n is for the order in small parameter h_0, and Ω is a linear operator. For n even, χ_n as a function of time is decomposed into Fourier components $e^{in\nu t}, e^{i(n-2)\nu t}, e^{i(n-4)\nu t},...$and a time independent part, say $\chi_{n,0}$. This time independent part is the solution of a linear equation of the form:

$$\Omega\chi_{n,0} = \text{(right hand side)}_0 \quad , \quad (32)$$

an ordinary differential equation in the variable x, and for the unknown function $\chi_{n,0}$. This equation cannot be solved as written, because Ω has a non trivial kernel for zero frequency functions (the explicit writing of this operator is given below). In technical terms, this means that it exists a non-zero function of x, $\chi_{n,k}$ such that $\Omega\chi_{n,k}=0$. Assuming now(and this can be made more precise, but at the price of a lot more work) that everything is like in ordinary algebra (i.e. that Ω is a numerical square matrix, and χ_n a vector, instead of differential operator and function repectively), the equation (32) has no solution in general, because the "matrix" Ω has zero as an eigenvalue. Before to consider in more details what to do next, I will (hopefully) make all the above considerations more precise by giving explicitly the form of the operator Ω when restricted to zero frequency perturbations. This is indeed nothing but the right hand side of (28), linearized near the Bloch wall solution χ_0:

$$\Omega\chi_{n,0} = \chi_{n,0} - 2|\chi_0|^2\chi_{n,0} - \chi_0^2\chi_{n,0}^* + \gamma\chi_{n,0}^* + \frac{d^2\chi_{n,0}}{dx^2},$$

and for the complex conjugate:

$$\Omega\chi^*_{n,0} = \chi^*_{n,0} - 2\,|\chi_0|^2\chi^*_{n,0} - \chi^{*2}_0\chi_{n,0} + \gamma\chi_{n,0} + \frac{d^2\chi^*}{dx^2}_{n,0}.$$

As announced this operator has a zero mode: putting in those two equations $\chi_{n,0} = \frac{d\chi_0}{dx}$, and the complex conjugate, one gets zero, because the corresponding expressions are nothing but the derivatives with respect to x of the r.h.s. of (28) and of its complex conjugate. This is a rather familiar phenomenon in applied mathematics and it occurs whenever one linearizes an autonomous nonlinear equation as (28).

The attentive reader will have noticed that the same solvability problem appears already at the *first* (instead of *second* at finite frequency) order in h_0, at zero frequency v. This is because then this first order solution depends already on the inversion of the operator Ω at zero frequency. Then the physics tells us what happens: a constant external magnetic field is imposed to the system and the wall (whatever its internal structure) drifts to replace continuously the domains antiparallel to this external magnetization by domains parallel to it. This means that the Bloch wall solution $\chi_0(x)$ becomes actually a moving wall solution(for h_0 small at least) of the form $\chi_0(x-ut)$. The velocity of the wall, u, is a free parameter till now, and it is small, as one expects the wall velocity to vanish at zero h. Indeed one expects the same in the other problem: time dependent h(t), and second order perturbation. Without going into all the details, that are quite tedious (but straightforward), one assumes in the time dependent problem that u is second order in h_0 and constant. This constant takes care of the solvability problel, that is of the non trivial kernel of Ω, by cancelling any component of the r.h.s. of (32) on the kernel.

This yields finally the following result, for the dominant contribution to u (again in the time dependent problem):

$$u = h_0^2 \frac{N}{D}, \text{ with:}$$

$$N = \int_{-\infty}^{+\infty} dx\{((\alpha_c\,\beta_c + \alpha_s\,\beta_s)\frac{d(X_0Y_0)}{dx} + \frac{1}{4}[3(\alpha_c^2 + \alpha_s^2) + \beta_c^2 + \beta_s^2]\frac{dX_0^2}{dx}\})$$

$$+ \int_{-\infty}^{+\infty}\frac{dx}{4}[3(\beta_c^2 + \beta_s^2) + \alpha_c^2 + \alpha_s^2]\,,$$

and

$$D = \int_{-\infty}^{+\infty} dx[(\frac{dX_0}{dx})^2 + (\frac{dY_0}{dx})^2].$$

190

In the small h_0 limit, the velocity u is proportional to the square of h_0 times a numerical function of v and γ. This function is such that u is odd with respect to v, in agreement with the qualitative considerations presented before: the motion of a corkscrew changes direction when the sense of rotation is changed. Moreover, by looking at the same expression for the Néel wall, without helicity (which can be done in the above expression for u by setting $Y_0 = 0$ and putting for X_0 its expression for a Néel wall) one finds by symmetry considerations that u is then equal to zero, as expected.

It is more delicate to show that u is of order v as it tends to zero. This is because in this low frequency limit the wall (Bloch or Néel) makes large excursions at first order in h_0: the external magnetic field is then quasistatic, and a constant magnetic field would give a constant velocity of order h_0: thus the amplitude of the displacement in a slow time-periodic external field is about of the order of this velocity times a period, that is of order $\frac{h_0}{v}$. Nevertheless one can show that, on top of those "large" first order(in h_0) fluctuations of a Bloch wall without mean drift, there is a mean drift at second order, proportional to v, that changes sign with the helicity of the Bloch wall.

All those predictions have been tested directly and successfully by solving numerically the model equation (30), with an equilibrium Bloch wall as initial condition. Concerning the possibility of an experimental observation of this displacement of Bloch walls in an externally rotating magnetic field, and without discussing this in any detail, I can only stress that such a displacement speed, of second order in an oscillating external field has been observed already by Schlömann and coll.[16], but with a geometry completely different of the one I have considered (in particular, the Bloch walls move in a direction independent on their handedness). However the connection with the problem I have considered here is that the velocity measured by Schlömann et al. [16] is of order v h_0^2, as here. Thus it seeems to be reasonable to expect that the one found here will be of the same order of magnitude for a real magnet as the one measured in [16] with the same h_0 and v, which leaves some hope to have it observable.

4. Phase dynamics in a constrained system[++]

In this section, I present an example of another fruitful approach to the study of nonequilibrium systems, that is the phase dynamics. The basic idea is very simple: consider a regular (or eventually not so regular, but with some kind of statistical homogeneity in space) steady structure. The archetypal example of this is the rolls of Rayleigh-Bénard convection. Now one looks at the slow large scale changes of this structure. This large scale dynamics is slow, because in the absence of bounadary conditions the system is translationally invariant, and then the equations have a neutral mode of zero frequency, representing the translations. Near this neutral mode, that is for nonuniform translations defined by a slowly space

---

[++] This is a part of a work done in collaboration with J. Prost and E. Guyon.

dependent vector, one finds [17] a phase diffusion mode. this mode is generic for dissipative systems, because of the irreversibility of the equations of motion. It may happen however that this mode becomes propagative near some degenerate situations, or indeed if the equations of motion are conservative.

This simple phase diffusion equation has a rather trivial behavior, and predicts relaxation to an uniform state. However, as shown below things may become much richer by adding the possibility of nonlinear effect in the phase gradients as well as the Pocheau-Croquette effect. This effect has been observed [18] in Rayleigh-Bénard convection and is rather subtle, in the sense that it puts out of equilibrium in some sense a roll structure that is already the result of a nonequilibrium process. This effect shows that a roll structure under a certain class of perturbation cannot stay in an uniform state: the roll wavelength becomes nonuniform. But, as shown below, this nonuniformity may create itself secondary bifurcations in the transverse direction (i.e. parallel to the roll axis) if this change in the roll wavelength exceeds a threshold. The remarkable result from the phase dynamics approach is that all tis can be described in an extended phase dynamics equations, using as only ingredients general assumptions on the symmetry of the basic equations. The resulting problem looks a bit like a buckling problem in classical elasticity, and everything can be paramerized by a single parameter, called C thereafter. The limit of a large C (=large deformations of the pattern) poses interesting mathematical questions.

(4.1) Phase equation in two space dimensions

Let u be the local displacement of a roll structure (this is its "phase"). When nonuniform, this is a priori a function of space (variable r) and time t. The dynamics of this phase in the most simple case (that is for small and one dimensional perturbations perpendicular to the roll axis) is given by the diffusion equation:

$$\frac{\partial u}{\partial t} = \frac{\partial^2 u}{\partial z^2} \, ,$$

where z is the coordinate perpendicular to the roll axis and the units have been chosen to set to 1 the parallel diffusion coefficient . As already said, I shall extend this equation in order to take into account effects nonlinear in the gradients of the displacement as well as its possible nonuniformity parallel to the roll axis, that is with respect to another Cartesian coordinate x. This can be done consistently by supposing that the underlying system is rotationally invariant and leads to a general equation of the form:

$$\frac{\partial u}{\partial t} = -\frac{\delta F}{\delta u}$$

where the right hand side represents the "elasticity" or phase diffusion effects through a Fréchet derivative of F. The form of the functional F[.] is obvious if one limits oneself to the linear diffusion term. We shall add more terms to it in order to represent the physical phenomena discussed before. This is accomplished with the following first choice for the "energy" functional F [17]:

$F[u] = \frac{1}{2}\int d\mathbf{r}\, E^2(u)$, where $E(u) = \frac{\partial u}{\partial z} + \frac{1}{2}(\nabla u)^2$. This form of E(u) is dictated by the requirement of global rotational invariance. Since the displacement u has the dimension of a length, E(u), as written is dimensionless. This is natural, since the phase approximation should be independent on the small length scale: this is very similar to the Hookian elasticity where, although the equations are for a regular lattice, they never make appear the mesh size. As we shall see below, the situation here is more complicated than the one of ordinary elasticity, because terms represented in E(u) can be at the end of the same order as terms formally of higher order in the gradient expansion and depending explicitly on the small length scale λ, that may be seen as the wavelength of the roll. At the very end, on can check that the gradient of the displacement is small enough to make the phase approximation consistent.

The term to be added to the integrand of F[.] and with this microscopic length is $\lambda^2(\Delta u)^2$, where Δ is for the Laplacian operator: $\Delta u = \frac{\partial^2 u}{\partial z^2} + \frac{\partial^2 u}{\partial x^2}$. Finally one has to add to the phase dynamics equation a term representing the external constraint. This constraint may be physically a constant flow perpendicular to the z-direction, and one can show [18] that its effect (at least at the dominant order) is to add a constant term to the phase equation, which takes the form:

$$\frac{\partial u}{\partial t} - V = -\frac{\delta F}{\delta u} \qquad , \qquad (33)$$

where $F[u] = \frac{1}{2}\int d\mathbf{r}\, [E^2(u) + \lambda^2(\Delta u)^2]$.

This is the equation we shall study now, that reads explicitly as:.

$$\frac{\partial u}{\partial t} - V = \frac{\partial^2 u}{\partial z^2} + \frac{1}{2}\frac{\partial}{\partial z}(\nabla u)^2 + \nabla(\frac{\partial u}{\partial z}\nabla u) + (1/2)\nabla(\nabla u(\nabla u)^2) - \lambda^2\nabla^4 u \quad . \qquad (34)$$

and may be written formally as:

$$\frac{\partial u}{\partial t} = -\frac{\delta G}{\delta u} \qquad \text{where } G = F + \int d\mathbf{r}\, V\, u \qquad (35)$$

However this does not necessarily imply relaxation to a steady state since the term linear in u in G may increase or decrease indefinitely with Neuman boundary conditions. This would not put a priori any restriction upon the value of u since F depends on the gradients of u only. In the case of Dirichlet b.c., G is uniformly bounded from below and one can look for stationary solutions.

(4.2) Linear stability against undulations of the constrained structure

With Dirichlet b.c. $u = 0$ at $z = O$ and $z = L$, the equation (34) has an elementary solution that reads:

$$u_0(z) = \frac{V}{2} z(L-z) \qquad , \qquad (36)$$

and we shall consider now its linear stability. Let w be a small perturbation to u_0. We shall deal with situations where the phase approximation makes sense. This is the "large box" limit $\lambda/L \ll 1$, meaning that they are many wavelength in the actual realization of the regular structure. Anticipating the typical length scale for the most unstable perturbations to be L along the z axis-as the basic solution itself- and $(\lambda L)^{1/2}$ along x, we obtain the relevant equation for w where we have omittted the terms of order λ/L smaller than the retained ones. Since we can always choose systems large enough, this limit is always meaningful.

$$\frac{\partial w}{\partial t} = \frac{\partial^2 w}{\partial z^2} + \frac{\partial u}{\partial z} \frac{\partial^2 w}{\partial x^2} - \lambda^2 \frac{\partial^4 w}{\partial x^4} \qquad . \qquad (37)$$

Although the approximation leading to this last equation is formally consistent, it neglects fourth order derivative in z. In such a situation, it seems likely that a boundary layer with a thickness of order λ will take care of the additional boundary condition due to this fourth derivative. We shall simply assume that this does not change, at the dominant order, the solutions, obtained with a second derivative in z only, far from the boundary.

Owing to the translational invariance in the x-direction and to the autonomous character of equation (36) with respect to time, one chooses to study perturbations of the form $w_0(z) \exp(iqx + \sigma t)$ so that σ becomes an eigenvalue of a linear differential operator with q as a parameter. From (37) this eigenvalue problem reads:

$$(\sigma - M_q)w_0 = 0 \quad , \qquad (38)$$

with the b.c. $w_0 = 0$ at $z = 0$ and L, and the following definition for the linear operator M_q:

$$M_q = \frac{\partial^2}{\partial z^2} + q^2 V(z - L/2) - \lambda^2 q^4. \qquad (39)$$

As equation (39) has a coefficient linear in z, it can be solved by Laplace's method. This yields Airy functions that become simple circular functions if the external constraint V is zero. Then the quantized σ values are :

$\sigma_n = -\{\pi^2[L(2n+1)]^{-2} + \lambda^2 q^4\}$, where n is a natural integer. Coming back now to the original problem, that is to perturbations around the basic solution given in (36), we shall use as a new variable $z' = z - z_0$ with $z_0 = L/2 + \frac{1}{V}(\lambda^2 q^2 + \frac{\sigma}{q^2})$. The coordinate z_0 corresponds to a turning point for the problem and bounds the spatial domain wherein the dilation instability takes place. This change of coordinates allows to recover the familiar Airy equation:

$$\frac{d^2w}{dz'^2}+bz'w=0,$$

with $b=Vq^2$. It has the general solution:

$$w=z'^{1/2}[\alpha\, J_{1/3}(2b^{1/2}z'^{3/2}/3)+\beta\, N_{1/3}(2b^{1/2}z'^{3/2}/3)],$$

where α and β are arbitrary constants for the moment and J and N are Bessel functions. Putting $z_1 = L-z_0$ one finds that the boundary conditions for w impose a transcendental equation to be satisfied by σ:

$$J_{1/3}(2b^{1/2}z_1^{3/2}/3)\, N_{1/3}[2b^{1/2}(-z_0)^{3/2}/3] = N_{1/3}(2b^{1/2}z_1^{3/2}/3)\, J_{1/3}[2b^{1/2}(-z_0)^{3/2}/3]\,.$$

$$(40)$$

Indeed, this last expression allows in principle to find numerically the unknown quantity σ. Let us consider for the moment the onset of instability. This is reached when $\sigma=0$ is a solution of (40). From this, a relation follows between L, V and q. The dimensionless number $C=\dfrac{VL^2}{\lambda}$ will be used in the following to characterize the relative efficiency of the driving force (it plays the role of a Peclet number for the problem). Thus, one can replace the equation (40) by:

$$J_{1/3}(Q_c^{1/2})N_{1/3}(-i\,Q_c^{1/2}) = J_{1/3}(-iQ_c^{1/2})N_{1/3}(Q_c^{1/2})\,, \qquad (41)$$

where Q_c is defined by $Q_c = q_c(VL^3)^{1/2}$. The pure number Q_c is the smallest root of (41), corresponding to the onset of linear instability, this implies that the instability sets in for $V \cong \dfrac{\pi\lambda}{L^2}$ with $q_c \cong (\pi/\lambda L)^{1/2}$. One expects that, if V is much bigger than this threshold value, a whole band of wavenumbers will become unstable.

Let us estimate the shape of those two borders. Let us assume first that the argument of the Bessel functions in (41) remains finite on one side of the band of unstable wavenumbers. This yields the long wavelength limit $q \cong Q_c\,(VL^3)^{-1/2}$ as obtained above.

The other side of the unstable band is reached when the two terms in z_0 and z_1 are of the same order of magnitude. This implies $q \cong (\dfrac{LV}{\lambda^2})^{1/2}$ and allows to use the asymptotic WKB-like expression for the Bessel functions in (40). However it is more transparent to come back to the original equations. With the scaling $Z=z'/L$, and $q=Q(\dfrac{LV}{\lambda^2})^{1/2}$, one transforms (37) into:

$$d^2w\,/\,dZ^2+Q^2(V/V_c)^2\,(Z-\tfrac{1}{2}-Q^2)w=0,$$

with the boundary conditions w=0 at Z=0,1. This has solutions at large $(V/V_c)^2$ if a "classical" region of the potential $Q^2(V/V_c)^2$ $(Z-1/2-Q^2)$ exists, that is if this potential is attractive somewhere for particles with energy zero. Otherwise one would have purely exponential (non oscillating) solutions which would not fit both boundary conditions. As the largest value of the potential is reached for $Z=1$, having a negative value of the potential imposes $Q^2 <1/2$. This condition defines completely the shortwavelength border of the stability domain at large C's with a variation of the critical wave vector as $V^{1/2}$.

This ends the linear stability analysis of the solution (38) of equation (37). Below we shall consider the domain of very strong nonlinearities, reached at large C's (but still respecting the inequality C<<L/λ).

(4.3) Undulation in the strongly nonlinear limit

Below we examine what happens very far above the linear threshold for the undulation instability. It turns out that this leads to a rather interesting mathematical structure that could have something to do with the famous von Karman conjecture on the buckling of plates for very large constraints [19].

The study of the nonlinear domain is made formally simpler by keeping the order of magnitudes coming from the above developments: $u \propto \lambda$, $x \propto (\lambda L)^{1/2}$, $z \propto L$. Then the strength of the constraint V is measured by the number C, and the nonlinear equation for the steady solution reads,after making a change of variables $u/\lambda \to u$, $x/(\lambda L)^{1/2} \to x$, $z/L \to z$:

$$- C = (u_z+u_x^2/2)_z + [u_x(u_z+u_x^2/2)]_x- u_{xxxx} \tag{42}$$

this has to be solved with the b.c. u=0 at z=0 and 1.We already established that a threshold value of C exists such that the x-independent solution of (37) bifurcates to a solution depending on x in a non trivial fashion. Moreover the equation (42) is the Euler-Lagrange condition expressing that the functional

$$G[u] = \frac{1}{2}\int dr \ \{[\frac{\partial u}{\partial z} + \frac{1}{2}(\frac{\partial u}{\partial x})^2]^2 +(\frac{\partial^2 u}{\partial x^2})^2- Cu\} \ . \tag{43}$$

is stationary. Thus it makes sense to look at the optimal solution with the lowest "energy". As G contains a term linear in u andmultiplied by the large quantity C, one expects that this optimal solution has the largest possible order of magnitude in C .

We can use a dimensional argument to evaluate the order of magnitude of the different terms in (42). More precisely we look for solutions of the form

$$u \propto C^{\alpha}, z \propto C^{\beta}, x \propto C^{\gamma} \tag{44}$$

If we look first for extended solutions along z, the scaling of the reduced z variable as 1 should correspond to $\beta = 0$. It is possible to match the first four terms in (42) with $u \propto C$,

$z \propto 1$ and $x \propto C^{1/2}$. This choice makes relatively negligible the last term on the right hand side, of order C^{-1}. It yields the parameterless equation, deduced from (42) with the scalings given by (44) but with the change of notations u in Cu, z in z, x in $C^{1/2}x$:

$$-1 = (u_z + u_x^2/2)_z + [u_x(u_z + u_x^2/2)]_x \quad , \tag{45}$$

with the b.c. u=0 at z=0 and 1. The scaling (44) indicates that the preferred wavelength of the optimal structure increases like $C^{1/2}$ at large C's, although the amplitude of the pertubation in u varies as C. This seems to be incompatible at first with the general assumptions of small gradients of u; in particular, the phase fluctuation has to be small in some sense in order to permit to write the equations in the coordinates of the unperturbed system. However this can be done consistently in the limit where C goes to infinity because there is a smallness parameter independent on C, the ratio $\frac{\lambda}{L}$. The tilt of the equiphase lines remains small, as required, if the dimensionless quantity $\frac{\partial u}{\partial x}$ (resp. $\frac{\partial u}{\partial z}$) is small. In the large C limit this gradient scales as $(\frac{C\lambda}{L})^{1/2}$ [resp $(C\lambda/L)$] and thus can remain small even with a large C as long as λ/L is smaller than C^{-1}.

It remains to prove that a nontrivial and smooth solution of (45) yields the lowest energy. This is a rather subtle question. First it can be guessed that the (trivial) solution u(z)=z(1-z)/2 does not represent the minimum of the energy by looking at fluctuations near this solution. Let again w be this fluctuation with the wavenumber q in the x direction. If neutral, this fluctuation has to be the solution of:

$$w_{zz} + \frac{1}{2}q^2 \, w \, (z-1/2) = 0 \text{ with } w=0 \text{ at } z=0 \text{ and } 1.$$

This Airy equation imposes quantized values of q^2 and those values exist, as can be shown in the WKB-limit (large q) for instance. Since the trivial solution u(z) is neutrally stable against some perturbations, a weakly nonlinear analysis would show that some small amplitude pertubation with a wavenumber close to the ones given by the above eigenvalue problem yields a solution with an energy less than the trivial one.

Let us come now to the smoothness of the solution of equation (45). This equation is the Euler-Lagrange condition for the stationarity of the functional:

$L[u] = \int dr \, \{-u + [u_z + \frac{1}{2}(\frac{du}{dx})^2]^2\}$, together with the b.c. u=0 at z=0 and 1. Suppose that one has found a smooth solution U(x,z) making the functional L stationnary and as negative as possible with our choice of sign. Consider now the second variation of L(u) around this solution. Let w(x,y) be the corresponding fluctuation of u near U, then this second variation has the form:

$$\delta^2 L[w] = -\int dr \, [w_z^2 + 2w_z \, w_x \, U_x + \frac{1}{2}w_x^2 \, (U_z + \frac{3}{2}U_x^2)],$$

This expression is quite remarquable in the sense that it is homogeneous as far as the derivation order of w is concerned. This implies that the sign of $\delta^2 L[w]$ is completely determined by the one of the quadratic form in the variable w_z and w_x that appears in the integrand of $\delta^2 L[w]$. This quadratic form may get positive values (and the initial solution may become unstable) if the determinant $-(U_z + \frac{1}{2} U^2_x)$ is positive somewhere in the physical domain.

Before going to the physical meaning of this last condition let us show that this determinant is certainly positive somewhere for a smooth solution U. Let us consider the following change of variables, kindly suggested by V. Hakim:

$$dz = d\tau \quad ; \quad dx = \frac{1}{2} d\tau . U_x$$

In term of the new variable τ, we can express the variation of U as

$$d U = U_z \, dz + U_x \, dx = d\tau \, (U_z + \frac{1}{2} U^2_x)$$

The variation of U between z =0, 1 is given in the new running variable τ as

$$\int \mathbf{grad} U . \, d\tau = 0$$

since the boundary conditions U=0 holds at both ends .

Thus, either U is zero everywhere, or it has a maximum in the interval. In such a case, the quantity $- (U_z + \frac{1}{2} U^2_x)$ must be positive somewhere in the interval z = 0,1. On the other hand, U cannot be zero everywhere on the line indexed by τ. Near z=0 (or z=1, but with a slight change of sign and of variable), the Taylor expansion of a solution of equation (43) should include a quadratic term ($-z^2/2$), and, as the line under consideration merges with the z=0 line at right angle, U cannot vanish everywhere on it.

The physical meaning of this result may be understood as follows: the determinant of the quadratic form is precisely equal to the local gradient of the wavelength expressed in intrisic coordinates. Thus the possibility of having locally an unstable fluctuation is another manifestation that this may increase the local wavenumber by a modulation in the transverse direction. Here we have the remarkable situation that this process may be continued down to arbitrary small scales. This implies that the equation (43) cannot be uniformly valid, because the original equation (42) had actually a built-in small space scale, represented by the fourth derivative, that has disappeared in (43) after the rescalings motivated by the large C limit. Indeed, this is the typical situation where a boundary layer has to exist in order to match the domains where the asymptotic equation [here equation (43)] joins formal singularities of its solution. Coming back to equation (42) and looking for the balance between all terms by using the scaling forms (44), we obtain the solution

$$\alpha= 0; \ \beta= -1/2; \ \gamma= -1/4.$$

This choice implies first that there is a boundary layer whose thickness along z goes to zero at large C as $C^{-1/2}$. More generally, the exponents β and γ define the scaling of lower cut-off lengths along the z and x direction. This scaling defines the smallest scales in the cascade of structures starting from an original large scale (zig-zag) solution as well as the adjustment of the boundary conditions at the wall. Much work remains to be done on this problem which could be approached from an asymptotic matching treatment.

However the above solution does not preclude that there are other solutions with an even smaller energy on average along the x-direction that were nonperiodic in y, but quasiperiodic or even chaotic. Let us simply notice that any solution of (42) has to have a Noether invariant in the x and z direction. Let us make this explicit for the x direction. Define the function K[u] as :

$$K[u]= \int_0^1 dz \ (u_x - \frac{1}{2}u_x^2 + \frac{1}{2}u_x^2 u_z + \frac{3}{16}u_x^4),$$

For a given u, this is a function of x only. This is a Noether invariant in the sense that, if u is a solution of (42) with the correct boundary condition, then dK[u]/dx=0.

This concludes our discussion of this application of phase approximation to a specific problem of periodic structure under constraint.

5. Conclusions and reflections

Through a discussion of three specific examples concerning the nonlinear dynamics of extended structures, I have tried to show the fascinating variety of problems that may arise in this field. As I was kindly asked by the organisers of the meeting to comment upon the existence or not of general principles, I would say here that, I believe (this is indeed a personal opinion only) that it is not yet time to draw such general principle from our limited knowledge of nonequilibrium systems, in the same sense that Thermodynamics model our thinking of equilibrium systems. This would require a better understanding of the statistical theory of those nonequilibrium systems than we have now. Hopefully more progress will be done in that direction in the coming years.

References

[1] Y.Pomeau; Physica **23D**,3 (1986).

[2] J.Smoller "Shock Waves and reaction-diffusion equations", Springer, Berlin(1983).

[3] A.C.Newell, J.Whitehead, J. of Fluid Mech.**38**, 79(1969); L.A.Segel, ibid; p.203.

[4] J. Hegseth, C.D. Andereck, F. Hayot and Y. Pomeau, Phys. Rev. Letters **62**, 257 (1989).

[5] O. Thual, S. Fauve, J. de Physique (Paris), **49**, 1829 (1988).

[6] V. Hakim, P. Jakobsen and Y. Pomeau, Europhys. letters **11**, 19 (1990).

[7] Ya.B.Zel'dovich, "Theory of combustion and detonation of gases", Moscow (1944).

[8] A.N.Kolmogorov, A.N.Petrovskii and N.S. Piskunov, Bulletin de l'Universite d'Etat a Moscou, Sec.A, Vol.**1**, Math. et Mec., 1 (1937).

[9] P.Tabeling, J. de Phys. Lettres **44**, 665(1983); P.Hall, Phys.Rev. **A29**, 2921(1984).

[10] I.J.Wygnansky and F.H. Champagne, J.of Fluid Mech. **59**,281 (1973).

[11] H.Chaté and P.Manneville, Phys.Rev. Lett. **58**,112 (1987).

[12] Y.Pomeau, in "Instabilities and nonequilibrium structures II", E. Tirapegui and D. Villaroel eds., Kluwer Academic press publishers (Boston, 1989).

[13] A.Pumir, P. Manneville and Y. Pomeau, J. of Fluid Mechanics, <u>135</u>, 27 (1983).
[14] A.C. Newell, "Solitons in Mathematics and Physics", SIAM pub. Philadelphia (1985).

[15] L.N. Bulaevskii and V.L. Ginzburg, Sov. Phys. JETP **18**, 530 (1964).

[16] E. Schlömann and J.D. Milne, IEEE Trans. Magn. **Mag-10**, 791 (1974); E. Schlömann, IEEE Trans. **Mag-11**, 1051 (1975).

[17] Chapter 10 in "Dissipative Structures and Weak Turbulence" by P. Manneville, Academic press, Boston (1990).

[18] A. Pocheau, Thèse Université Paris 6, unpublished (1987); A. Pocheau, V.

Croquette and P. Le Gal, Physical Review Letters, **55**, 10 (1985); A. Pocheau, J. de Phys. **50**, 25 (1989).

[19] see for instance D. Schaeffer and M. Golubitsky, Comm. on Math. Phys. **69**, 209 (1979), and references therein.

GROWTH PATTERNS : FROM STABLE CURVED FRONTS TO

FRACTAL STRUCTURES

Yves Couder

Laboratoire de Physique Statistique*, Ecole Normale Supérieure
24 rue Lhomond, 75231 Paris Cedex 05, France

* associé au C.N.R.S. et aux Universités Paris 6 et Paris 7

I Introduction : The Stefan problem

The present lectures intend to review some recent experiments on the properties of patterns obtained in growth processes. This covers a very large range of situations and I will limit myself here to three systems which can be considered as archetypes demonstrating some of the effects at work. The three systems, Saffman Taylor viscous fingering [1-37], dendritic crystal growth [38-50], and diffusion limited aggregation [51-68] have been widely studied and reviews on these three problems can be found respectively in references 3-6, 39-42, and 52-55. Though they are related to still other pattern forming systems these have been chosen here for their (relative) simplicity. All three are close to an ideal problem which is intrinsically mathematical.

Ideally this problem is known as a Stefan problem[69-71]. We will describe it in the situation sketched on Figure 1. An interface separates two regions of an infinite plane respectively labeled 1 and 2. In this plane exists a scalar field P. We assume that P is constant in region 1 and varies in region 2. The field P obeys the Laplace law :

$$\Delta P = 0 \tag{1}$$

At a given time the determination of the field depends on the boundary conditions defined at the interface and at infinity. As a function of time a physical process makes the interface move with a normal velocity proportional to the local gradient of P.

$$V_n \propto \nabla P \tag{2}$$

This displacement of the interface, because it changes the location of the boundaries, modifies the field of P, thus in turn the interface velocity. The problem is : what shapes of the interface will result from processes of this type.

If we consider the situation drawn on figure 1 two cases can exist: if region (1) shrinks, the interface is stable and we have the Stefan problem which reduces to a (difficult) topological problem concerning the connexity of the shrinking domain. If region (2) expands we have what could be called the "inverse" Stefan problem in which the interface is unstable and produces patterns. There is no known solution to this problem, but it was shown that ideally, in the absence of any stabilizing factor, the instability of such an interface would lead to the formation of singularities in a finite time[69-71]. However in the experimental or numerical situations an additional effect is present at the interface which creates a limiting small scale under which the interface is stable. In the experimental system that we will

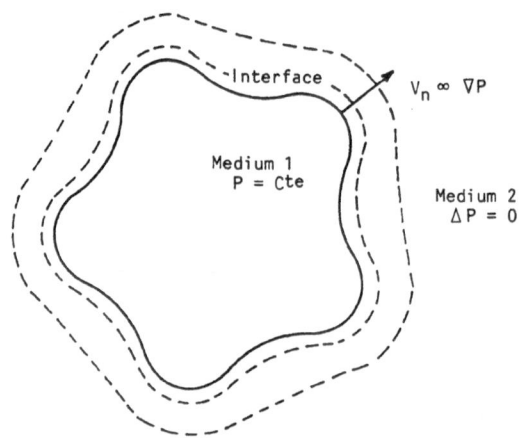

Figure 1. Scheme of the Stefan problem

describe it is due to the interfacial capillary effects, in the numerical system it is related to the computing grid mesh size.

A more general family of problems of the same type is obtained in diffusive systems ; equation (1) is then replaced by a diffusion law

$$\frac{\partial P}{\partial t} = D \, \Delta P \qquad (3)$$

Near the interface moving at a velocity V, P varies in a layer limited by a diffusion length $l_d = D/V$. When this diffusion length is smaller than the characteristic length scales of the domain the growth is independent of the boundaries. We will be interested here in the other limit, where the diffusion length is so large that the system is quasi Laplacian.

Returning now to the Laplacian systems, a general characteristic is their sensitivity to the boundary conditions. For these problems the imposed geometry is thus very important. In the most generic geometry, at the initial time, region 1 is reduced to one central point and region 2 occupies all the rest of the plane. With time region 1 expands radially into region 2. This configuration is very attractive because it corresponds to a situation often met in physics, chemistry or biology where the growth starts from a seed. This radial configuration has been widely investigated in the case of diffusion limited aggregation. It turns out to create patterns of increasing complexity, having a fractal structure. It is in my opinion the most difficult geometry for reasons that will be discussed below. Fortunately there are variants of this problem in which the growth occurs only in a part of the plane. These situations are easier for the analysis of the processes at work.

For instance we can choose an initial condition in which each region occupies half of the plane, separated by a linear front. This configuration lends itself to a linear analysis of the stability which describes the initial destabilization. We will describe in part II the result of such analyses for two of the physical systems at hand. The experimentally observed destabilization of a planar front is given by the results of these linear analyses, Their validity however is limited to the initial times when the perturbations are still of very small amplitude. When the amplitude of the disturbances increases the different protrusions in the front interact non-locally and the fastest growing ones tend to screen off the others so that they stop growing. Simultaneously these fastest protrusions have more space to grow so that they become unstable either by tip splitting or by side branching. The structure is thus submitted to two antagonist trends which will give it, when it has grown sufficiently, a fractal structure. The screening-off is responsible for the formation of larger and larger, increasingly widely spaced trees. This process is often called coarsening. Typically the spacing of the surviving trees is of the order of the front thickness. It is clear that this process generates ever larger scales to the growing fractal. Simultaneously the side branching keeps generating constantly the small scales of the fractal. The overall characteristics of this process have received no general theoretical treatment as yet. For these reasons it is useful to turn to even more confined experimental configurations limiting the freedom of the system.

In these geometries, the space occupied by regions 1 and 2 is bounded by walls which form a channel. In this way we have only a partial Stefan problem because only one of the boundaries is moving while the others are fixed. We will see that in this way we can force the system to have steady regimes of growth (if the geometry has translational invariance), or also self similar growth. The resulting patterns can be either stable curved fronts described in part III or fractal structures described in part IV. These two parts will thus present general properties of the non linear growth in these confined geometries. Emphasis will be put on three determinant factors in the growth :

- *The geometry.* For Laplacian systems grown in a part of the plane only, the geometry of the fixed boundaries define families of possible analytical solution. Because the geometry forces the curvature of the front the selection due to isotropic or non isotropic surface tension is affected by it.

- *The isotropy or the non isotropy*. Whenever the interface is isotropic and homogeneous, a specific selection is induced which can be destroyed by either anisotropy or localized disturbances of the front or of its surroundings.

- *The length scales.* The instability mechanism creates a typical instability length scale. The growth itself determines a second length scale, either the size of the pattern in the case of the growth in an infinite plane, or sizes imposed by the boundaries for the growth in channels. This second length scale in all the cases of interest here is larger than the first. Roughly, if it is only slightly larger we obtain stable curved fronts, if it is much larger the pattern is fractal in the range between these two scales.

II Introduction to the physical characteristics of three experimental systems : the instability of their linear fronts

II-1 Saffman-Taylor fingering

The instability first introduced by Saffman and Taylor (1958) [1] occurs in Hele-Shaw cells formed of two solid plates placed at a very small distance b from each other. The flow of a fluid between these plates, dominated by the friction on the plates is a Poiseuille flow. A mean velocity across the cell's thickness can be defined, governed by a potential law:

$$\vec{V} = - \frac{b^2}{12\,\mu} \vec{\nabla}p \tag{4}$$

Where μ is the fluid viscosity. The fluid being incompressible the pressure p obeys a Laplace law :

$$\Delta p = 0 \tag{5}$$

The Saffman-Taylor instability occurs at the interface between two fluids of very different viscosities when the fluid of low viscosity forces the other fluid to recede. To obtain a situation as close as possible to the ideal defined in part I, we will assume that the viscosity of one of the fluids is zero. A good approximation is obtained in practice when air pushes a high viscosity oil. The problem is completed by the interfacial forces due to capillarity ; they create across the interface a pressure jump δp :

$$\delta p = \gamma\,\kappa \tag{6}$$

Where γ is the surface tension and κ the local curvature of the meniscus in the plane of the cell. It is important to note that the analysis of the Saffman-Taylor instability relies on a two dimensional modeling of a really three dimensional situation. In particular the assumption of a constant curvature of the meniscus across the cell's thickness is not always valid as was shown experimentally [10] and theoretically [11,12]. Tanveer[12] has recently treated the complete three dimensional system and shown in which limits the two-dimensional assumption is valid.

Within this approximation the problem at hand is exactly of the type described in the introduction. A typical length scale of the instability is given by the linear stability analysis of a plane front (Chuoke et $al.$ 1959) [2]. The interface is assumed to be deformed by a sinusoidal deformation (Fig. 2a) so that its position is given by :

$$\zeta = Vt + \varepsilon\, e^{\sigma t} \sin k_y y \qquad (7)$$

The amplification rate σ of this deformation is given by:

$$\sigma = V|k|\left(1 - \frac{b^2\gamma}{12\,\mu V}k^2\right) \qquad (8)$$

Figure (2b) shows the graph $\sigma(k)$. The front is unstable at all velocities, and the wavelength l_c corresponding to the wavevector k_c of maximum amplification rate (at velocity V) is :

$$l_c = \pi b \sqrt{\frac{\gamma}{\mu V}} \qquad (9)$$

In the following we will call l_c this capillary length scale of the system. We can note that no disturbance with a wavelength smaller than $l_m = l_c/\sqrt{3}$ is unstable (Fig 2b).

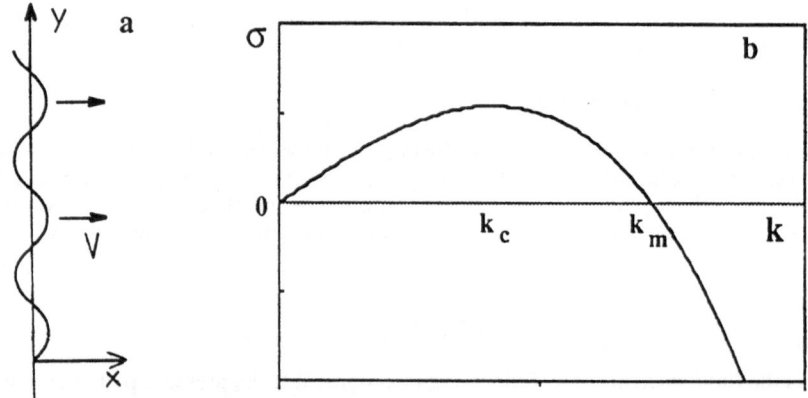

Figure 2.   a) Sketch of the moving plane interface.
                  b) The wave-vector dependence of the amplification rate predicted by the linear analysis

II-2 Solidification instability

The growth of a crystal in a pure melt occurs when the system is put out of equilibrium by undercooling [38,43]. As the solid grows into the liquid, it releases latent heat : a diffusive field of temperature is formed in front of the crystallization front. It is controlled by equations of the type of equation (3) where P is replaced by the temperature T and D by the heat diffusion coefficient D_T. The linear analysis of the stability of a plane solidification front in a pure melt (Mullins and Sekerka (1964) [38]) shows that it is unstable at all velocities but that small scales are stabilized by capillary effects through the Gibbs-Thomson effect. The amplification rate of a perturbation of wave-vector k is :

$$\sigma = V|k|\left(1 - \frac{D_T}{V}d_o k^2\right) \qquad (10)$$

a result very similar to that of Chuoke et al. [2] (equation 8). The maximum instability of a front growing at velocity V occurs at a length scale :

$$l_c^s = 2\pi \sqrt{3} \sqrt{\frac{D_T d_o}{V}} \qquad (11)$$

l_c^s is proportional to the geometrical mean of the diffusion length scale D_T/V and a capillary length scale d_o which is the size of the smallest crystal in equilibrium in the melt at this temperature. It is related to L the latent heat of solidification, γ the surface tension and c_p the heat capacity :

$$d_o = \frac{\gamma c_p T}{L^2} \qquad (12)$$

In most real cases however the growth of a crystal is dominated by the diffusion of impurities (this is naturally the case for growth in solutions, the solvent being then the impurity). The impurities rejected by the crystal form a diffusive field in front of the crystal. Now it is C the concentration of impurities which is determined by an equation of the type of equation (3), the diffusion constant being D_C rather than D_T. The latter being much larger than the former, most crystallisation processes are controlled by the concentration diffusion field. It is possible, for impurity controlled fronts, to do a linear stability analysis [39] of the Mullins and Sekerka type. The situation differs in that, while the diffusion of heat is somewhat symmetrical in the solid and the liquid, the diffusion of impurities is large in the solution and small in the solid. The results are however similar and the maximum instability occurs at a length scale l_c^{sc} proportional to the geometrical mean between the impurity diffusion length D_C/V and a chemical capillary length d'_o.

II-3   Diffusion limited aggregation

This numerical model system was introduced by Witten and Sander (1981)[51]. In the initial model, random walkers usually moving on a square lattice are emitted far away from a central seed. If a walker visits any of the sites neighbouring this seed, it sticks there. The process is repeated a large number of times, each walker sticking whenever it visits a site neighbouring a site already occupied. The growth is therefore radial and results in a much investigated fractal aggregate. It was shown by Paterson[56] that P the probability of visit by the random walker of a given site of the lattice obeys a Laplacian law :

$$\Delta P = 0 \qquad (13)$$

The average speed at which a region of the aggregate grows is:

$$V_n \propto \nabla P \qquad (14)$$

In the standard model there is no surface tension and the process is unstable at all scales. However the numerical technique itself, because it always uses either a lattice or walkers of finite size, introduces a length scale l_u : the lattice mesh or the walker's size. Its role is evident in a variant of the classical DLA model in which the punctual seed is replaced by a linear base line[52,55]. In this model the instability of a plane front is observed : the first layer deposited on the base line is lacunary at a scale of the order of l_u. Several works [57-62] have been devoted to variations in which the rules of the game are changed so as to introduce an effective surface tension in DLA. The small length scale is then larger than l_u.

III   The stable curved fronts

III-1   Isotropic case

Isotropic fronts are obtained in the case of Saffman Taylor instability and in DLA.

However the latter being violently unstable it is only in the former case that stable curved fronts have been studied. Viscous fingering has classically been investigated mainly in three geometries.

In the first, introduced in their initial work by Saffman and Taylor [1], the two fluids move in a Hele Shaw cell in the shape of a long channel closed on both lateral sides and of width W.

In the second geometry introduced recently by Thomé et al.[7] the fluids are bounded by two walls at an angle θ_0 with each other. In these sector-shaped cells the front can move either in a divergent direction ($\theta_0 > 0$) or in a convergent one ($\theta_0 < 0$).

The third geometry which corresponds to the radial growth in an infinite plane, was introduced in the case of viscous fingering by Bataille[8] and investigated further by Paterson[9]. The viscous fluid is enclosed between two large circular plates and air is injected at the center of these plates.

The interest of the first two geometries is that they lead to the formation of well defined curved fronts

III-1-1 Linear channel

This configuration has translational invariance along the cell and generates a well known steady solution : the Saffman-Taylor finger. The experiments and their two-dimensional numerical simulations show that the state of the front is controlled by a parameter B which is usually chosen as :

$$B = \frac{\gamma}{12 \, \mu V} \left(\frac{b}{W}\right)^2 = 8.45 \; 10^{-3} \left(\frac{l_c}{W}\right)^2 \tag{15}$$

B is proportional to the square of the ratio of the small length scale to the large length scale of the system.

For moderate values of B one single finger is observed moving steadily along the cell with a well defined width and shape. This situation persists[10] down to values of $B \sim 1.4 \; 10^{-4}$ (where $W \sim 8 \, l_c$). In this range as $B \rightarrow 0$ the ratio λ of the finger's width to the channel's width tends to a limit value $\lambda = 0.5$ (Fig.3a). For very small values of B however the finger becomes unstable (See part IV-1-1). Saffman and Taylor had shown that if surface tension was neglected it was possible to find, by conformal transform techniques, a family of analytical smooth shapes for the interface.

$$x = \frac{W(1-\lambda)}{2\pi} \ln\left[\frac{1}{2}\left(1 + \cos\frac{2\pi y}{\lambda W}\right)\right] \tag{16}$$

These fingers are parametrized by their relative width λ which can take any value from 0 to 1 (Fig. 3b). The shape of the solution of width 0.5 is in very good agreement with the experimentally observed finger (Fig. 3a). However in this analysis there was no hint as to why this particular solution was selected.

Theoretically the answer to the selection problem came much later after numerical investigations by McLean and Saffman [13] and Vanden-Broeck[14] and analytic works by Combescot et al.[15a and b], Hong and Langer [16], and Shraiman [17]. They showed that surface tension could not be treated as a perturbation because it introduced higher order terms in the integro-differential equation defining the interface. When the transcendentally small terms due to surface tension are taken into account, a solvability condition is introduced for the finger to be smooth at the tip. For a given value of B there is only a discrete set of solutions for which this condition is satisfied ; their width λ_n all tend towards 0.5 for vanishing B. In the limit of very small B the λ_n dependance on B is given analytically by

$$\left(\lambda_n - \frac{1}{2}\right) = \frac{\left(16 \, \pi^2\right)^{\frac{2}{3}}}{8} a_n \; B^{\frac{2}{3}}$$

where the coefficients a_n are approximately given by $a_n = 2(n+4/7)^2$. This dependance can also be found numerically; it is shown on Figure (3d) and in good agreement with the theoretical prediction [18a]. The stability of the various branches was investigated by Bensimon *et al*.[18] and Tanveer[18b]: only the lower branch n=0 is stable, the others are unstable by tip splitting.

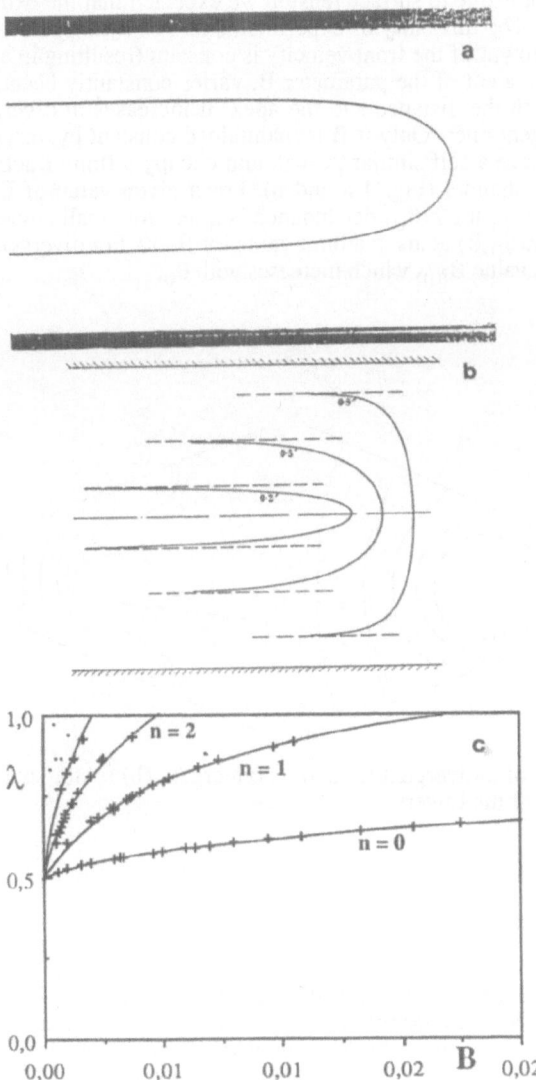

Figure 3. a) The experimental Saffman Taylor finger of width λ≈0.5 at B~2 10⁻⁴ (isotropic situation)

b) Three finger profiles of the continuous family given by equation (16) (from Ref 1).

c) The first levels of the discrete set selected by surface tension in the isotropic case (courtesy M. Ben Amar). The level n=0 is stable and corresponds to the experimentally observed finger.

Experimentally we showed at about the same time (Couder et al. [25a] and Rabaud et al.[25b]) that this selection could be lifted by various types of localized disturbances of the finger tip. We thus obtained narrow fingers that we called anomalous fingers. This was a direct experimental confirmation of the singular role ascribed by the theory to the tip, and will be discussed in more details in section III-2.

III-1-2 Sector shaped cells

The basic idea which led us to the investigation of sector shaped cells [7] is that in these configurations the geometry itself creates a curvature. Far away from the growing front the isobars are arcs of a circle centered on the apex of the cell (Fig. 4). As the selection of the Saffman Taylor finger is due to surface tension we expected that the extra curvature would modify the selection. The difficulty of experiments in sector shaped cells comes from the unsteadiness of the growth. If the front velocity is constant (resulting in a constant capillary length scale) the equivalent of the parameter B, varies constantly because the local width $W(r) = \theta_0 r$ varies with the distance r to the apex. It increases in divergent channels and decreases in convergent ones. Only if B is maintained constant by varying the velocity as r^{-2} will the fingers have a self similar growth and occupy a finite fraction $\lambda(\theta_0,B)$ of the angular width of the channel (Fig. 4 a and b). For a given value of B this fraction is a function of the angle θ_0 ; the $\lambda(\theta_0)$ dependance is linear for small angles (Fig.5d). For all convergent channels $\lambda(\theta_0,B)$ tends to a limit value for $B\to 0$. For divergent angles the finger becomes unstable at a value B_{lim} which increases with θ_0.

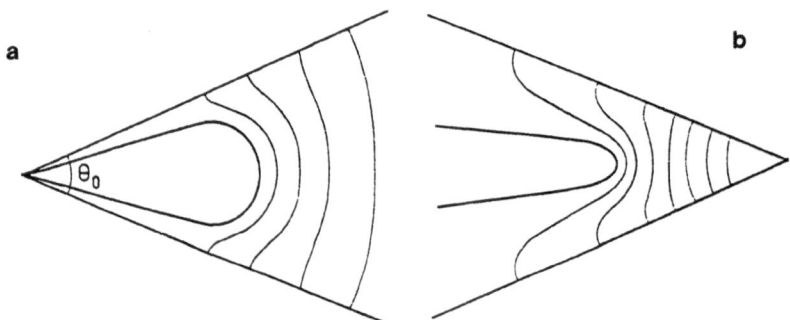

Figure 4. Sketch of a divergent (a) and a convergent (b) sector shaped cell and of the isobars ahead of the fingers.

From the theoretical point of view V. Hakim[7] first showed that a self similar set of analytical solutions could be found in the absence of surface tension for sector shaped cells with an angle $\theta_0=90°$. Self similar solutions of this type exist for all angles, they have been found recently by M. Ben Amar[19]. She showed that in these cases the free boundary problem can be transformed in a non-linear Ricatti differential equation. For each value of the angle θ_0 there is a family of solutions given by hypergeometric functions parametrized by their width λ. Fig 5a shows three profiles of this type for $\theta_0=60°$. The expression of these solutions can be simplified only in the particular cases $\theta_0=0°$, 90° and -90° ; in these cases the previously known solutions are recovered. The finger shapes obtained experimentally at small B are well fitted by the theoretical profiles of corresponding λ.

The selection of the solution in the presence of isotropic surface tension has been investigated both numerically and analytically (M. Ben Amar et al [20a, 20b]. As in the linear cell, for a given value of the control parameter B there is a discrete set of solutions and the observed finger corresponds to the solution of smallest width. Relatively to the set selected for $\theta=0$, these solutions have widths λ shifted to smaller values for $\theta_0<0$, and to larger

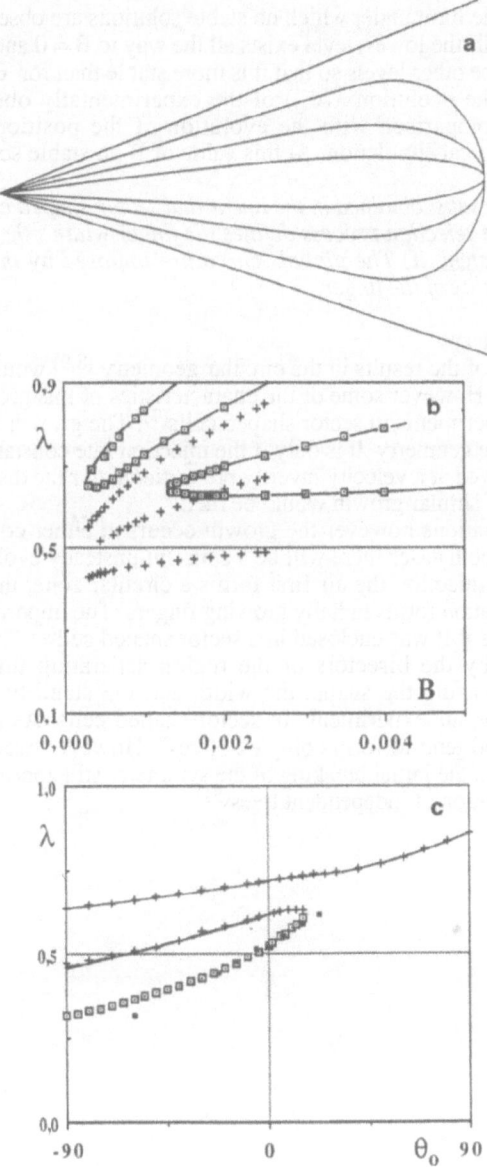

Figure 5. (a) Three finger profiles of the analytical solutions[19] in the absence of surface tension for $\theta_0 = 40°$
(b) The B dependance [20a] of the first λ_n of the discrete set of widths selected by surface tension. □ $\theta_0 = 20°$, + $\theta_0 = -20°$.
(c) Finger widths obtained by numerical simulation at a fixed value $B = 1 \ 10^{-3}$ as a function of the cell's angle θ_0; □ Level λ_0, + Levels λ_1 and λ_2. The experimentally observed fingers (■) correspond to the first level

values for $\theta_0 > 0$. But the most remarkable feature is that these solutions do not exist at all values of B. In the case of $\theta_0 > 0$ the $\lambda(B)$ dependance (Fig. 5b) shows that at a finite value of B the levels n=0 and n=1 coalesce so that they form a loop. Similarly n=2 and n=3 will form a loop at smaller B. As will be discussed below (III-3) the value B_l where there is a loop corresponds to the limit under which no stable solutions are observed. In contrast, in the case of convergent cells the lowest level exists all the way to B = 0 and at finite values of B is well separated from the other levels so that it is more stable than for $\theta_0 = 0$ stable.

Fig.5c shows the evolution $\lambda(\theta_0)$ of the experimentally observed finger-width at B= $2 \ 10^{-3}$ and its comparison with the evolution of the position of the lower branch observed in the numerical simulation. At this value of B no stable solution exists for angles larger than ~20°

Altogether the results obtained in the linear and sector-shaped cells show that : i) for an isotropic interface the selection process defines the finger width : the selection occurs at the large scale of the system. ii) The global curvature imposed by the geometry affects the selection and the stability of the finger.

III-1-3 Circular geometry

The discussion of the results in the circular geometry [8,9] would go beyond the limits of the present article. However some of the characteristics of this growth can be understood in the light of the experiments in sector shaped cells[7]. The growth is likewise intrinsically unsteady, owing to the geometry. It is only if the injection rate constantly decreased in such a way that the front moved at a velocity invertly proportional to r the distance to the center, that the conditions for self similar growth would be met.

In practical situations however the growth occurs at either constant injection rate or constant velocity. In both cases there will be a constant unsteady evolution of the pattern. At the beginning of the injection the air first forms a circular zone, then the axisymmetry is broken and destabilization forms radially growing fingers. The important point is that each of these fingers grows as if it was enclosed in a sector shaped cell[7]. The walls of this cell are virtual and formed by the bisectors of the region separating this finger from its two neighbours. With this rule the shape, the width and the stability of each finger can be predicted by the previous experiments in sector shaped cells. As they keep growing the fingers destabilize and tend to form complex "trees". However even when the growth has formed a fractal object, the initial breaking of the symmetry still appears to persist and to be at the origin of the formation of independent trees[51].

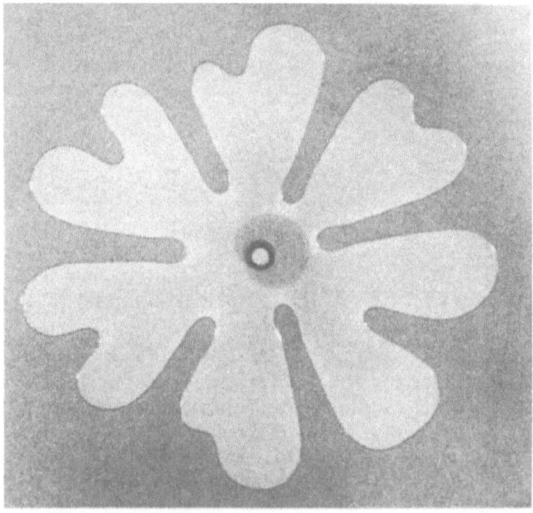

Figure 6. A pattern obtained in the circular geometry

III-2 Non isotropic situation

Non isotropic fronts can be obtained in the three types of growth described here but they are unstable for DLA. We will first describe stable anisotropic fronts in the case of anomalous Saffman Taylor fingering because the difference in the selective process is particularly clear by contrast to the previously described isotropic case. We will then recall some of the results concerning the dendritic growth.

III-2-1 Linear channel, the anomalous Saffman Taylor finger

Experiments [21-27,29], numerical simulations [26] and theoretical calculations [28,30-32] have now shown that the selection of the Saffman Taylor finger is completely changed when several types of disturbance are applied. The first experiment giving a hint of this effect was done by Ben Jacob et al. [21]. in the circular geometry and showed qualitatively that the morphology of fingers was affected if the system was made globally anisotropic by replacing the plates of the Hele-Shaw cell by plates deeply engraved with periodic grooves. In this case dendritic looking structures were observed. It was then shown (Couder et al.[22]) that one single localized disturbance at the finger tip was enough to obtain this type of finger so that global anisotropy was only one of the means by which the tip could be disturbed. But, as noted above, the circular geometry does not lend itself to a quantitative analysis of the processes involved. For this reason systematic investigations were done [25-29] in the linear geometry.

In the presence of either anisotropy or of a localized disturbance, the fingers obtained in a linear channel have shapes that are still well fitted by Saffman Taylor analytical solutions (equation 16) but have a width much smaller than 0.5 (Fig.7a). Couder et al. [25a] and Rabaud et al. [25b] investigated the selection of these anomalous fingers ; it turned out to be completely different from the isotropic case. Experiments done with the same disturbance in various channels showed that at a given velocity it was the radius of curvature at the tip which was selected. As the finger shapes are given by equation 16 this radius of curvature ρ can be deduced from the measured λ :

$$\rho = \frac{\lambda^2 W}{\pi (1-\lambda)} \tag{17}$$

For all the anomalous fingers grown in various cells at the same velocity with the same disturbance we find ρ to be the same. Varying the velocity we found that ρ remains proportional to the capillary length scale (Fig.7b) :

$$\rho = \alpha \, l_c \tag{18}$$

The coefficient α was found experimentally [25b] to be $\alpha \sim 0.85$ (another value is given in Ref.26 because of a different choice of capillary length). This value of α is dependent on the strength of the applied disturbance, but this point has not yet been quantitatively investigated. This regime however is observed in a limited domain of values of the parameter B. At very low or very high velocities there are two crossovers. A very slow finger will be only weakly affected by anisotropy and its width will be close to that of the isotropic one: the scaling on W dominates (Fig 7c). (This cross-over investigated by Dorsey and Martin [28] is similar to that investigated for crystals grown in a narrow channel [44b]).

Finally, for very fast fingers the two dimensional approximation breaks down when the tip radius of curvature becomes of the order of b the thickness of the cell (when $\rho \sim 3b$), value at which a saturation is observed [25b] (Fig 7b).

Finally let us note that we have shown above that the selection process of normal Saffman Taylor fingers is lifted by a disturbance which affects the isotropy or the homogeneity of the interface; this is the most common natural situation and the more relevant to an analogy with crystal growth. However this selection being due to a delicate relation between an isotropic surface tension and a hydrodynamic flow can also be lifted by a disturbance of the flow only, the interface remaining isotropic. This was suggested by a

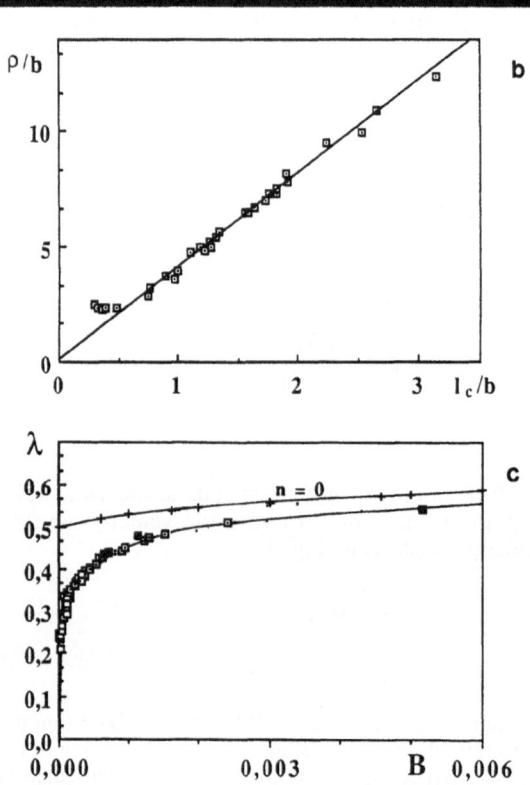

Figure 7. a) An anomalous finger [25b] obtained with a groove etched in one of the glass plates (Its shape is given by the narrow profiles of Fig. 3b)

b) Graph ρ/W versus l_c/W showing the fit by relation (18). The saturation at small B occurs when two dimensionality breaks down.

c) The anomalous finger's width obtained with the same disturbance in cells of various width. For small B, relation 18 is satisfied and λ tends to 0. For large B the influence of the cell's width dominates and a crossover is observed : the anomalous finger then hardly differs from the isotropic one (the level n=0 in the isotropic situation (cf. Fig3c) is recalled (+).

214

theoretical analysis of Combescot *et al.* [32]. in their interpretation of the previous experiments[22] in which a small bubble was present at the finger tip. This result was quantitatively confirmed by Thomé *et al* .[29] in experiments in which the flow is disturbed by a small disk moving ahead of the finger.

III-2-2 Open geometry: the dendrite

This type of selection is interesting in that it is similar to the selection which determines the shape of a dendritic crystal. It is well beyond the scope of the present course to present all the characteristics of dendritic crystal growth. The problem is more complex because it is a finite range diffusive process. We will limit ourselves here to the case which is nearest to the Laplacian inverse Stefan problem. This occurs when there is only diffusion on one side of the interface (one-sided model) a condition well satisfied in the case of impurity diffusion. We will also limit ourselves to cases where the diffusion length is very large compared to the crystal size (i.e. small Peclet numbers) and where the boundaries are also far away so that they have no influence on the growth (the medium is considered infinite). Generally it was shown by Ivantsov [43] that, in the absence of surface tension, parabolas were stationary solutions of the shape of the interface. As in the Saffman Taylor problem the argument gave no hint as to the selection of the experimentally observed solution. In the case of dendrites the introduction of isotropic surface tension is not enough to solve the problem, there is still no selection [44a,45]. However crystals are anisotropic and the surface tension γ of the interface is a function of the angle Θ of the normal to the interface with the main axis of the crystal. For a cubic crystal :

$$\gamma = \gamma_o (1 + \varepsilon \cos \Theta) \tag{19}$$

If this anisotropy of the surface tension is taken into account, for a given undercooling a discrete set of parabolas are selected [46]. The observed solution is the fastest one. It is defined by its radius of curvature which is proportional to $l_c{}^s$. The coefficient of proportionality is a function of the anisotropy ε. In the limit of very weak anisotropy the theoretical result can be written [46] :

$$\rho = 0.5 \, \varepsilon^{-7/8} \, l_c^{sc} \tag{20}$$

This type of selection had been observed by Dougherty *et al* .[47] and Maurer *et al* .[48] but the experimental measurement of ε is very difficult so that the dependence on ε is still uncertain. It is worth mentioning that Honjo *et al.* [49] have shown that the presence of very strong external noise could destroy the effect of anisotropy ; so that no dendritic growth of the crystal is observed and the growth looks isotropic.

Altogether the results of both the anomalous Saffman Taylor fingers and the dendritic case show that either a local disturbance or anisotropy result in a selective process which defines the finger's radius of curvature at the tip. The selection occurs at the small scale of the system.

In both cases considered above the coefficient of proportionality between ρ and l_c is a number of order unity so that the two lengths are of the same order of magnitude. This explains that for large velocities the tip of these structures always remains stable (as discussed in III-3-2). We will see in Part IV-2 that this is not necessarily the case [50,79]. It is possible in DLA to obtain patterns of weak anisotropy for which ρ is quite different from the lattice mesh l_u. As we will see this results in ρ forming a third length scale for the system. Then, even though the growth is of a non isotropic type, the tip can be unstable.

III-3 The stability of the smooth fronts

The stability of curved fronts was first investigated by Zel'dovich[72] in the case of flame fronts. A review on the stability problems can be found in P. Pelcé [6,73]. The previous results have shown that the selective mechanisms at work respectively in the isotropic case and in the anisotropic case lead to the tip being of small or strong curvature respectively. For this reason the type of instability observed in the two cases is very different.

III-3-1 Isotropic case

In the case of isotropic fronts (e.g. normal Saffman Taylor fingers) the finger is selected to be on the scale of the large scale (e.g. half of the channel in the parallel geometry). Decreasing B means that the finger tip becomes wider and wider in relation to l_c. The front of small curvature is more and more similar to a straight front and will be unstable at the capillary length scale l_c corresponding to the local normal component of the velocity. This will result in a tip splitting type of instability (Fig. 8). The complete theoretical analysis in the parallel channel gives a more subtle result [18]; the lower branch n=0 on Fig.3c is stable and the other branches are unstable. Experimentally however for B values smaller than 1.4 10^{-4} the observed finger is unstable. This can be ascribed to the fact that as B tends to zero the branches become more closely spaced and the natural noise is enough to hide the quantization of the solutions: the finger destabilizes by tip splitting.

For sector-shaped cells the experiments[7] show that for divergent channels the finger destabilizes more and more easily by tip splitting for increasing angles. In contrast, for convergent fingers tip-splitting tends to be inhibited; the finger becomes increasingly stable when the angle is large.

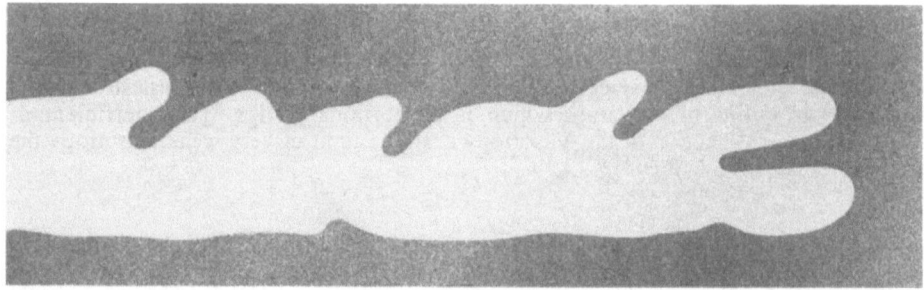

Figure 8. The side branching and tip splitting instabilities of an isotropic Saffman Taylor finger in a linear channel.

There is an intuitive way of understanding this result. On figure 4 are sketched the isobars in front of fingers moving in a divergent and a convergent channel respectively. Near the finger tip these isobars have in both cases a curvature of the same sign as that of the interface. Far away from the finger tip the isobars are arcs of circles centered on the apex of the sector. The point is that in convergent channels the isobars near and far from the finger have curvatures of opposite signs, while in divergent channels they are of the same sign. In convergent channels the pressure gradient is thus enhanced along the cell's axis by a focusing effect which opposes tip splitting. In reverse in divergent cells the gradient of pressure is spread over a large width of the finger tip so that tip splitting becomes easier. More generally this illustrates the role of a mean curvature imposed by the geometry which is also responsible for the shift in the selected λ.

The results of the numerical simulation[20] shown on Fig 5c puts this intuition on a more quantitative basis; it shows that in sector-shaped cells withor $\theta_0 > 0$, the lower level n=0 ceases to exist when it forms a loop with the second level n=1 at a finite value B_l. Though the stability analysis of the solutions in sector shaped cells was not yet done, the experimental results are coherent with the assumption that, as in the linear cells, only the lowest level is stable and the others are unstable by tip splitting. In divergent cells the instability is thus intrinsic; the finger destabilizes when the lowest level ceases to exist. As the value of B at which the loop occurs increases with the angle, so does the value at which the finger destabilizes.

This result is important to analyse the growth of fingers moving at constant velocity in sector shaped cells or in the circular geometry. The observed behaviour shows that the finger adapts "adiabatically" to the evolution of a local parameter B defined in the tip region. This B decreases with the increasing width of the cell. The finger is at first stable and destabilizes by tip splitting when B reaches the limit value B_l. The pattern shown on Fig 6 exhibits several tip-splittings of that type

For fingers growing in a convergent geometry ($\theta_0 < 0$) the lowest level n=0 exists for all values of B, furthermore Fig 5c shows that, at finite values of B, this level is shifted to lower λ more than the others and is thus further apart from them. It is only at smaller values of B that the noise will destabilize the finger; this is coherent with the enhanced stability observed experimentally.

III-3-2 Non isotropic case

In contrast for the non-isotropic fronts (e.g. anomalous Saffman Taylor fingers [25]) the finger tip is defined by the small scale. Decreasing B the tip curvature always remains in the same ratio to the capillary length scale ; thus the tip has no reason to become unstable. The same is not true of the lateral sides. As the scale of ρ moves away from the scale of W the finger's relative width λ decreases and the finger's profile becomes parabolic in a larger and larger region behind the tip. As a result a lateral instability grows (Fig.9a). It is identical in nature to the lateral instability of crystalline dendrites (Fig. 9c).

The growth of an instability advected along a curved front was investigated theoretically by Zel'dovich et al.[72], Pelcé et al. [6,73] and Caroli et al .[74] and numerically by Kessler et al. [75]. They put forward two characteristic processes. Firstly [43,75] the growing wave being advected along a curved front is stretched by a kinematic effect due to the growing tangential velocity. Secondly[74] each region of the front is unstable by a process described by the linear stability analysis of equation (8) or (10). Along the curved fronts the normal velocity decreases constantly away from the tip. As a result there is a continuous shift of the maximum amplification rate towards larger wavelengths. Both effects tend to increase the wavelength away from the tip. The most important point which underlies these arguments is that the side branching is an instability of a convective type. This means that it has a spatial growth rather than a temporal one (for a discussion of absolute and convective instabilities see Huerre et al. [76]). In both the dendrites and the anomalous Saffman Taylor fingers the characteristics of convective instabilities have been observed. Dougherty et al. [47] showed experimentally that the side branches of dendrites result from selective amplification of the noise incident at the tip. As for anomalous fingers Rabaud et al. [25b] gave a sudden local disturbance to the tip of an anomalous finger and showed that it resulted in the growth of a wave packet which, in the finger's frame of reference, was advected away. No wave remained in the front region so that any later disturbance was uncorrelated from an earlier one. This is precisely the characteristic of a convective instability. Furthermore this experiment[25b] permitted the direct observation of the wavelength change along the profile due to both the Zel'dovich stretching and the shift in wavelength selection. A similar experiment was done by Quian et al. [77] applying a heat pulse at the tip of a dendrite.

In convective instabilities noise can be replaced by a periodic forcing of larger amplitude. The instability is thus made strictly periodic at the imposed frequency. This experiment was done by Rabaud et al.[25b]. Periodic side branches (Fig.9b) are obtained when the forcing frequency is in the amplified frequency range. More recently a similar forcing was applied to dendritic growth by Bouissou et al.[78], a case where it is more difficult to modulate the growth. They achieved it by applying a modulated flow around the tip of the growing dendrite and obtained the periodic side branching shown on Figure 9d.

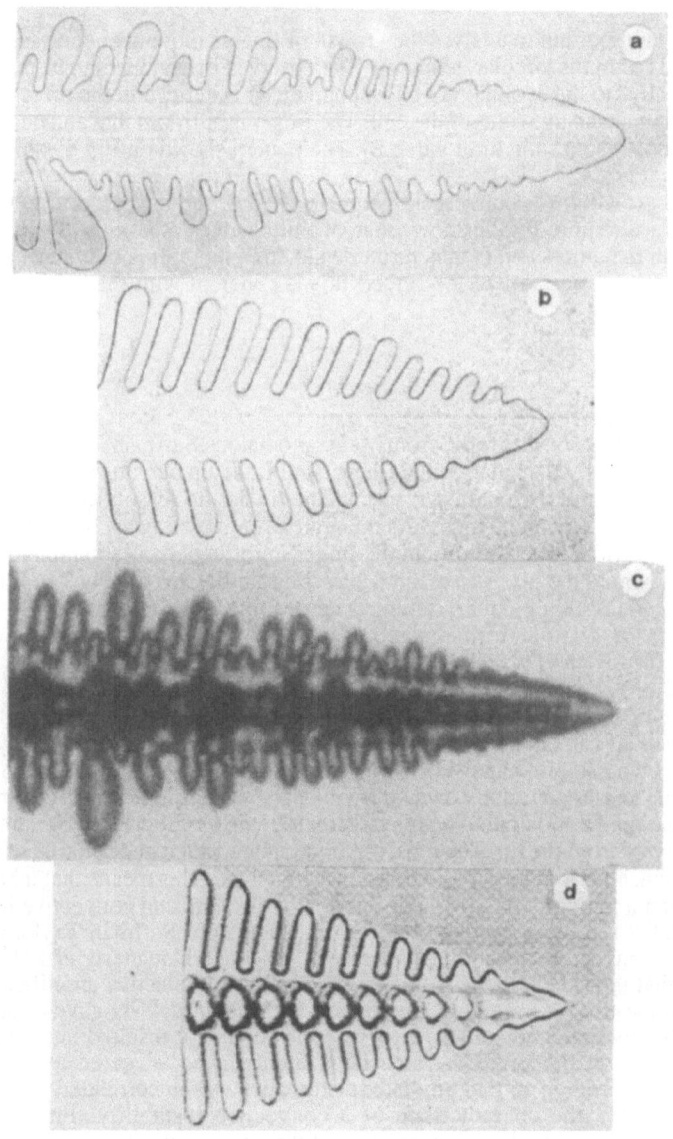

Figure 9. a) The natural instability of an anomalous Saffman Taylor finger grown in a wide channel.
b) An anomalous Saffman Taylor finger forced by a periodic modulation [25b].
c) The natural destabilization of the side of a dendrite.
d) A dendrite forced by a periodic modulation (courtesy P. Bouissou et al [78])

IV The fractal fronts

The question we will address now concerns the structure of very unstable patterns obtained when the two characteristic length scales of the problem (e.g. W and l_c) are very far apart from each other. In particular what are the statistical properties of these patterns and are they somewhat related to the stable curved fronts. Two points are important in this investigation.

-We will limit ourselves here to very unstable patterns obtained in the geometries in which stable smooth fronts are known : the linear and the sector shaped geometries for isotropic growth, the linear and the opened geometry for anisotropic growth.

-We will constantly compare the patterns obtained in Saffman Taylor fingering and crystal growth with those resulting from numerical growth processes of the DLA type. As was discussed in II-3 except for the absence of surface tension, this random growth is also an inverse Stefan problem . Works have been devoted to the comparison of DLA patterns to viscous fingers and dendrites, others to a comparison with the reverse perspective.

Many variants of the DLA model have pointed out these similarities. By introducing specific rules [36] for the sticking of particles on the aggregate it is possible to introduce an equivalent to surface tension. Kadanoff [60] and Liang [61] have shown that it was thus possible to recover the stable isotropic Saffman Taylor finger in a linear channel. It is also possible to obtain anisotropic growth. This is done by either revealing the anisotropy of the underlying lattice by a noise reduction (Nittman and Stanley [62], Kertèsz and Vicsek [63] and Meakin [64-65]) or to introduce anisotropy of a different symmetry by specific rules Eckmann et al.[68]). The resulting patterns have the general structure of dendrites.

In reverse the similarity of the fractal structure of viscous fingering (at very high constraint) with DLA has been a more controversial matter. This fractal structure was observed early in the case of porous cells (Paterson [56], Nittman et al.[80], Jing-Den Chen et al. [81]) or with non-newtonian fluids (Lenormand et al. [82] and Van Damme et al.[83]). The case of pure Saffman Taylor fingers is more decisive because all the characteristics of the process are better known. But the first experiments done in the radial geometry lead to divergent conclusions. Ben Jacob et al. [33] concluded that the pattern reached a space filling shape that they call dense branching morphology (this term is also used to describe regimes of growth in electrodeposition). Rauseo et al. [34] and Couder et al. [35] working in the same geometry concluded to a fractal growth of the DLA type. The DLA-like growth is only obtained in experimental situations as near as possible to the ideal, where wetting by the viscous fluid is perfect, where there is no flexion of the glass plates due to the high applied pressure, and finally where the viscosity contrast between the two fluids is practically infinite. Whenever one of these conditions is not met there is a cross-over to a dense morphology.

IV-1 The isotropic case

IV-1-1 Linear channel

The structure of very unstable fingers in a linear channel was first studied experimentally by Kopf Sill et al.[36]. They measured the average portion of the cell's width occupied by the fingers and the mean width of each of the branches. They found the former to be related to the control parameter by a power law, while the latter was on the scale of the capillary length.

We undertook (Arnéodo et al.[37]) a set of experiments, having in mind to investigate *statistically* the rate of occupation of the channel. For this purpose we performed a series of N identical runs of the same experiment. We sought to reach very small values of B. For dimensional reasons (discussed in ref.(7)), it is best, in order to reduce l_c, to use a cell of very small thickness and an oil of very large viscosity. We thus reached values of B ranging from 10^{-4} to 10^{-6}. A typical pattern is shown on (Fig. 10a). We chose a section of the cell where the pattern had finished its evolution. We built an histogram of the occupancy by air in all the runs. A division by N gave the mean occupancy r in this section. In both experiment and simulation we wanted, after a large number of independent runs, to measure the mean occupancy of each site of the channel. The most extensive analysis was carried out in the DLA case. In a given strip we grew N (N~512) aggregates with the same total number M of

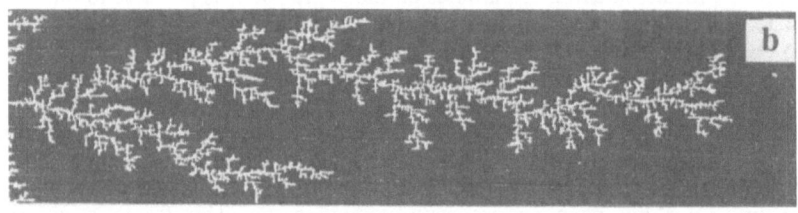

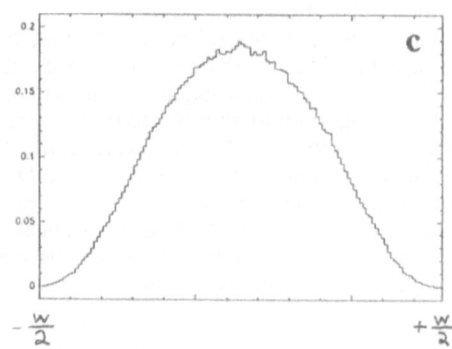

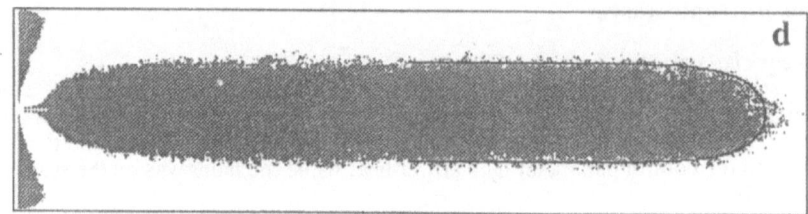

Figure 10. a) A very unstable Saffman Taylor finger.
b) A DLA aggregate grown in a strip.
c) The profile of the mean occupancy rate across the strip.
d) The region of the channel with an occupancy rate above average[37].

particles (Fig. 10b). We then counted for each point of the grid how many times it had been occupied by a particle of an aggregate. This number, divided by N, gave r(x,y), the mean occupancy of this point. In viscous fingering our smallest scale is larger than in DLA so we did not need such a large number of runs ; N ranged from 50 to 120. We limited ourselves to the measurement of the mean occupancy across the cell.

The histogram of the occupancy along the axis of the linear strip obtained from the DLA simulations shows that, except in the initial region and in the tip region, r is independant of x. This means that in the regions where the growth has ceased, the cell translational invariance imposes itself to the occupancy profile. We grew aggregates with M from 1000 to 6000 in strips of W = 32, 64 and 128. Scaled on W, the mean length of the cluster x_t/W is proportional to its mass. The fall-off of r in the tip region, of width Δx_t, corresponds to the active part of each pattern and to the dispersion of the tip position. Δx_t is also scaled on W and increases slowly[16] with x_t. Across the linear cell, all transverse occupancy profiles have a maximum value r_{max} at the center (y=0) and decrease to zero at the walls (y=±W/2). For viscous fingers, the profiles (a sharp step-shaped profile for stable fingers) become smoother with decreasing B and the width of the regions which are never visited (along the walls) reduces. The limiting profile of the histogram obtained for DLA far from the tip (in the region where the evolution is finished) is surprisingly well fitted (Fig. 10c) by

$$r(x,y) = r_{max}\cos^2(\pi y/W) \tag{21}$$

We obtain an average finger shape by seeking for each x the value y_m of y such that

$$y_m = \frac{1}{r_{max}}\int_0^\infty r(x,y)dy \tag{22}$$

The important result is that the profile defined by the $y_m(x)$ has (with some statistical fluctuation) the shape of the Saffman Taylor finger of width $\lambda= 0.5$ (Fig. 10d). In other terms, going from the stable finger to unstable patterns, the occupancy rate becomes smeared out but the stable solution remains underlying.

Two questions arise from this result.

The first is relative to the fractality. What relation is there between the fractal structure of DLA aggregates and the steady averages obtained here? Confined in the cell the mean growth has become translationnaly invariant along the channel. Across the channel the patterns have retained their fractality on all scales between W and l_u (or l_c). By varying either W (in the DLA experiments) or lc (in the Saffman Taylor experiments) we found that the mean occupancy of the cell's width varies as $r_{max} \propto B^{0.19} = (W/l_u)^{-0.38}$. If the pattern is a fractal of dimension d_f its section by a line will have a dimension d_f -1 and a mean density along this line scaling as d_f-2. The experimental exponent -0.38 corresponds to d_f =1.62±0,02, a value in good agreement with the usual dimension of DLA clusters.

The second is relative to the selection of the mean profile. This result is very surprising in that all the arguments that are used in the analysis of the selection of the stable Saffman Taylor finger rely on the stability of the profile and the smoothness of its tip. Therefore they all seem to have lost their validity here. It could result from a mere coincidence if the results summarized below which concern unstable growth in sector shaped cells and unstable anisotropic growth did not make it a well documented general result.

IV-1-2 Sector shaped cells

A similar result is obtained in sector shaped cells. Using the same procedure we measured for each angle the relative angular width $\lambda_m(\theta_0)$ of the profile given by $y_m(x)$ (Fig.11). Comparison with Fig.5c shows that for all convergent cells and for weakly divergent ones, $\lambda_m(\theta_0)$ coincides remarkably well with the limiting width of the stable fingers obtained in these cells at small B. For $\theta_0 > 20°$, the width $\lambda_m(\theta_0)$ follows the trend

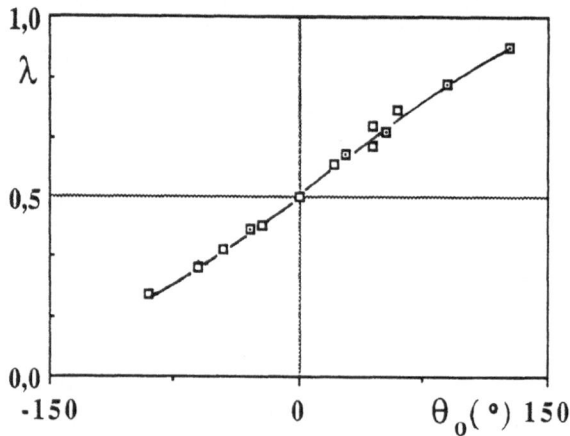

Figure 11. The width of the region of large occupancy in sector shaped cells as a function of the cell's angle
□ Widths obtained averaging unstable Saffman Taylor fingers
▣ Widths obtained averaging DLA aggregates

observed for smaller angles and appears to correspond to the extrapolation at B=0 of the level n=0 which is only observed as a stable solution at finite B (as it disappears due to the loop).

Thus, in all the isotropic cases in which the smooth profiles are known, the mean occupancy profile of very unstable patterns reflects both their shape and their selection .

IV-2   Non isotropic case

IV-2-1 Linear channel, the unstable anomalous Saffman Taylor finger
The next thing to do is to investigate with this averaging procedure the non isotropic case (Couder et al [50] and Arnéodo et al [79]). Here we wish the anisotropy to affect the whole cell. For Saffman-Taylor fingers this is obtained as in Ref. 21 by one of the plates having a periodic square structure of small scale. For DLA the on-lattice procedure has a natural, but weak, anisotropy which can be revealed by a noise reduction procedure [62-65]. In practice we require for a given site of the cluster's border to be visited m times before sticking a particle onto it. Figure 12a shows an aggregate of 2000 particles grown with a value m=5 of the noise reduction parameter. This pattern is markedly narrower than the isotropic pattern shown on Figure 10b. Again 516 aggregates with the same number of particles are grown and the repartition of the mean occupancy in the cell is sought. The transverse profiles differ from those obtained in the isotropic case (relation 21) they are now narrower and sharper : for large m the value of r_{max} becomes independant of W. Profiles obtained with the same m but different W only differ by their lateral extension of their wings (Arnéodo et al [79]). Even though the profiles are different the procedure defined by equation (22) is applied to define regions of maximum occupancy ; Fig.12b shows such a region in the case of m= 3 and W=128.

Again these regions are in the shape of Saffman Taylor fingers but being narrow they appear analogous to the stable anomalous fingers (see § III-2-1). As the profile given by $y_m(x)$ is well fitted by equation 16, we can deduce from the observed λ the tip radius of curvature ρ (relation 17). For a given m the mean profiles obtained in cells of different W have different widths λ but the same ρ. The radius of curvature is naturally scaled on the lattice scale l_u, the coefficient of proportionality depending on the anisotropy (which is fixed by m). We find :

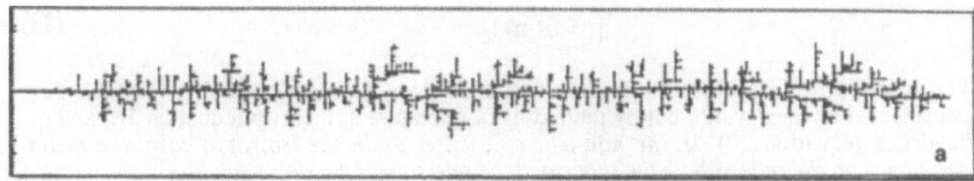

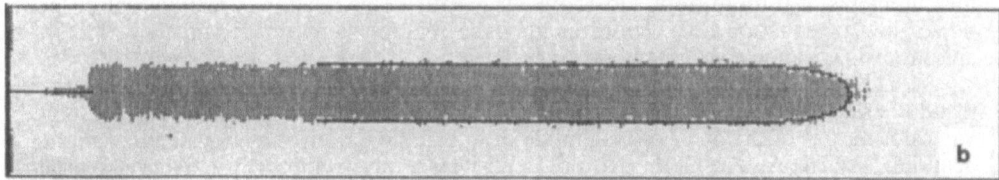

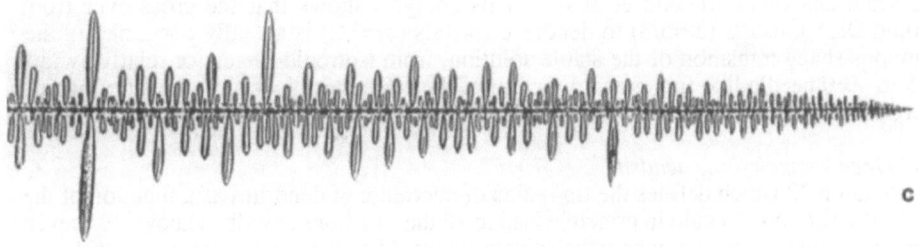

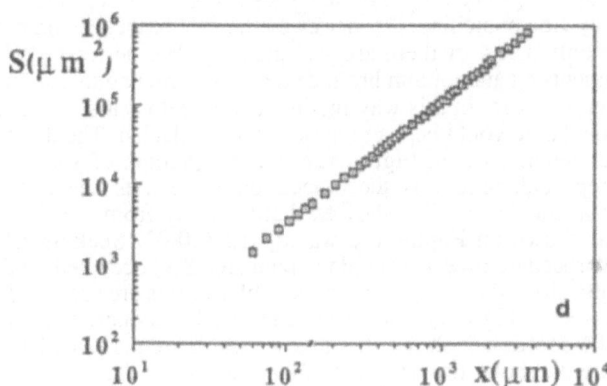

Figure 12. a) An anisotropic DLA cluster obtained with a noise reduction factor m=15 in a channel of width W=64

b) The mean occupancy rate[50] across the channel for anisotropic DLA clusters with m=3

c) The general structure of the large scale growth of a dendrite. The photograph shows a length of 1.25mm of a dendrite with a tip radius of curvature ρ=2.5μm

d) Logarithmic plot of the area of this dendrite [50] as a function of the distance to the tip

$$\rho = \alpha(m) \, l_u \qquad (23)$$

This is precisely the characteristic of the stable anomalous Saffman Taylor fingers which are not selected by their width, but by the radius of curvature at their tip (equation 18) and of dendrites (equation 20). *In the non-isotropic case, as in the isotropic case, the mean occupancy profile reflects both the shape and the selection of the stable solution.*

Inspection of the patterns shows qualitatively that increasing m increases the anisotropy. There is an empirical relation (Arnéodo et al[79]) between α and m : measurements of ρ for m from 2 to 15 give $\alpha \propto m^{-1,5}$. This relation is reminiscent of the power law of equation (20). Unfortunately the relation between m and the effective anisotropy is not known quantitatively.

Finally we can remark that for small values of m the radius of curvature is much larger than l_u. As a result on a given realization of the aggregate the tip is unstable. But for large values of m, ρ is of the order of magnitude of l_u and the tip of the aggregate, as that of a dendrite, is stable. As a result the axis of a dendrite is always compact. The transverse mean occupancy profiles are thus very different from the istropic one (shown on Fig. 10c) because on the axis the maximum occupancy $r_{max} = 1$. Going from a narrow channel to a wide one the profiles will differ by their lateral extension. It is possible to apply to these profiles a finite scale analysis (Arnéodo et al [79]). This analysis shows that the cross-over from isotropic DLA clusters (d_f=5/3) to dendritic fractals (d_f=3/2) is actually contained in the continuous shape transition of the stable solution, from isotropic fingers of relative width λ=0.5 to λ=0 needle like finger as a function of the effective anisotropy (transition which has been discused above in III-2-1).

IV-2-2 Open geometry, the dendrite

Relation 20 which defines the tip radius of curvature of dendrites as a function of the crystalline anisotropy ϵ could in principle lead to all the situations described above. However, to our knowledge, a situation where the anisotropy would be very small and ρ very different from $l_c{}^S$ has not been observed yet in crystal growth. For the crystals usually investigated ϵ is such that ρ is of the order of $l_c{}^S$ and the tip of dendrites is observed as a stable parabola. However it destabilizes by side branching. Further away the side branches compete so that far enough from the tip only a few of them are still growing. We performed (Couder et al.[50]) a series of experiments on ammonium bromide dendrites growing in a solution in the limit of very small Peclet numbers. In this way the interaction between side branches was long range and similar to what it would be, had the field been laplacian. The dendrites were grown in a quasi two dimensional cell. Figure 12c is a photograph of such a dendrite. Though the crystal is compact along its axis, thecompetition between its side branches gives it the structure of a self affine fractal. We showed that far away from the tip the fractal dimension of the crystal shown on Figure 12c was d_f=1.58±0.03. Seeking an average profile we used a simple procedure : we measured the total area S(x) occupied by the pattern from the tip to a distance x along the axis. A logarithmic plot of this area as a function of x shows a power law $S(x) \propto x^{1,5}$ (Fig.12d) : the area occupied by the dendrite is that of dense parabola. Furthermore the coefficient shows that this virtual dense parabola has the same radius of curvature as the parabolic tip. The total area is the same that would have been obtained, had the dendrite remained stable. If we had made a statistic on identical dendrites and applied the rule defined by equation (22) we would also have found the same result. The reason for which the average property is apparent on one single realization is due to the stability of the tip which creates a rate of occupancy $r_{max} = 1$ along the axis of the dendrite (a situation also obtained in DLA for large m). An averaging on many realizations is thus not necessary.

V-Conclusions

Several general conclusions can be drawn from this survey concerning the Laplacian patterns.

The growth in the inverse Stefan problem is unstable. In the experimental or numerical reality there is always a lower limit to the size of the structures emerging from the growth. In the two physical examples that we examined this small length scale can be obtained by the linear analysis of the stability of a plane front. Under this scale the front, stabilized by surface tension, is smooth. We have been interested here in situations with weak surface tension where the length of the front is either slightly larger or much larger than this scale.

Up to now the problems which have been solved are partial Stefan problems in which the growth occurs in cells with fixed boundaries. The successive steps are the following. When the existence of surface tension is neglected possible shapes of curved fronts can be found, depending on the geometry of the cell. In this limit, families of exact analytical solutions were found for two geometries : the linear cells and the sector shaped cells. In the first case the solutions are translationally invariant, in the second they are self similar. Then the selection and stability of these solutions was investigated in the presence of surface tension. The growth of stable curved fronts in the case where the two length scales of the problem are not too far from each other was first examined. *In all these cases, whether the growth is isotropic or not, the actual front is selected amongst the family of solutions found neglecting surface tension.* In the isotropic growth the selection provides a finger defined by its width ; the selection occurs at the cell's width scale. In the non-isotropic case the selection occurs at a scale which is usually of the order of the small scale.

When the two length scales are far from each other the observed patterns are very unstable. The realizations of the patterns resulting from identical experiments are all different and have a fractal structure. It would thus seem that all the characteristics of the smooth front are lost. However investigation of the mean occupancy of the cell when many runs of the same experiment are done show that *the region of mean occupancy reflects both the shape of the corresponding stable solution and its selection.* This result about unstable patterns appears to be general and has been extended to all the investigated cases. It was shown [84] that these results could not be interpreted using the normal mean field equations [51]. Very recently Brenner et al [85] found that average patterns of the type we observed could be found, provided an ad-hoc modification was introduced in the mean field equation. I believe more experimental, numerical and theoretical work is still needed to reach a full understanding of the fractal structures.

Many of the results that I have reported here have been obtained by the collaboration of a group of experimentalists and theorists. It is the place here to express the pleasure I had working with F. Argoul, A. Arnéodo, M. Ben Amar, R. Combescot, G. Grasseau, V. Hakim, J. Maurer, M. Rabaud and H. Thomé.

References

(1) P. G. Saffman and G. I. Taylor, Proc. Roy. Soc. London , Ser. A, **245**, 312 (1958)
(2) R. L. Chuoke, P. Van Meurs and C. Van der Pol, Tr . AIME **216**, 188 (1959)
(3) D. Bensimon, L. P. Kadanoff, S. Liang, B.I. Shraiman and Chao Tang, Rev. Mod. Phys. **58,** 977 (1986)
(4) P. G. Saffman, J. Fluid Mech. **173**, 73 (1986)
(5) G. M. Homsy, Ann. Rev. Fluid Mech. **19**, 271 (1987)
(6) P. Pelcé, *Dynamics of Curved Fronts* (Academic Press, Orlando, 1988)
(7) H. Thomé, M. Rabaud, V. Hakim and Y. Couder, Phys. Fluids **A1**, 224 (1989)
(8) J. Bataille, Revue Inst. Pétrole **23**, 1349 (1968)
(9) L. Paterson, J. Fluid Mech. **113**, 513 (1981)
(10) P. Tabeling, G. Zocchi and A. Libchaber, J. Fluid Mech. **177**, 67 (1987)
(11) D.A. Reinelt, Phys. Fluids, **30**, 2617 (1987)
(12) S. Tanveer, Preprint
(13) J. W. Mc Lean and P. G. Saffman, J. Fluid Mech. **102**, 455 (1981)
(14) J. M. Vanden-Broeck, Phys. of Fluids **26**, 2033 (1983)
(15) (a) R. Combescot, T. Dombre, V. Hakim, Y. Pomeau and A. Pumir, Phys. Rev. Lett. **58**, 2036 (1986) and (b) Phys. Rev. A **37**, 1270 (1988)
(16) D. C. Hong and J. Langer, Phys. Rev. Lett. **56**, 2032 (1986)
(17) B. Shraiman, Phys. Rev. Lett. **56**, 2028 (1986)
(18) (a) D. Bensimon, P. Pelcé, B.I. Shraiman J. Phys. **48**, 2081 (1987)
     (b) S. Tanveer, Phys. of Fluids, **30**, 2318 (1987)

(19) M. Ben Amar, To appear in Phys. Rev. A, in May 1991
(20) (a) M. Ben Amar, V. Hakim, M. Mashaal and Y. Couder, to appear in Phys. of Fluids, July 1991
(b) R. Combescot and M. Ben Amar, To appear in Phys. Rev. Lett.
(21) E. Ben Jacob, R. Godbey, N. D. Goldenfeld, J. Koplik, H. Levine, T. Mueller and L.M. Sander, Phys. Rev. Lett. **55**, 1315 (1985)
(22) Y. Couder, 0. Cardoso, D. Dupuy, P. Tavernier and W. Thom, Europhys.Lett. **2**, 437 (1986)
(23) A. Buka, J. Kertesz and T. Vicsek, Nature **323**, 424 (1986)
(24) S.K. Sarkar and D. Jasnow, Phys. Rev. A **39**, 5299 (1989)
(25) (a) Y. Couder, N. Gerard and M. Rabaud, Phys. Rev. A **34**, 5175 (1986)
(b) M. Rabaud, Y.Couder and N. Gerard, Phys. Rev. A **37**, 935 (1988)
(26) G. Zocchi, B.E. Shaw, A. Libchaber and L.P. Kadanoff, Phys. Rev. A **36**, 1894 (1987)
(27) A.R. Kopf-Sill and G.M.Homsy, Phys. Fluids **30**, 2607 (1987)
(28) A. T. Dorsey and O. Martin, Phys. Rev. A **35**, 3989, (1987)
(29) H. Thomé, R. Combescot and Y. Couder, Phys. Rev. A **41**, 5739 (1990)
(30) D. C. Hong and J. Langer, Phys. Rev. Lett. **56**, 2032 (1986)
(31) D. C. Hong and J. Langer, Phys. Rev. A **36**, 2325 (1987)
(32) R. Combescot and T. Dombre, Rev. A **39**, 3525 (1989)
(33) E. Ben Jacob, G. Deutscher, P. Garik, N. D. Goldenfeld and Y. Lereah, Phys. Rev. Lett. **57**, 1903 (1986)
(34) S. N. Rauseo, P. D. Barnes and J. V. Maher, Phys. Rev. A **35**, 1245 (1987)
(35) Y. Couder, in *Random Fluctuations and Pattern Growth*, edited by H.E. Stanley and N.Ostrowsky (Kluwer Academic Publisher, Dordrecht, 1988)
(36) A.R. Kopf-Sill and G.M. Homsy, Phys. Fluids **31**, 242 (1988) and E. Meiburg and G.M. Homsy, Phys. Fluids **31**, 429 (1988)
(37) A. Arneodo, Y. Couder, G. Grasseau, V. Hakim and M. Rabaud, Phys. Rev. Lett. **63**, 984 (1989)
(38) W.W. Mullins and R.F. Sekerka, J. of App. Phys. **35**, 444 (1964)
(39) J. S. Langer, Rev. of Mod. Phys. **52**, 1 (1980)
(40) C. Huang and M. E. Glicksman, Acta Metall. **29**, 701 (1982)
(41) J. S. Langer, in *Chance and Matter*, edited by J. Souletie, J. Vanimenus and R. Stora (North-Holland, Amsterdam, 1987)
(42) D. A. Kessler, J. Koplik and H. Levine, Advances in Physics **37**, 255 (1988)
(43) G. P. Ivantsov, Dokl. Acad. Nauk. **58**, 567 (1947)
(44) (a) D. Kessler, J. Koplik and H. Levine Phys. Rev. A **33,** 3352 (1986)
(b) D. Kessler, J. Koplik and H. Levine, Phys. Rev. A **34**, 4980 (1986)
(45) B. Caroli, C. Caroli, C. Misbah and B. Roulet, J. de Phys.**48**, 547 (1987)
(46) M. Ben Amar, Phys. Rev. A. **41**, 2080 (1990)
(47) A. Dougherty and J. P. Gollub, Phys. Rev. A. **38**, 3043 (1988).
(48) J. Maurer, P. Bouissou, B. Perrin and P. Tabeling, Europhys. Lett. **6**, 67 (1989)
(49) H. Honjo, S. Ohta and M. Matsushita, J. of the Phys. Soc. Japan **55**, 2487 (1986)
(50) Y. Couder, F. Argoul, A. Arnéodo, J. Maurer and M. Rabaud, Phys. Rev. A **42**, 3499 (1990)
(51) T. Witten and L. M. Sander, Phys. Rev. Lett. **47**, 1400 (1981) and Phys. Rev. B **27**, 5686 (1983)
(52) *Random Fluctuations and Pattern Growth*, edited by H.E. Stanley and N. Ostrowsky (Kluwer Academic Publisher, Dordrecht, 1988)
(53) P. Meakin, in *Phase Transitions and Critical Phenomena*, Vol. 12, Edited by C. Domb and J.L. Lebowitz (Academic Press, Orlando, 1988)
(54) D. A. Kessler, J. Koplik and H. Levine, Advances in Physics **37**, 255 (1988)
(55) T. Vicsek, *Fractal Growth Phenomena* (World Scientific, Singapore, 1989)
(56) L. Paterson, Phys. Rev. Lett. **52**, 1621 (1984)
(57) T. Vicsek, Phys. Rev. Lett. **53**, 2281 (1984)
(58) S. K. Sarkar, Phys. Rev. A **32**, 3114 (1985)
(59) L.M. Sander, P. Ramanlal and E. Ben Jacob, Phys. Rev. A **32**, 3160 (1985)
(60) L.P.Kadanoff, J. of Stat. Phys. **39**, 267 (1985)
(61) S. Liang, Phys. Rev. A **33**, 2663 (1986)
(62) J. Nittmann and H. E. Stanley, Nature **321**, 663 (1986)

(63) J. Kertèsz and T. Vicsek, J. Phys. A **19**, L257 (1986)

(64) P. Meakin, Phys. Rev. A **36**,332 (1987)

(65) P. Meakin, J. Kertesz and T. Viczek, J. Phys. A **21**, 1271 (1988)

(66) H. Kondo, M. Matsushita, and S. Ohnishi. *Proceedings of the first symposium for science on form*, Editor S. Ishizaka, (KTK Scientific publishers, Tokyo, 1986)

(67) R. Julien, M. Kolb and M. Botet, J. de Phys. **45**, 395 (1984)

(68) J.P. Eckmann, P. Meakin, I. Procaccia and R. Zeitak, Phys. Rev.A **29**, 3185 (1989)

(69) B. Shraiman and D. Bensimon, Phys. Rev.A **30**, 2840 (1984)

(70) S. D. Howison, J. Fluid Mech. **167**, 439 (1986)

(71) D. Bensimon and P. Pelcé, Phys. Rev. A **33**, 4477 (1986)

(72) Ya. B. Zel'dovich, A. G. Istratov, N. I. Kidin and V. B. Librovich, Combust. Science Technol. **24**, 1 (1980)

(73) P. Pelcé, Thése d'état, and P. Pelcé and P. Clavin, Europhys. Lett. **3**, 907 (1987)

(74) B. Caroli, C. Caroli and B. Roulet, J. Phys. **48**, 1423 (1987)

(75) D. Kessler and H. Levine, Europhys. Lett. **4**, 215 (1987)

(76) P.Huerre and P.A. Monkewitz, J.Fluid Mech. **159**, 151 (1985)

(77) X. W. Quian and H. Z. Cummins, Phys. Rev. Lett. **64**, 3038 (1990)

(78) P. Bouissou, A. Chiffaudel, B. Perrin and P. Tabeling, Europhys. Lett. **13**, 89 (1990)

(79) A. Arnéodo, F. Argoul, Y. Couder and M. Rabaud, Phys. Rev. Lett. **66**, 2332 (1991)

(80) J. Nittmann, G. Daccord and H. E. Stanley, Nature **314**, 141 (1985), and G. Daccord, J. Nittmann and H. E. Stanley, Phys. Rev. Lett. **56**, 336 (1986)

(81) Jing-Den Chen and D. Wilkinson, Phys. Rev. Lett. **55**, 1892 (1985)

(82) R. Lenormand, E. Touboul and Zarcone, J. Fluid Mech.**189**, 165 (1988)

(83) H.Van Damme, E. Alsac, C. Laroche, and L. Gatineau, Europhys. Lett. **5**, 25 (1988)

(84) V. Hakim, Private communication

(85) E. Brener, H. Levine and Y. Tu, Phys. Rev. Lett. **66,** 1878 (1991)

Simple and Complex Patterns in Coupled Map Lattices

L. Bunimovich †

Institute for Scientific Interchange, 10133, Torino, Italy
†On leave from Institute of Oceanology, 117218, Moscow, USSR

Coupled map lattices appear mainly as a tool to understand some features of space-time motions of nonlinear spatially distributed systems and especially of turbulent flows in fluids. The last ones demonstrate two basic phenomena, i. e. space-time chaos and coherent structures. The general questions that arise in the field are:

1. What does it mean that the motion of a dynamic system is space-time chaotic?
2. What does it mean that in the motion of a dynamic system one faces coherent structures?
3. How could coherent structures appear from chaos?
4. What are the types of coexistence of these phenomena (intermittency)?

Experimentalists have in fact answers to these questions and those are the pictures where all these phenomena appear [CH]. For the theorists the natural way is to try to understand what happens in some relatively simple models. Coupled map lattices serve in fact as such simple (but far non trivial) models. These models were studied numerically by many authors (see, for instance, reviews [CK], [K], where an extensive set of references can be found) and some phenomenology was developed.

The aim of this lectures is to give a short survey of rigorous results obtained on this subject. We want to stress that such results seem to be especially important in this field where the first problem that one meets is to derive reasonable definitions of such rather vague motions as space - time chaos, intermittency and coherent structures.

A coupled map lattice (CML) is a dynamical system with a discrete time that acts in the phase space $\hat{M} = M^{\mathbf{Z}^k}$ where $\mathbf{Z}^k$ is a k-dimensional lattice and M is a (local) phase space that is attributed to each site of the lattice. Hence at each site $i = (i_1, ..., i_k) \in \mathbf{Z}^k$ we have a variable $x_i \in M$. A point of the phase space $M^{\mathbf{Z}^k}$ has the form

$$X = \{x_i\}, \quad i \in \mathbf{Z}^k, \quad x_i \in M$$

A local dynamical system is defined by some map $f: \quad M \to M$.

Chaos, Order, and Patterns, Edited by R. Artuso *et al.*
Plenum Press, New York, 1991

The dynamics on $M^{\mathbf{Z}^k}$ is generated by a map $\Phi : M^{\mathbf{Z}^k} \to M^{\mathbf{Z}^k}$, where Φ is the convolution of two maps

$$F : \quad \hat{M} \to \hat{M}, \quad (Fx)_i = f(x_i), \quad i \in \mathbf{Z}^k \qquad (1)$$

and

$$A : \quad \hat{M} \to \hat{M} \qquad (2)$$

here F corresponds to the independent and simultaneous action of local dynamical system f at all sites of the lattice $\mathbf{Z}^k$ and A defines space interactions. Moreover the group of space translations acts in $M^{\mathbf{Z}^k}$. It is easy to see that the action of this group commutes with the dynamics generated by Φ.

In case when $k = 1$ and A is a diffusive coupling with nearest neighbors interaction Φ has the following form

$$(\Phi x^{(n)})_i = x_i^{(n+1)} = f(x_i^{(n+1)}) + \frac{\epsilon}{2}(f(x_{i-1}^{(n)}) - 2f(x_i^{(n)}) + f(x_{i+1}^{(n)})) \qquad (3)$$

Then the phase space $\hat{M}$ consists of all doubly infinite sequences $\{x_i\}_{-\infty}^{+\infty}$, $x_i \in M$.

Let us first assume that $A = Id$, i.e. there are no spatial interactions. Then $\Phi = F$ and local map f acts independently at each site. Therefore if f is chaotic then the infinite-dimensional dynamical system under consideration obviously exhibits space-time chaos. So the most chaotic motion in CML is generated if the local maps are chaotic and evolve without interactions, because space correlations are identically equal to zero in that case.

It is natural to assume that space-time chaos in such CML preserves if spatial interactions are sufficiently weak. The corresponding statement was proven in [BC]. We shall formulate now this result that allows us to give the definition of space-time chaos (at least for some infinite-dimensional spatially extended dynamical systems, i.e. for CMLs).

Consider a function $f : \quad [0,1] \to R^1$ such that

(1) $f(0) = 0$; $f(1) = d$, where $d \geq 2$ is an integer.
(2) the derivative f' of f satisfies Hölder condition with some positive exponent γ;
(3) $f'(y) \geq \lambda > 1$ for all $y \in [0,1]$.

Let also f be the one-dimensional expanding map $f : \quad x \to \{f(x)\}$, where $\{..\}$ denote the fractional part of a real number. It is known that this class of one-dimensional transformations has the most strong stochastic properties ([CE], [R]), i.e. f has a unique invariant measure which is absolutely continuous. With respect to this measure it is ergodic, mixing and the corresponding time correlation functions decay exponentially fast.

We introduce now a slightly modified spatial interaction. Take a function $\alpha(y)$ on $I = [0,1]$ such that

(i) $\alpha(y) = \epsilon$ for $\delta \leq y \leq 1 - \delta$
(ii) $\alpha(0) = \alpha(1) = 0$

230

(iii) $\alpha'(y) \geq 0$ for $0 \leq y \leq \delta$; $\alpha'(y) \leq 0$ for $1 - \delta \leq y \leq 1$
(iv) $\alpha(y)$ has continuous second derivative.

Take now a space averaging operator A in the form

$$(Ax)_i \ = \ (1 - \alpha(x_i))x_i + \frac{\alpha(x_i)}{2}(x_{i-1} + x_{i+1}) \tag{4}$$

If $\alpha(y) \equiv \epsilon$ then (4) reduces to (3).

The result of [BS] says, that if ϵ and $\max_{y \in [0,1]} |\alpha'(y)|$ are sufficiently small, then on $[M]^{\mathbf{Z}}$ there exists an unique measure μ such that

(1) μ is invariant under Φ and under spaceshift S;
(2) for any integers N_1 and N_2 the induced measure on the space of finite sequences $\{x_i\}$, is absolutely continuous with respect to Lebesgue measure;
(3) the dynamical system on M with the measure μ generated by the group (Φ, S) of space and time translations is mixing.

It is natural to take the last property (3) as the definition of space-time chaos. It means that correlations in the system under consideration decay in time as well as in space, i.e. two sites of the lattice $\mathbf{Z}^1$ sufficiently far apart evolve almost independently.

The result of [BS] on space-time chaos is valid for all expanding one-dimensional transformations $f: [0,1] \rightarrow [0,1]$ such that

$$[0,1] \ = \ \bigcup_{i=1}^{m} [\alpha_i, \alpha_{i+1}], \quad f([\alpha_i, \alpha_{i+1}]) \ = \ [\alpha_i, \alpha_{i+1}]$$

where $m < \infty$. Besides it holds not only for nearest neighbor interactions but for those ones decaying sufficiently fast. For instance, one such condition is given by the relation

$$|(A^{-1}x)_i \ - \ (A^{-1}y)_i| \ < \ const \sum_{j=-\infty}^{j} (const \ \epsilon)^{|i-j|}|x_j \ - \ y_j| \tag{5}$$

Recently V.L.Volevich [V] proved that in the CLM considered in [BS] nonequilibrium initial distributions tend, as time goes to infinity, to the equilibrium Gibbs state μ. We have to mention that the smoothening of spatial interactions via the introduction of the function $\alpha(y)$ in due to special features of one-dimensional transformations with chaotic behavior. (For instance, these are noninvertable ones.) The same result as in [BS] can be proved without introducing this smoothing, if the local map is a smooth uniformly hyperbolic dynamical system and the corresponding CMLs have weak space interactions (see [PS]).

The main point in the proof of these results is the construction of thermodynamic formalism for the CML. The usual thermodynamic formalism which has proved very useful in the theory of dynamical systems with hyperbolic structure reduces the study of such dynamical systems to the investigation of a one-dimensional lattice model of statistical mechanics with fast decaying interactions (see eg. [S], [R]). In such models there are no phase transitions. On the contrary CML with the dynamics defined by

(3) and (4) are naturally described by two-dimensional lattice models of statistical mechanics. In these models one direction corresponds to the space translation while the other corresponds to the time direction generated by the dynamics. From this point the result of [BS] is analogous to absence of phase transitions for high temperatures in statistical mechanics. The high-temperature region corresponds, according to the thermodynamic formalism, to the region of weak space interactions in CML.

Having in mind the analogy with two-dimensional statistical mechanics, in [BS] it was formulated the conjecture that the low-temperature domain of parameters (i.e. $\epsilon \sim 1$) may lead to phase transitions where the emerging phases can be interpreted as coherent structures (CS). The results, that we discuss below, seem to support this conjecture. (We want to stress that some of them were obtained numerically.)

In this context the appearance of new ground states can be interpreted as the arising of coherent structures from chaos in systems with an infinite number of degrees of freedom. It gives the possibility to provide a reasonable definition of coherent structures (at least for some CML). The general opinion which has now emerged considers CS as new types of bifurcations. We discuss below some results obtained on this subject for CML. They were obtained for coupled logistic maps, i.e. for $f(x) = ax(1 - x)$, where $0 \le a \le 4$. If $a = 4$ then this CML could be reduced to one considered in [BS] via the famous von Neumann – Ulam change of variables [V].

First, some computer simulations [BLL] have been perfomed to find new stable states of the system for strong spatial interactions ($\epsilon \sim 1$). We used $a = 3.6$, where the local dynamical system is known to be chaotic one [CE]. It was shown that there exists an interval $(\epsilon_{cr_1}, \epsilon_{cr_2})$, $\epsilon_{cr_1} \approx 0.860$; $\epsilon_{cr_2} \approx 0.884$ where there are two stable states of the lattice. These states are period-two standing waves in space with even and odd points of the lattice fixed at different values. These standing waves are the simplest CS in space for the given value of a. (It could be shown [BLL], [RR] that homogeneous state of the lattice is unstable in that case.) At the value $\epsilon = \epsilon_{cr_2}$ a bifurcation takes place and, instead of the standing waves described before, two new states appear such that the point at odd and even sites oscillate with an amplitude that tends to zero as $\epsilon - \epsilon_{cr_2} \to 0$ (Fig.1). These phenomena have been studied then in some details in [RR]. But before to go to that let's make some general comments on CS (or simple patterns)in coupled map lattices.

First we want to stress that complex patterns (patterns generated by space-time chaos) are rather robust and do not depend on precise analytical expression for the spatial interactions. This follows from [BS], [PS] and was in fact discussed above. On the contrary the shape of coherent structures depend on subtle features of operator A.

As an example let us mention the following simple property of CMLs with diffusive coupling of the type defined in (3). Let $\Phi = \Phi_\epsilon$ has a solution $C_{2,m}$, that is a cycle with period 2 in space and period m in time. (The standing wave mentioned above is a $C_{2,1}$ – cycle.) Then it is easy to verify that $\Phi_{1-\epsilon}$ has a cycle $C_{2,n}$, where $n = m(2m)$ if m is an even (odd) number. Thus knowing a cycle of Φ_ϵ it is possible to construct a new cycle of $\Phi_{1-\epsilon}$. It allows one to design simple solutions of local dynamics which are usually known. Consider a set C_i, $i = 1, ..., N$, of cycles with periods p_i of f. Then take a sequence $\{x = x_i\}$, that is periodic with period N and $x_i, 1 \le i \le N$, is some point of a cycle C_i. It is easy to see that this point belongs to the cycle $C_{N,m}$ of the corresponding CML for $\epsilon = 0$, where m is the least common multiple of the numbers p_i, $i = 1, ..., N$. Then by using the theorem on implicit function one can check if this cycle exists for $\epsilon > 0$ and investigate its stability. Making use of the duality mentioned above one can extract the same information in the vicinity of $\epsilon =$). This

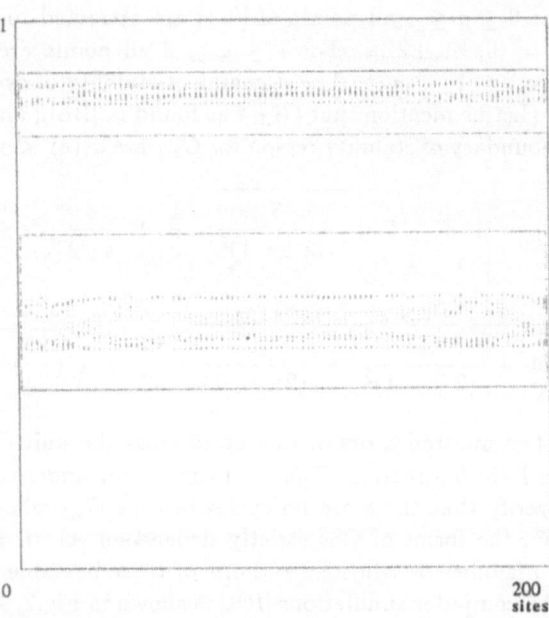

Fig.1. Plot of 20 iterations after a transient of 100 iterations for the logistic map (horizontal axis: $0 \leq i \leq 200$; vertical axis: $0 \leq x \leq 1$), $a = 3.6$, $\epsilon = 0.96$ with random initial conditions for even sites of the lattice in I_1 and odd sites in I_2, where $f(I_1) = I_2$, $f(I_2) = I_1$ and $I_1 \cup I_2$ is the support of absolutely continuous invariant measure for f.

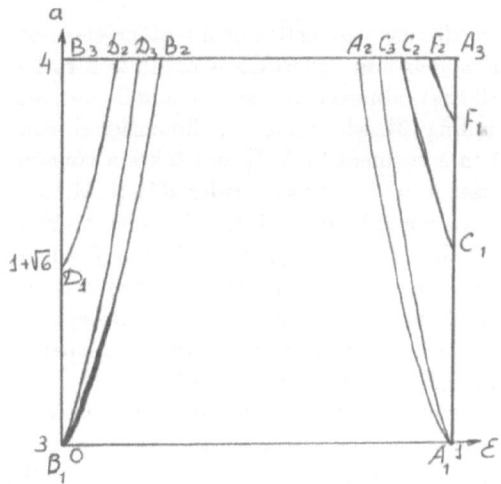

Fig.2. A_1, A_2, A_3, (B_1, B_2, B_3) – the region of existence of the cycle $C_{2,1}$ $(C_{2,2})$, C_1, C_2, C_3, (D_1, D_2, D_3) – the region, where it is stable. F_1, F_2, F_3 – the region of stability of the cycle $C_{3,1}$.

233

approach was derived in [RR] where some analytical and computer – assisted proofs where obtained. For $0 \leq a \leq 1$ all points of $[0,1]$ are attracted to the homogeneous state $x_i = 0$, $i = 0, \pm 1, \pm 2, \dots$. For $1 \leq a \leq 3$ all points are attracted to the homogeneous state $x_i = 1 - a^{-1}$. The domains of stability of cycles $C_{2,1}$ and $C_{2,2}$ are shown in Fig.2. (Let us mention that $C_{2,1}$ was found in [BLL] and $C_{2,2}$ was known earlier [CK].) The boundary of stability region for $C_{2,1}$ are $a_1(\epsilon) < a < a_2(\epsilon)$, where

$$a_1(\epsilon) = 1 + \sqrt{1 + \frac{3}{(2\epsilon - 1)^2}}; \quad \frac{1}{2} + \frac{8}{4\sqrt{2}} \leq \epsilon \leq 1$$

$$a_2(\epsilon) = 1 + \sqrt{1 + \frac{4\epsilon}{(2\epsilon - 1)^2} + \frac{1}{(2\epsilon - 1)}}; \quad \frac{19}{32} + \sqrt{(\frac{19}{32})^2 - \frac{9}{32}} \leq \epsilon \leq 1$$

In the curve $a_2(\epsilon)$ two multiplicators of this cycle cross the unit circle. So it seems that here one meets Hopf bifurcation. This problem is now under consideration.

It is easy to verify that there are no cycles of type $C_{3,1}$ where all three points are different. (Again the forms of CSs strictly depend on A). If $x_1 = x_2 \neq x_3$ then there exist two families of solutions and one of these is stable. The region of its stability, obtained by computer simulations [RR] is shown in Fig.2. In [RR] there were constructed also some cycles of Φ with larger periods.

Computer experiments with two-dimensional lattice of logistic maps give that there exists interval $0 < \epsilon_1 < \epsilon_2 < 1$ where a cycle $C_{2,1}$ is stable [BBL]. This state has the form of chess – board, i.e. $x_{i,j} = b_1(a, \epsilon)$ ($x_{i,j} = b_2(a, \epsilon)$) if $i + j$ is an even (odd) number. The interesting phenomenon observed here is that, for random initial conditions, an asymptotic state does not form kinks (as for small ϵ [CK]) but instead is nearly homogeneous in space (Fig.3). Also in this case we can show the presence of the long order in the system for $\epsilon \sim 1$, but the trivial spatially homogeneous states are unstable (for such a that one-dimensional transformation has no stable cycles).

As a last issue we discuss properties of an other class of CML that seems to be one of the simplest models and nevertheless exhibits a rather rich behavior. The advantage of it is as well that (almost) all results could be derived rigorously. Consider CML with diffusive coupling (3), where the one-dimensional transformation f is linear with the slope $\lambda > 1$ in a segment $[0, \lambda^{-1}]$ and takes a constant value $a < \lambda^{-1}$ in $(\lambda^{-1}, 1]$ (Fig.4). (The case $a = \lambda^{-1}$ was considered in [L,M,R].) The following results were obtained in collaboration with R. Livi, G. Martinez-Mekler and S.Ruffo.

It is easy to show that for one-dimensional map f all points except $y = 0$ tend to the periodic trajectory of the point $y = a$ which is superstable one. It can be proved that for ϵ small enough the asymptotic (in time) motion is restricted to n nonoverlapping segments of the unit interval, where n is the period of the point $y = a$. The widths of these intervals tend to zero as $\epsilon \to 0$.

In what follows we shall consider for simplicity the case $n = 2$. The first critical value appears when these strips begin to overlap. Then we get kinks that are stable in time. By further increasing ϵ one meets travelling waves that correspond to the motion of kinks boundaries. Then there are two possibilities depending whether $\gamma < \frac{1}{2}(a + f(a)) < \lambda^{-1}$ or not. Let us mention first that for $\epsilon = 1$ there is a state similar to that one found in [BLL] (i.e. period-two standing wave). Moreover homogeneous states are present for all values of ϵ. If $\gamma > \lambda^{-1}$ then the only attractive states are clusters of consecutive sites in the lattice states with homogeneous or periodic-two states. Boundaries between these clusters don't move. When $\gamma < \lambda^{-1}$, then these

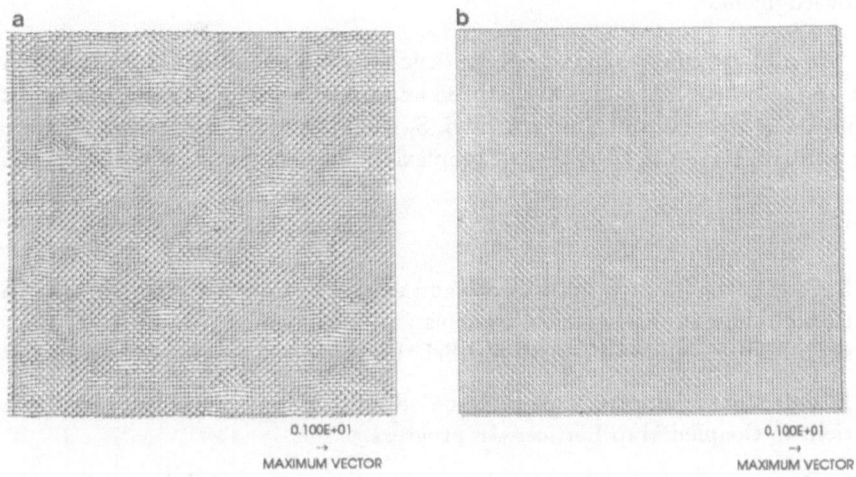

0.100E+01

MAXIMUM VECTOR

0.100E+01

MAXIMUM VECTOR

Fig.3. $a = 3.6$, $\epsilon = 0.980$ with random initial conditions at all sites. Instant states after 499 (a) and 500 (b) iterations. Directions of arrows correspond to the value $x_{i,j}$ at the site (i,j) of the lattice.

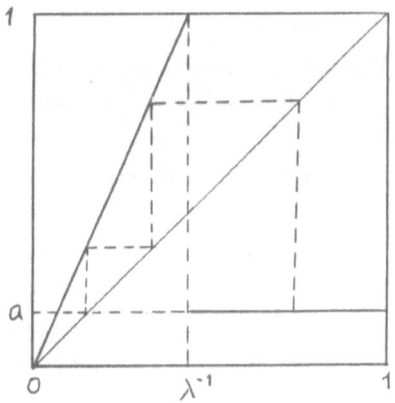

Fig.4. One-dimensional map with superstable **periodic point of period 3**.

235

boundaries can move and travelling waves appear. The difference with the travelling waves discussed above is that those correspond to moving fronts between clusters with homogeneous states. So one can see that variety of different patterns appear even in this simple CML. And the relevant bifurcation parameter is again the strength value of spatial interactions.

Acknowledgments

The author expresses his sinsere gratitude to ISI Foundation for the kind hospitality at Torino, where this paper was written and some mentioned results were obtained. I would like to thank also R. Lima, R. Livi, S. Ruffo, G. Martinez-Mekler, A. Politi and other participants of the Workshop "Complexity and Evolution" for useful discussions.

References

[B] L. Bunimovich. Space-Time Chaos and arising of Coherent Structures in Lattices of Maps In: "Dynamical Systems and Turbulence" (ed. by A. N. Sharcovsky), Kiev, Publ. Math. Inst. Ac. Sci. UkSSR, 1989.

[BBL] L. Battiston, L. Bunimovich, R. Lima. Robust Quasihomogeneous Configurations in Coupled Map Lattices. in progress.

[BLL] L. Bunimovich, A. Lambert, R. Lima. The Emergence of Coherent Structures in Coupled Map Lattices. J. of Statistical Physics, 61, (1990).

[BS] L. A. Bunimovich, Ya. G. Sinai. Space-Time Chaos in the Coupled Map Lattices. Nonlinearity. 1, (1988).

[CE] P. Collet, J.-P. Eckmann. Iterated Maps on the Interval as Dynamical Systems. Birckhäuser, Boston, 1985.

[CH] "New Trends in Nonlinear Dynamics and Pattern Forming Phenomena" (ed. by P. Coullet, P. Huerre), Plenum, NY, 1990.

[CK] J. P. Crutchfield, K. Kaneko. Phenomenology of Spatial-Temporal Chaos. In: "Directions in Chaos". (ed. by Hao Bai-lin), Singapore, World Scientific, 1987, 272-353.

[K] K. Kaneko. Pattern Dynamics In Spatial-Temporal Chaos. Physica D 34 (1986).

[LMR] R. Livi, G. Martinez-Meckler, S. Ruffo. Periodic Orbits and Long Transients in Coupled Map Lattices. Physica D, (1990), to be published.

[RR] A. I. Rakhmanov, N. K. Rakhmanova. On one Dynamical System with Spatial Interactions. Preprint, Keldysh Inst. for Applied Mathem., Moscow, 1990.

[R] D.Ruelle. Thermodynamic Formalism. London, Addison-Wesley, 1978.

[S] Ya. G. Sinai. Gibbs Measure in Ergodic Theory. Russian Mathem. Surveys 27, (1982).

[V] V. L. Volevich. On some Coupled Map Lattices. 1990 to be published.

COUPLED MAP LATTICE

Kunihiko Kaneko

University of Tokyo, Komaba, Meguro-ku, Tokyo 153, JAPAN

1 Introduction

In the first two lectures of the summer school, I have reported the studies of spatiotemporal chaos with the use of coupled map lattices (CML). On the first lecture, I have shown some qualitative results of CMLs.

Since some of the inovating features in CML modelling are not recognized, some questions and answers are listed up here, starting from the most trivial one.

What is CML?

It is a dynamical system with discrete time ("map"), discrete space ("lattice"), and a continuous state. It usually consists of dynamical elements on a lattice interacting ("coupled") among suitably chosen sets of other elements [1-19,24-26,29-33].

Why is CML introduced?

It has been introduced to study spatiotemporal chaos. Spatiotemporal chaos (STC) covers the turbulent phenomena in general, not only in fluids, but also in plasma, solid-state physics, optics, chemical reaction, and so on, and also in biological information processing with nonlinear elements like neural dynamics. STC is high-dimensional chaos whose dimension increases with the system size.

Otto Rössler has introduced the term "hyperchaos" as chaos with the number of positive Lyapunov exponents more than 1. Extending his idea, the above STC, in general, may be described as (hyper)$^\infty$-chaos in space. The simplest modelling for this is to put nonlinear maps on a lattice with some interaction among them. This is the original picture that the author has adopted 9 years ago [1,2].

In other words, it is interesting to examine if it is possible to construct a model for turbulence starting from a direct product of many chaotic systems. This construction means the fusion of Landau's picture of turbulence and Rössler's hyperchaos [16]. Landau's picture is to regard turbulence as a direct product of periodic states (a quasiperiodic state with many incommensurate frequencies). This direct product state, however, is unstable and is locked to a lower-dimensional torus, or attracted to nearby strange attractors [21,2]. On the other hand, a turbulence model as a direct product of chaos ((hyper)$^\infty$chaos) may be structurally stable. This picture has recently been confirmed at least in the fully developed spatiotemporal chaos in CML [18].

How is a CML constructed?

The essence of CML lies in the reductionism in procedure, not in each elementary process. Physicists are often trained to adopt the reductionism in its exact sense, starting from a microscopic level. We do not take such a simple viewpoint. Starting from a suitably coarse-grained description, we introduce the reductionism in a macroscopic level. This procedural reductionism has been appreciated not in physics, but in biology and in artificial intelligence, as a functional module.

Let us now stop this abstract speculation and come back to physics. As an example, assume that you have a phenomenon in fluids, created by a local chaotic process and diffusion. A simple reductionist might start from a microscopic model, like molecular dynamics or lattice gas cellular automata. Some others may believe only in equations in a coarse-grained level like Navier-Stokes equation. In CML we take a different approach from both of the two; reductionism in procedure. We reduce the phenomena into local chaos and diffusion processes. Then we choose a suitable lattice model in a semi-coarse grained level for each process. As the simplest choice we can adopt a local logistic map for chaos, and a discrete Laplacian operator for the diffusion.

The former process is given by

$$x'_n(i) = f(x_n(i)) \tag{1}$$

where $x_n(i)$ is a variable at time n and lattice site i, and $x'_n(i)$ is introduced as the intermediate value. For the logistic map $f(x)$ is chosen to be $1 - ax^2$. The discrete Laplacian operator for diffusion is given by

$$x_{n+1}(i) = (1 - \epsilon)x'_n(i) + \epsilon/2(x'_n(i+1) + x'_n(i-1)) \tag{2}$$

Combining the above two processes, our dynamics is given by

$$x_{n+1}(i) = (1 - \epsilon)f(x_n(i)) + \epsilon/2(f(x_n(i+1)) + f(x_n(i-1))) \tag{3}$$

The above CML has extensively been investigated as a standard model for spatiotemporal chaos. We note that the local chaos and diffusion processes are carried out separately in the above, which is one of the merits of our CML.

What are the merits of CML for the study of spatially extended dynamical systems?

One important feature in CML lies in its semi-macroscopic description. Each variable at a lattice point represents not a microscopic but a semi-macroscopic state. This is in contrast with cellular automata (CA) or spin systems of Ising type. In order to simulate physical problems, CA requires a huge number of cells, while a moderate numer is sufficient for CML.

Another merit in CML is applicability of dynamical systems and statistical physics. In CA, the former application is rather difficult, since there is no continuity in states, while the application of statistical physics is not easy in the description by partial differential equations.

The separation of procedures makes the applications easier, since they are rather straightforward for each process. Perron-Frobenius operator for dynamical systems, for example, can be easily applied to our lattice system [14].

238

How is our CML approach justified?

There is superstition in physics that the nature should be described by continuos time and space. In fact, this is just one of the viewpoints of the nature. We can choose any others if they are more relevant to an object of our study. CML offers such a novel viewpoint:

Nature viewed from qualitative universality

Conventionally, a model equation in physics is believed to have one-to-one correspondence with a phenomenon in concern. For example, if one tries to model turbulence, one often adopts molecular dynamics or a PDE which has a fair basis on the phenomena at the level of microscopic process. If one succeeds in reproducing the phenomena from the model equation, then what can one learn? One dangerous trap in computational physics is that one may be still in the same level as the direct observation of the complex phenomenon itself. What one might obtain is just that the equations are correct or reasonable, without any **understanding** of the complex phenomena. Some of the simulations with Navier-Stokes equation, for example, might have a danger to fall in this trap when it is too successful.

Let us recall a lesson from low-dimensional chaos. If one makes a splendid numerical simulation on sets of equations with velocity field and temperature (e.g., Navier-Stokes with buoyancy and heat), one possibly can get the same oscillatory behavior of rolls as in experiments. Does this success give any better intuition on the origin of this strange oscillation than that a simple chaotic dynamical system provides? After chaos revolution in science, I am sure that most people agree with the negative answer. One of the most important lessons from chaos lies in that it has opened a road to qualitative dynamical viewpoint.

In the studies of CML, we would put forward this point of view. Without bothering the details of phenomenology, we search for a novel qualitative universality class. Through this approach we may succeed in understanding how such phenomenology appears, in what class it is commonly seen, and what the essence of the phenomena is.

This standpoint of CML is essentially based on the universality.

In physics, quantitative universality is justfied with the use of renormalization group (RG), as is seen in critical phenomena, onset of chaos, and field theory. A familiar discrete model, lattice gauge theory, indeed, is justified by the approach to continuum based on RG. Similar justification may be possible for CML, and the search for it is strongly recommended.

Even without this quantitative basis, however, our CML approach can have a basis in a wider context; qualitative universality. The qualitative universality is loosely related with the science as metaphor. This direction of science has been put forward by Prigogine and his group in the study of dissipative structure [23]. To develop further our scope of physics to dynamical systems, this qualitative picture is important. Mathematical foundation of qualitative universality is left for the future.

2 Brief Review of Pattern Dynamcis

In the lecture, I have discussed the following phenomena in CML, which are briefly summarized here. Two topics will be discussed in later sections. For the following phenomenology (I)-(IV), diffusively coupled logistic lattice (eq. (3)) are mainly used, although the phenomena have large universality class for different choices of local dynamics and couplings.

(I) Period-doubling of kinks with scalings [1,5]

If the parameter is chosen so that a single (logistic) map shows the period-doubling bifurcation, corresponding CML exhibits doublings of a kink structure. A kink separates two domains with different phases of oscillations. There appear kinks with smaller and smaller scales, through the period-doubling. The kink structure has self-similarity as is seen in Figs.11-12 of [5]. We note that the self-similarity can be understood as an extension of RG approach for the period-doubling bifurcation [24,25].

(II) transition among pattern dynamics; frozen random pattern, pattern selection with suppression of chaos, and spatiotemporal intermittency transition to fully developed spatiotemporal chaos (FDSTC) [4]

These successive transitions have been intensivley studied in [4]. The frozen random pattern leads to spatial bifurcation. Even if the model is homogeneous in space, attractors can have strong spatial dependence. At some point the motion is fully chaotic, while it is almost periodic with relatively short periods at other points. In the pattern selection, such patterns that suppress chaos are selected as attractors. Among the above transitions of pattern dynamics, spatiotemporal intermittency has extensively been studied in various contexts, which will be discussed in §4.

(III) (Structural) stability of FDSTC sustained by the supertransients [18,19]

In FDSTC, we have not observed any window structures, for almost all initial conditions. This is rather mysterious, since the homogeneous state with the stable cycle for the window is linearly stable also in a coupled system.

Of course, if we start from the vicinity of a homogeneous solution, our system is attracted into the homogeneous state within few time steps. Indeed, there is a set of initial conditions which are attracted into the homogeneous periodic state within few time steps. The volume of this set of initial conditions, however, decreases exponentially with the system size. If the size is not too small, our system enters into FDSTC state for almost all initial conditions.

After a long time step, our state can escape from FDSTC and hit the homogeneous attractor. We have examined if the FDSTC is really the ultimate attractor for our logistic lattice at the parameter for the period-3 window. If the system size is not too large, there is always an escape from a chaotic state to the homogeneous periodic state. This transient time for the escape, however, diverges with the system size exponentially. If the system size is large, thus, it is practically impossible to see the homogeneous attractor. The stability of FDSTC is sustained by this supertransient.

(IV) Two types of supertransients: Type-I: decaying turbulence; the transient length grows with some power to the system size: Type -II: quasistationary transients with exponential (or faster) growth of transients with system size [11,19]

The above supertransient is often seen in spatiotemporal chaos. There are two types of these supertransients.

In type-I transients, strength of chaos decays with time. There is no quasistationarity in the transient regime. In the type-I, the length of transient time increases with some power to the system size. A typical example is seen in the Brownian motion of chaotic defects [4].

In type-II, there is a quasistationary measure in the transient regime. The transient time diverges exponentially or faster with the system size. For type-II, it is almost impossible to see the ultimate attractor if the system size is large. Furthermore, the quasistationarity makes almost impossible to distinguish transients from attractors. It is an important question to examine if real turbulence is an attractor or a transient of this type.

For the logistic lattice at the period-3 window, we have seen successive transitions from type-I to type-II as the coupling ϵ is increased [19].

(V) Soliton turbulence in coupled circle lattice [5]

In lattices of circle maps $f(x) = x + \Omega + (K/2\pi)sin(2\pi x)$, there is a kink structure which propagates with a constant velocity. If the nonlinearity K is small, interactions of kinks make them to pair-annihilate or to pass through.

If the nonlinearity is large, interactions of kinks can create turbulent bursts or a nucleus emitting kinks. The motion is turbulent, but it consists of propagation of kinks and their interactions. This type of phenomena is called soliton turbulence, which is seen in a turbulent system with propagating structures. See [28] for the soliton turbulence in cellular automata.

(VI) Spatial bifurcation in open-flow model [6,3]

To study a system with spatial flow as in a fluid pipe flow, it is useful to introduce spatial asymmetry in our model. The extreme limit is a one-way coupling model given by

$$x_{n+1}(i) = (1 - \epsilon)f(x_n(i)) + \epsilon f(x_n(i)), \tag{4}$$

which has been introduced as a model for the open fluid flow.

The model (4) exhibits the spatial period-doubling in space [6,7] and selective amplification of noise. Dynamical state changes from the fixed point, period-2, period-4, $\cdots$ sucessively as the lattice point goes downflow. After some doublings, our system goes to a turbulent state. Spatial bifurcation to downflow is also studied with the use of Lyapunov exponents and dimension. How the dimension grows or approaches a constant is an important problem for future experimental and theoretical studies.

In an open-flow system, we have to distinguish absolute instability from convective one. If a small perturbation against a reference state grows in a stationary frame, it is called "absolute instability", while if the perturbation grows only in some frames with finite velocities, it is called "convective" one". In the above model, if the coupling is large, our system shows only the convective instability. Only within a certain band of velocity $v_L < v < v_t$, the co-moving Lyapunov exponent (§5 (5))takes a positive value [7].

Aronson et al. [26] have investigated the spatial period-doubling in open flow, using the renormalization group approach [27]. In an absolute instability case, they have calculated the scaling relation of the spatial interval in the period-doubling. If the coupling is larger, the instability which causes the period-doubling is not absolutely but *convectively* unstable. In this case, a small noise is enhanced as it propagates to downflow. The scaling of doubling is involved both with the noise strength and a lattice site.

As the nonlinearity is increased, successive changes among (i) flow of randomly chosen patterns, (ii) flow with selected patterns, (iii) transmission of defects, (iv) spatiotemporal intermittency, and (v) fully developed spatiotemporal chaos are observed.

Our results can be seen in a class of CMLs with other choices of couplings and local dynamics. Change from the frozen random pattern to the pattern selection is commonly seen in a CML with a local map with a 2-band oscillation. Spatiotemporal intermittency is a common route to FDSTC, as will be discussed in §4. Soliton turbulence is seen in general as a route to FDSTC in a system with local propagating structures like kinks. Supertransient is a generic behavior in a system with local topological chaos (without a local chaotic attractor). These examples give a support for our qualitative universality.

3   Spatial Dimension

The spatial dimension plays an importatnt role as in most of statistical mechanical problems.

Extension of our diffusively coupled map lattice to the higher dimension is straightforward. For example, a 2-dimensional logistic lattice is given by [30] (see also [31]).
$$x_{n+1}(i,j) = (1-\epsilon)f(x_n(i,j))+$$

$$\frac{\epsilon}{4}[f(x_n(i+1,j)) + f(x_n(i-1,j)) + f(x_n(i,j+1)) + f(x_n(i,j-1))] \tag{5}$$

where n is a discrete time step and i,j denotes a 2-dimensional lattice point ($i,j = 1,2,\cdots,N$=system size) with a periodic boundary condition.

In a 2-dimensional system with a weak coupling, we have again observed the same transition sequence as in the 1-dimensional lattice ((II) in §2); frozen random pattern, pattern selection, spatiotemporal intermittency transition, and FDSTC [30].

If the coupling is larger ($\epsilon > .35$), however, neither a frozen random pattern nor a pattern selection is observed. This is explained through a surface tension effect [30].

The diffusive coupling tries to destroy the domain boundary between two frozen patterns ("surface tension"). To destroy a domain, the lattice point at its boundary has to pass through an unstable state (e.g., the unstable fixed point of the logistic map), which brings about some barrier. If this barrier is large, the domain is preserved with the pinning of the boundary. The surface tension is estimated to be $\epsilon M^{(1-1/d)}$ for a domain of length M, while the barrier is roughly independent of M, if M is large enough. Thus, a smaller domain has smaller stability for $d > 1$. For a lattice with a dimension ≥ 2, it is expected that there is an upperbound on the coupling strength beyond which any frozen domain pattern loses its stability.

Indeed, even in a 1-dimension lattice, we often encounter with a floating domain [32] if the coupling is very large.

In a higher dimensional lattice, it is expected that the frozen pattern is much harder to be sustained in a short-ranged model, although no detailed simulations are carried out as yet. In an infinite dimensional lattice, i.e., in a mean-field coupling model, we have a novel class of dynamical transitions; **Clustering** [1]. As long as the correlation length is finite, this exact clustering may be seen only in an infinite dimension. It is an open question if there is a critical dimension for the mean-field behavior in our CML.

---

[1]Globally coupled maps have been discussed in the third lecture. The lecture note is not reproduced here. See [17]

4 Spatiotemporal Intermittency Revisited

Transition from an ordered pattern to FDSTC occurs via spatiotemporal intermittency (STI). In STI there are laminar motion and tubulent bursts in spacetime. Since the introduction of STI in 1984 [1], studies on STI have been growing both experimentally and theoretically.

So far, there are two classes of STI. As the zeroth order approximation, the spacetime region can be classified into laminar (L) and bursts (B).

type-I STI [1,10,19]

There is no spontaneous creation of bursts locally. Therefore, in the above 2-state representation for nearest-neighbor interaction, the next state of a site is L if the present state of it and of nearest neighbors are L (symbolically written as LLL → L). The ground state before the onset of STI has a spatially homogeneous state, given by a stable periodic point for the local map. This STI can be attributed as the supertransient in §2 [19].

Possible relationships of type-I STI with directed percolation have been intensively investigated. Qualitatively results are similar, but there seems to exist a quantitative difference [10].

type-II STI [9,4]

There is spontaneous creation of turbulent bursts, at least in a two-state representation. In other words, the probability for $LLL \rightarrow B$ does not vanish. Making k partitions of the original L state, it might be possible to construct a representation with more states as $(L_0, L_1, \cdots, L_k), B$ in which the probability for $L_0 L_0 L_0 \rightarrow B$ vanishes. (Or it might require infinite state partition, $k = \infty$).

This STI is observed with the transition with a spatial pattern. The ground state before the onset of STI has a spatially periodic structure. So far, this type of STI is observed as a transition from local chaos to global chaos.

Statisitical mechanical studies of type-II STI are not developed as yet. One quantitative feature here is the existence of selective-flicker noise; The dynamical form factor $P(k, \omega)$ (power of Fourier transform of the spacetime pattern $x_n(i)$) exhibits the $\omega^{-\beta}$ noise, only for the wavenumber $k \approx k_p$, the wavenumber of selected pattern [4].

Experimental observation of STI

Among the above two types, only the latter has been observed so far. The first experimental investigation on STI is due to Cilibeto and Biggazzi [34]. They have studied Benard convection in an annulus with a large aspect ratio. They have found STI transition (i) from a roll pattern selection to FDSTC. The transition here is (ii) a chaos-chaos transition, i.e.. from local to global chaos; (iii) there is spontaneous creation of turbulent bursts from a laminar region, as far as the reduction to two states is adopted. These three observations clearly show that their STI is type-II. Furthermore, their roll selection before the onset of STI has similarity with our pattern selection in the logistic lattice.

Daviaud, Dubois and Berge [36] have also studied the Benard convection with periodic and fixed boundaries. They have clarified the importance of a spontaneous creation of turbulent states.

Nasuno, Sano, and Sawada [35] have studied 2-dimensional electric convection of liquid crystal. They have found STI as the collapse of selected checkerboard pattern. This again is a chaos-chaos transition and includes spontaneous creation of turbulent states from a laminar region. In the Faraday instability of surface wave, STI has also been observed [37].

In the summer school, Couder [38] has reported a novel discovery of STI in viscous rotating fluid. This STI again appears through the competition of selected patterns, and includes spontaneous creation of turbulent states. These two features are coincident with our type-II STI. Indeed, his spacetime pattern is quite similar to our observation in the logistic latice. One difference, however, lies in that his transition is not chaos-chaos but that from a fixed pattern to chaos. This is a novel feature in the type-II STI.

It is important to study other experimental systems for STI to examine the universality and to search for the possible existence of other types.

5   Quantification of Spatiotemporal Chaos

Some dynamical and statistical quantifiers are introduced and investigated in CML, to characterize the pattern dynamics, flow of information in spacetime, and strength of spatiotemporal chaos. Among those, the following quantifiers are extensively studied:

(1) Traditional power spectrum in space and time [4]; useful to characterize the change of patterns in space-time; power-law-behaviors at the transition are extracted from the spectra.

(2) Distribution of patterns and pattern entropy; useful to characterize the change of pattern dynamics [4].

(3) Lyapunov spectrum and corresponding eigen vectors [1,3,4]: The spectrum changes its shape from step-like (at frozen random pattern), to concave (at intermittency), and then convex one (at FDSTC) as the nonlinearity is increased. The convex shape of spectra is in strong contrast with the spectra for Kuramoto-Sivashinsky equation [39] and for Gledzer model for turbulence [40].

(4) Density of Kolmogorov-Sinai entropy and Lyapunov dimension obtained from (3) [4]: As has been noted [3], the entropy and density are extensive quantifiers. The density gives a size-independent measure for strength of chaos.

(5) Co-moving Lyapunov spectra [7,3,13]: They are the spectra in a Galilean frame with a finite velocity. The spectra are especially important in the convective chaos [7] and also useful in the analysis of the flow of information.

(6) Propagation speed of disturbance [3]: How a small disturbance at one lattice point propagates to other lattice points are related with Lyapunov analysis. Three patterns of propagation; localization, tunneling, and constant spreading are discovered [3]. The speed of propagation is calculated with the positivity of co-moving Lyapunov exponent.

(7) Sub-space-time Lyapunov spectra [18]: They are introduced as the extension of Lyapunov analysis to local space-time patches. They are powerful to distinguish local chaotic regions from laminar ones. The distribution function of these subspace-time Lyapunov exponents clearly characterizes the change of pattern dynamics [18].

(8) Mutual information in spacetime [15]: It is calculated from conditionl probability of variables of two distant points, $x_n(i)$ and $x_{n+\ell}(i+m)$. It is an effective measure to

see the correlation in space and time. In FDSTC, this quantity decays exponentially in time and space.

(9) Co-moving mutual information flow [3]: It is a measure how information flows in space-time even in turbulent media. In soliton turbulence, the propagation of information by solitons can be confirmed through this quantifier. Even in FDSTC, there remains some finite information flow [33].

(10) Dimension density [13] related with (5): Dimension algorithm has been a standard diagnosis technique for low-dimensional chaos. In our system, the density of dimension is important. There are some proposals for the extraction of dimension density from experimental data [13], but practical applications are not developed as yet.

Theoretical formulation for these quantifiers has just been started, and most problems are left for the future. Relationships of Lyapunov exponents with other quantifiers are discussed. The localization of Lyapunov vectors are pointed out in [3] in connection with Anderson localization.

Bunimovich and Sinai [12] have constructed a statistical mechanical formulation for CML. The application of their theory is so far restricted to CMLs with complete hyperbolicity. Another trial for the construction of thermodynamics is the use of Perron-Frobenius operator with self-consistent approximation [14,18]. The applicability of this approach, however, is limited to FDSTC so far. We are still far from thermodynamics of CML.

Acknowledgements

This manuscirpt is partly written during my stay at Institute for Scientific Interchange at Torino. I would like to thank their support and hospitality. I am also grateful to Professor S. P. Kuznetsov for sending me an English translation of [24] by himself. This work is partially supported by a Grant-in-Aid for Scientific Research from the Ministry of Education, Science, and Culture of Japan.

References

[1] K. Kaneko, Prog. Theor. Phys. 72 (1984) 480, 74 (1985) 1033; in *Dynamical Problems in Soliton Systems* (Springer, 1985, ed. S. Takeno) 272-277

[2] K. Kaneko, Ph. D. Thesis, *Collapse of Tori and Genesis of Chaos in Dissipative Systems*, 1983 (enlarged version is published by World Sci. Pub., 1986)

[3] K. Kaneko, Physica 23D (1986) 436

[4] K. Kaneko, Physica 34D (1989) 1; Europhys. Lett. 6 (1988) 193; Phys. Lett. 125 A (1987) 25

[5] J. P. Crutchfield and K. Kaneko, "Phenomenology of Spatiotemporal Chaos", in *Directions in Chaos* (World Scientific, 1987) 272

[6] K. Kaneko, Phys. Lett. 111A (1985) 321

[7] R.J. Deissler and K. Kaneko, Phys. Lett. 119A (1987) 397

[8] R. J. Deissler, Phys. Lett. 120A (1984) 334: I. Waller and R. Kapral, Phys. Rev. 30A (1984) 2047: Y.Oono and S. Puri, Phys. Rev. Lett. 58 (1986)836; T. Bohr and O.B Christensen, Phys. Rev. Lett. 63 (1989) 2161

[9] J. D. Keeler and J. D. Farmer, Physica 23D (1986) 413

[10] H. Chate and P. Manneville, Europhys. Lett. 6 (1988) 59; Physica 32D (1988) 409

[11] J.P. Crutchfield and K. Kaneko, Phys. Rev. Lett. 60 (1988) 2715

[12] L.A. Bunimovich and Ya.G. Sinai, Nonlinearity 1 (1989) 491

[13] G. Mayer-Kress and K. Kaneko, J. Stat. Phys., 54 (1989) 1489

[14] K. Kaneko, Phys. Lett. 139 A (1989) 47; J.M. Houlrik, I. Webman, and M. H. Jensen, Phys. Rev. 41A (1990) 4210

[15] K. Kaneko, "Simulating Physics with Coupled Map Lattices" in *Formation, Dynamics, and Statistics of Patterns*, 1 (ed. K. Kawasaki, A. Onuki, and M. Suzuki, World. Sci. 1990)

[16] K. Kaneko "Climbing Up Dynamical Hierarchy", in *Chaotic Hierarchy* (ed. G. Baier and M. Klien, World. Sci. 1990)

[17] K. Kaneko, Phys. Rev. Lett. 63 (1989)219; Physica 41 D (1990) 137; Phys. Rev. Lett. 65 (1990) 1391

[18] K. Kaneko, Prog. Theor. Phys. Suppl. 99 (1989) 263

[19] K. Kaneko, Phys. Lett. A (1990) in press

[20] L.D. Landau and E.M. Lifshitz, *Fluid Mechanics* (Pergamon, London, 1959) Chapt.3

[21] D. Ruelle and F. Takens, Comm. Math. Phys. 20 (1971) 167; 23 (1971) 343

[22] O.E. Rössler, Phys. Lett. 71 A (1979) 155; in *Lecture Notes in Applied Math.*, 17 (1979) 141-156; Z. Naturforschung 38a (1983) 788

[23] G. Nicolis and I. Prigogine, *Self-Organization in Nonequilibirum Systems*, (John Wiley adnd Sons, 1977)

[24] S.P. Kuznetsov, Radiophysics 29 (1986) 888 (in Russian)

[25] H. Kook, F.H. Ling and G.Schmidt, preprint (1990)

[26] I.S. Aronson, A.V. Gaponov-Grekhov, and M.I. Rabinovich, Physica 33D (1988) 1

[27] M.J. Feigenbaum, J. Stat. Phys. 21 (1979) 669

[28] Y. Aizawa, I. Nishikawa, and K. Kaneko, Physica D, in press (1990)

[29] See M.H. Jensen, Phys. Rev. Lett. 62 (1989) 1361 for an extension of open flow model in 2-dimensional intermittency, with a possible relation with Benard convection

[30] K. Kaneko, Physica 37D (1989) 60

[31] R. Kapral, Phys. Rev. 31A (1985) 3868

[32] P. Grassberger, in one of the lectures given at the summer school (Como, 1990)

[33] T. Schreiber, J. Phys. 23 A (1990) 393

[34] S. Ciliberto and P. Bigazzi, Phys. Rev. Lett. 60 (1988) 286

[35] S. Nasuno, M. Sano, Y. Sawada, in Cooperative Dynamics in Complex Systems, (ed. H. Takayama, Springer, 1989)

[36] F. Daviaud, M. Dubois, P. Berge, Europhys. Lett. 9 (1989) 441

[37] J. Gollub and R. Ramshankar, in *New Perspectives in Turbulence* (eds. S. Orszag and L. Sirovich, Springer)

[38] Y. Couder, in one of the lectures given at the summer school (Como, 1990)

[39] P. Manneville, in *Macroscopic Modelling of Turbulent Flows*, (eds. U. Frisch et al., Lecture Notes in Physics 230; Springer, 1985)

[40] M. Yamada and K. Ohkitani, Phys. Rev. Lett. 60 (1988) 983

GLOBAL IMPLICATIONS OF THE IMPLICIT FUNCTION THEOREM

Kathleen T. Alligood[1]

Department of Mathematics
George Mason University
Fairfax, Virginia 22030

James A. Yorke[1,2]

Institute for Physical Science and Technology
University of Maryland
College Park, Maryland 20742

INTRODUCTION

Sensitivity analysis has long been an important technique in science: How sensitive is the international price equilibrium to the availability of oil? How does the strain on a bridge depend upon the weight carried? Or, how does the incidence of an infectious desease depend upon the fraction of the population that is checked daily? These questions can be answered by determining how solutions of systems of equations depend on a parameter. Does the solution continue as the parameter is varied? If it does, how far through the parameter range does it extend? Formally, let $F: R^{n+1} \rightarrow R^n$ be a C^1 map. (In practice, f is usually defined on a subset of R^{n+1}.) We denote by (x,α) a point in $R^{n+1} = R^n \times R$, where $x \in R^n$ and $\alpha \in R$ (a scalar parameter), and let

$$C = \{(x,\alpha) \in R^{r+1} : f(x,\alpha) = 0\}$$

be the set of zeroes of f. When the implicit function theorem is applied at a point $(\bar{x},\bar{\alpha}) \in C$, one learns about the local structure of C near $(\bar{x},\bar{\alpha})$, and thus how small changes in the parameter α affect the set of zeroes. The conclusion of this theorem, however, says nothing about the

---

[1]Research partially supported by the National Science Foundation.
[2]Research partially supported by the Air Force Office of Scientific Research.

Chaos, Order, and Patterns, Edited by R. Artuso *et al.*
Plenum Press, New York, 1991

set of zeroes globally (i.e., over the entire parameter range). Concentrating on the (connected) component of C that contains $(\bar{x},\bar{\alpha})$, we show that the hypotheses of the implicit function theorem imply certain global facts about this component.

In the first three sections of this paper, we familiarize the reader with some of the definitions, ideas, and applications of continuation theory. In the fourth section, we use these ideas to prove a global extension of the implicit function theorem.

1. THE IMPLICIT FUNCTION THEOREM AND REGULAR VALUES

For $(\bar{x},\bar{\alpha}) \in C$, let $D_{(x,\alpha)}f(\bar{x},\bar{\alpha})$ denote the $(n+1) \times n$ matrix of partial derivatives of f evaluated at $(\bar{x},\bar{\alpha})$, and let $D_x f(\bar{x},\bar{\alpha})$ denote the $n \times n$ matrix of partials of f with respect to the first n variables at $(\bar{x},\bar{\alpha})$. The implicit function theorem says that if $D_x f(\bar{x},\bar{\alpha})$ is nonsingular, then there is an $(n+1)$-dimensional neighborhood V of $(\bar{x},\bar{\alpha})$ such that $C \cap V$ is a 1-dimensional path (in α); i.e., there is an open interval I about $\bar{\alpha}$ in R and a C^1 function $h: I \to V$ such that for $(x,\alpha) \in C \cap V$, $x = h(\alpha)$. Of course, there is no need to emphasize the last coordinate α. In general, if $D_{(x,\alpha)}f(\bar{x},\bar{\alpha})$ has full rank (rank n), then there exists a neighborhood $\tilde{V}$ of $(\bar{x},\bar{\alpha})$ such that $C \cap \tilde{V}$ is a (1-dimensional) path. If, when the kth row ($1 \le k \le n+1$) of $D_{(x,\alpha)}f(\bar{x},\bar{\alpha})$ is deleted, the resulting $n \times n$ matrix is nonsingular, then the path can be parametrized by x_k.

We say that the zero $(\bar{x},\bar{\alpha})$ is _locally_ _continuable_ if $D_{(x,\alpha)}f(\bar{x},\bar{\alpha})$ has full rank. Recall that 0 is a regular value of f if and only if either 0 is not in the image of f or $D_{(x,\alpha)}f(\bar{x},\bar{\alpha})$ has full rank for _every_ $(\bar{x},\bar{\alpha}) \in C$. Hence 0 is a regular value if an only if each point in C is locally continuable. In other words, C is an embedded 1-manifold. In this case, each component of C is homeomorphic to the real line R or the circle S^1 (see, for example, the classification of 1-manifolds in [15]). We illustrate these ideas with a simple example.

Example. Let $f: R^3 \to R^2$ be given by

$$f(x_1,x_2,\alpha) = (x_1+x_2, 8-x_1^2-x_2^2-\alpha).$$

The set C of zeroes of f is shown in Figure 1. For $(\bar{x}_1,\bar{x}_2,\bar{\alpha}) \in C$.

$$D_{(x,\alpha)}f(\bar{x}_1,\bar{x}_2,\bar{\alpha}) = \begin{bmatrix} 1 & -2\bar{x}_1 \\ 1 & -2\bar{x}_2 \\ 0 & -1 \end{bmatrix} .$$

Since $D_{(x_1,x_2)}f(\bar{x}_1,\bar{x}_2,\bar{\alpha})$ is nonsingular for $\bar{x}_1 \neq \bar{x}_2$, the implicit function theorem implies there is a 1-dimensional path that can be parametrized by α through each point in C, except $(0,0,8)$. For example, taking $\bar{x} = (2,-2)$ and $\bar{\alpha} = 0$, we can solve for the solution $x = h(\alpha)$ for α up to 8: namely,

$$x_1 = \left(\frac{8-\alpha}{2}\right)^{\frac{1}{2}}$$

$$x_2 = -\left(\frac{8-\alpha}{2}\right)^{\frac{1}{2}}.$$

Moreover, viewing C as a path of solutions in R^3, the parameter value $\alpha = 4$ is not a termination but rather a turning point of the path. Since $D_{(x_1,\alpha)}f(0,0,8)$ is nonsingular, the path through $(0,0,8)$ can be parametrized by x_2. (It can also be expressed as a function of x_1, since $D_{(x_2,\alpha)}f(0,0,8)$ is nonsingular.) Each point in C is locally continuable; hence, 0 is a regular value.

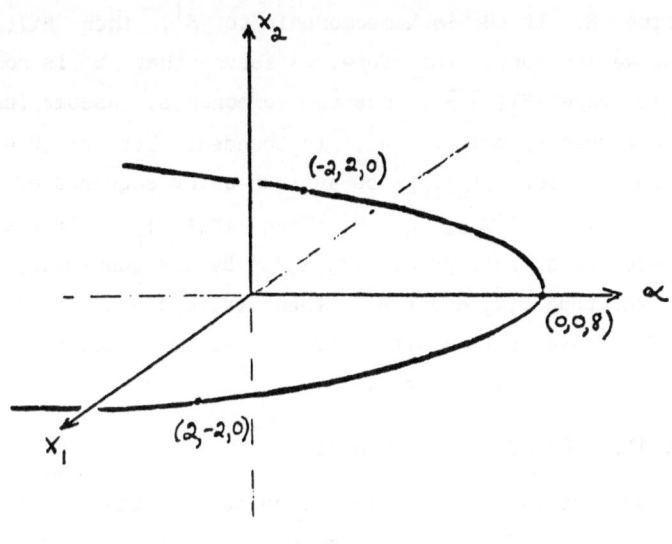

Figure 1

2. A GLOBAL CHARACTERIZATION: THE REGULAR VALUE CASE

Before continuing with the characterization of the zeroes of f, we want to stress here the importance of studying regular values. From looking at the definition of regular value, one might expect that it is a rare coincidence when 0 is a regular value. In fact, Sard's Theorem tells us that this case is very likely. According to Sard's Theorem (see, for example, [15]), the set of regular values of a C^1 map $g: R^m \to R^n$ ($m \geq 1$) includes almost every point (in the sense of Lebesgue measure) in R^n. If, instead of studying f, we look at the map $f - c$, where c is a constant in R^n, then for almost every choice of c, 0 is a regular value. (We will exploit this observation in the application to numerical analysis in Section 3.) Furthermore, understanding the case in which 0 is a regular value will help us understand the general case.

Generally, we canot expect C to be connected, as in the example given. Restricting the analysis, let B be the component of C containing $(\bar{x}, \bar{\alpha})$. In the following proposition we give a global characterization of B, when 0 is a regular value.

Proposition. Suppose 0 is a regular value of f. Then either

    (i) $B \setminus \{(\bar{x}, \bar{\alpha})\}$ is connected, or

    (ii) $B \setminus \{(\bar{x}, \bar{\alpha})\}$ has two components, each of which is unbounded in R^{n+1}.

Proof. Since 0 is a regular value, B is an embedded 1-manifold.

It is also connected and hence is homeomorphic either to the circle S^1 or to the real line R. If B is homeomorphic to S^1, then $B \setminus \{(\bar{x}, \bar{\alpha})\}$ is connected, and we are done. Therefore, we assume that B is homeomorphic to R, in which case $B \setminus \{(\bar{x}, \bar{\alpha})\}$ has two components. Assume further that one of these components, call it B^+, is bounded. Let $h: (0, \infty) \to B^+$ be a homeomorphism, and let $\{t_n\}_{n \in N}$ be an increasing sequence of positive real numbers such that $\lim_{n \to \infty} t_n = \infty$. Then $\{h(t_n)\}_{n \in N}$ is a bounded sequence and hence has a limit point (x_*, α_*). By the continuity of f, $f(x_*, \alpha_*) = 0$ and thus $(x_*, \alpha_*) \in B^+$. Then $(x_*, \alpha_*) = h(t_*)$ for some t_* in $(0, \infty)$, and we have $h(t) \to h(t_*)$ as $t \to \infty$, contradicting the fact that B^+ is an embedded 1-manifold.

3. AN APPLICATION TO NUMERICAL ANALYSIS

One obvious, but important, observation here is that if the component B satisfies case (ii) of the above lemma, and if it is uniformly bounded in x-space for all α, then the path of solutions through $(\bar{x}, \bar{\alpha})$ must have $|\alpha|$ unbounded. In particular, if we follow a path B from a zero of $f(\cdot, 0)$ at $\alpha = 0$) into the right half-space $(\alpha > 0)$, then either B returns to $\alpha = 0$ at a different zero of $f(\cdot, 0)$, or it extends to $\alpha = 1$ at a zero of $f(\cdot, 1)$. This aspect of the lemma has an important application to numerical analysis, which we will briefly describe.

Consider a C^1 map h defined on the n-dimensional disk D^n. By the Brouwer Fixed Point Theorem, h has at least one fixed point. Observe that the set of fixed points of h equals the set of zeroes of the map $k = h - id$, (i.e., $k(x) = h(x) - x$, for $x \in D^n$). Among the many proofs of this theorem, those that depend on the continuation ideas discussed so far yield a particularly useful approach. M. Hirsch [11] was the first to publish a proof based on regular values. (Aee also the exposition in [15].) His was a proof by contradiction, but it was shown in [12] that a variant of the proof provided a computer implementable method for finding fixed points. We will describe a related technique which is the basis of the Homotopy Continuation Method, as originally developed in [6] and [14] for finding fixed points (or zeroes). B. C. Eaves and H. Scarf [9] used similar ideas for simplicial (piecewise linear) maps.

We start with a map of the disk for which the fixed points are known—say, for example, the constant map $h_a(x) = a$, for all $x \in D^n$.

Construct a homotopy f from $k_a = h_a - id$ to the map in question, $k = h - id$:

$$f(x,\alpha) = (1-\alpha)k_a(x) + \alpha k(x)$$

for $x \in D^n$, $0 \le \alpha \le 1$. If 0 is a regular value of f, then each component of the set C of zeroes in $D^n \times I$ is homeomorphic to a circle or to a line segment with each of two endpoints on the boundary of $D^n \times I$. See Figure 2. We require that f has exactly one zero at $\alpha = 0$ (as in the case $f(\cdot,0) = k_a$) and that C does not intersect the boundary of D^n. Then the component of C which begins with the unique zero at $\alpha = 0$ must have as its other end point a zero of f at $\alpha = 1$, (i.e., a fixed point of h). This path can be followed numerically. While sophisticated techniques of differential equations can be used to follow the path, we describe the simplest approach. For $(\bar{x},\bar{\alpha}) \in C$, if $D_x f(\bar{x},\bar{\alpha})$ is nonsingular, then $(\bar{x},\bar{\alpha})$ is locally continuable in α, and a new zero n-dimensional x-space by Newton's method; i.e., begin Newton's method in D^n at $(\bar{x},\bar{\alpha} \pm \epsilon)$. We do not necessarily always move monotonically in α, since the path can turn around. (Of course, eventually it ends at $\alpha = 1$.) At points where the path changes direction in α, a different variable assumes the role played by α. As long as $D_{(x,\alpha)} f(\bar{x},\bar{\alpha})$ has full rank, the path through $(\bar{x},\bar{\alpha})$ can be followed by incrementing an appropriate coordinate and correcting (again using Newton's Method) in the n-dimensional hyperspace perpendicular to this coordinate axis. The advantage of this method over, say, Newton's method is that we stay close to the solution set at all times. Larger steps can be taken by calculating the tangent vectors to the curve and incrementing in that direction.

Unfortunately, this discussion is hypothetical, since we have no way of determining whether 0 is a regular value of f. For the particular problem of finding fixed points of D^n, however, we can virtually overcome this difficulty with the help of Sard's Theorem. Recall that fixed points of h are found by numerically following solutions of

$$f_a(x,\alpha) = (1-\alpha)(a-x) + \alpha(h(x)-x) = 0 \qquad (3.1)$$

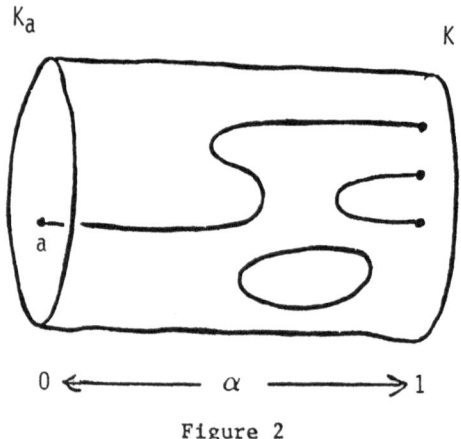

K_a K

$$0 \longleftarrow \alpha \longrightarrow 1$$

Figure 2

for some constant $a \in D^n$, from $\alpha = 0$ to $\alpha = 1$. Equivalently, solving (3.1) for a, we have

$$a = \frac{x-\alpha h(x)}{1-\alpha}, \qquad \text{for} \quad \alpha \in [0,1).$$

Let $\xi: D^n \times [0,1) \to D^n$ be given by

$$\xi(x,\alpha) = \frac{x-\alpha h(x)}{1-\alpha}.$$

Since ξ is C^1 and maps onto D^n, by Sard's Theorem, almost every choice of $a \in D^n$ is a regular value of ξ. If $a \in D_n$ is a regular value of ξ, then the set $C_a = \{(x,\alpha) \in R^{n+1}: \xi(x,\alpha) = a\}$ is a 1-manifold. But C_a can equivalently be characterized as the set of zeroes of f_a (given by (3.1)) for $\alpha \in [0,1)$. Thus for almost every choice of constant $a \in D^n$, a smooth path Γ_a of zeroes of f_a begins with $x = a$ at $\alpha = 0$ and continues through increasing α. As discussed previously, the path Γ_a eventually leaves each bounded set $D^n \times [0,\alpha]$, for $0 < \alpha < 1$, and thus must pass through each $\alpha < 1$. Hence Γ_α has limit points at $\alpha = 1$. By the continuity of f_a, all the limit points of Γ_a at $\alpha = 1$ are zeroes of $f(\cdot,1)$ and thus fixed points of h. In other words, the fixed points of h can be approximated as closely as desired by following Γ_a as close as one wants to $\alpha = 1$. It was shown in [13] that if f_a is (real) analytic in $n + 1$ variables, then Γ_a has exactly one limit points at $\alpha = 1$, giving more rapid convergence. In [12] fixed point of 20 dimensional problems were found by this method with 1 second of computer time.

For other problems, more subtle theory is required to implement the technique in the general c^1 case. We direct interested readers to an excellent survey (complete with algorithms) by E. Allgower and R. Georg in [3]. As a final note on this subject, we should mention that the only practical method of finding numerically all solutions of a system of n complex polynomials in n variables utilizes the Homotopy Continuation Method [8] (see also, the exposition in [7]).

4. A GLOBAL CHARACTERIZATION: THE GENERAL CASE

Returning to the original problem, we again ask: what is the behavior of solutions for large changes in the parameter? The proposition in Sec. 2 gives an answer in the special case where 0 is a regular value. If 0 is not a regular value, i.e., if $D_{(x,\alpha)}f(\bar{x},\bar{\alpha})$ does not have full rank for some $(\bar{x},\bar{\alpha}) \in C$, then we have no nice characterization of the topology of C near $(\bar{x},\bar{\alpha})$. We might imagine something like the solution set shown in Figure 3 occuring. Surprisingly, however, the type of configuration shown there cannot occur for a c^1 map, as we will show. Suppose again that B is the component of C containing $(\bar{x},\bar{\alpha})$. Following the characterization of the proposition, we say that the zero $(\bar{x},\bar{\alpha})$ is <u>globally</u> <u>continuable</u> if either

    (i) $B\backslash\{(\bar{x},\bar{\alpha})\}$ is connected, or

    (ii) $B\backslash\{(\bar{x},\bar{\alpha})\}$ has two components, each of which is unbounded in R^{n+1}.

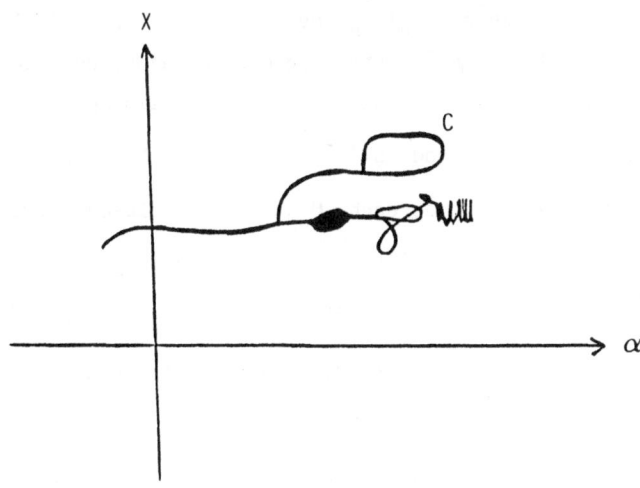

Figure 3

The following theorem shows that the local continuability of $(\bar{x},\bar{\alpha})$ in fact implies its global continuability. This result is a special case of a theorem originally proved in [2]. We give a new, simpler proof. This proof depends on Sard's Theorem and some point-set topology which relates the structure of the component B to that of nearby smooth curves which map to regular values of f.

Theorem. Let $f: R^n \times R \to R^n$ be a C^1 map, and let $(\bar{x},\bar{\alpha})$ be a point in $R^n \times R$ such that $f(\bar{x},\bar{\alpha}) = 0$. If $D_x f(\bar{x},\bar{\alpha})$ is nonsingular then $(\bar{x},\bar{\alpha})$ is globablly continuable.

The theorem is proved with the aid of the topological properties given in the following lemma. (Proofs are left as exercises.)

Lemma. Let $\{K_n\}_{n \in N}$ be a nested family of connected, closed sets such that $K = \bigcap_{n \in N} K_n$ is non-empty. Then

(i) either K is connected, or each component of K is unbounded;

(ii) if K_n is unbounded, for each n, then K is unbounded.

Proof of Theorem. Since $D_x f(\bar{x},\bar{\alpha})$ is non-singular, and f is C^1, there is an n-dimensional neighborhood U of $(\bar{x},\bar{\alpha})$ in x-space such that for $(x,\bar{\alpha}) \in U$, $D_x f(x,\bar{\alpha})$ is non-singular. In addition, by the Inverse Function Theorem, there is an n-dimensional neighborhood V of $(\bar{x},\bar{\alpha})$ in x-space such that $f|_V$ is a homeomorphism. Let $W = U \cap V$.

We assume, for the sake of contradiction, that $(\bar{x},\bar{\alpha})$ is not globally continuable; i.e., that $B\backslash\{(\bar{x},\bar{\alpha})\}$ is not connected and that one component B^+ of $B\backslash\{(\bar{x},\bar{\alpha})\}$ is bounded. (In the following, for sets P and Q, we write $P\backslash Q$ for $P\backslash P \cap Q$.) For $\epsilon > 0$, let

$$N_\epsilon = \{(x,\alpha): |f(x,\alpha)| \leq \epsilon\}.$$

Choose a decreasing sequence $\{\epsilon_n\}_{n \in N}$ such that $\epsilon_n > 0$ and $\lim_{n \to \infty} \epsilon_n = 0$. Notice that $\{N_{\epsilon_n}\}_{n \in N}$ is a nested sequence of sets, and that $C = \bigcap_{n \in N} N_{\epsilon_n}$. Let M_{ϵ_n} be the component of N_{ϵ_n} containing B, for each $n \in N$. Then $B = \bigcap_{n \in N} M_{\epsilon_n}$, and $B\backslash\{(\bar{x},\bar{\alpha})\} = \bigcap_{n \in N} (M_{\epsilon_n}\backslash W)$. We are almost set up to use the lemma, except that each $M_{\epsilon_n}\backslash M$ is not closed. Thus we modify the construction by endowing $R^{n+1}\backslash W$ with the "arc-length" metric. Namely, for $x_1, x_2 \in R^{n+1}\backslash W$, let d, the distance from x_1 to x_2, be given by

$$d(x_1,x_2) = \min\{\text{arc length of } \gamma, \text{ where } \gamma \text{ is an arc}$$
$$\text{from } x_1 \text{ to } x_2 \text{ and } \gamma \subset R^{n+1}\backslash W\}.$$

When we complete this metric space, the closure $M_{\epsilon_n}\backslash M$ denoted $(\overline{M_{\epsilon_n}\backslash W})_d$ remains disconnected if $M_{\epsilon_n}\backslash W$ is disconnected. That is, $(\overline{M_{\epsilon_n}\backslash W})_d$ and

$M_{\varepsilon_n} \backslash W$ are either both connected or both not connected, for each $n \in N$. Hence, since $B \backslash \{(\bar{x},\bar{\alpha})\}$ is not connected, $\overline{(B \backslash \{(\bar{x},\bar{\alpha})\})}_d$ is not connected and $\overline{(B^+)}_d$ is one of its components. In addition, since B^+ s bounded, $\overline{(B^+)}_d$ is bounded. Observe that $\overline{(B \backslash \{(\bar{x},\bar{\alpha})\})}_d = \bigcap_{n \in N} \overline{(M_{\varepsilon_n} \backslash W)}_d$ By the lemma, there exists $j \in N$ such that $\overline{(M_{\varepsilon_j} \backslash W)}_d$ and therefore $M_{\varepsilon_j} \backslash W$, is not connected. For $\varepsilon < \varepsilon_j$, let Q_ε be the component of $M_\varepsilon \backslash W$ containing B^+. Then $\overline{(B^+)}_d = \bigcap_{n > j} \overline{(Q_{\varepsilon_n})}_d$. Again, by the lemma, there exists $k \in N$, $k > j$, such that $\overline{(Q_{\varepsilon_k})}_d$, and therefore Q_{ε_k}, is bounded.

Since f is a homeomorphism on W, by Sard's Theorem we can choose a regular value ν in $f(W)$ such that $|\nu| < \varepsilon_k$. Let $(\bar{y},\bar{\alpha})$ be the unique pre-image of ν in $M_{\varepsilon_k} \cap W$. In addition, let

$$\Gamma = \{(x,\alpha): f(x,\alpha) = \nu\},$$

and let β be the component of Γ containing $(\bar{y},\bar{\alpha})$. Notice that β is contained in M_{ε_k}. By the proposition, either $B \backslash \{(\bar{y},\bar{\alpha})\}$ is connected, or each component is unbounded. Since $M_{\varepsilon_k} \backslash W$ is not connected and $B \backslash \{(\bar{y},\bar{\alpha})\}$ is contained in $M_{\varepsilon_k} \backslash W$, $\beta \backslash \{(\bar{y},\bar{\alpha})\}$ is not connected. Let β^+ be the component of $\beta \backslash \{(\bar{y},\bar{\alpha})\}$ in Q_{ε_k}. Then β^+ is bounded, contradicting the proposition. Therefore, $(\bar{x},\bar{\alpha})$ is globally continuable. $\square$

We have given some basic ideas involved in continuation theory. For underlying topological considerations, [1] and [4] investigate when continuation methods are applicable. Aside from the application given here (in Section 3), continuation ideas have considerable ramifications in the fields of ordinary differential equations (see, e.g., [5]) and partial differential equations (see, e.g., [16]), with applications to biology (see, e.g., [10]), economics (see, e.g., [17]), and numerical engineering (see, e.g., [18]).

REFERENCES

1. J.Alexander, Topological theory of an embedding method, in Continuation Methods, H. Wacker, ed., Academic Press N.Y., 1978: proceedings of a symposium at the University of Linz, Austria, 1977; pp. 37-69.
2. J. Alexander and J. Yorke, The implicit function theorem and the global methods of cohomology, J. Funct. Anal. 21 (1976), 330-339.
3. E. Allgower and K. Georg, Simplicial and continuation methods for approximating fixed points and solutions to systems of equations, SIAM Rev. 22 (1980), No. 1, 28-85.
4. K. Alligood, Topological conditions for the continuation of fixed points, in Fixed Point Theory, E. Fadell and G. Fournier, ed.,

points, in Fixed Point Theory, E. Fadell and G. Fournier, ed., Springer Lecture Notes in Math #886 (1981).

5. K. Alligood and J. Yorke, Families of periodic orbits: virtual periods and global continuation, J. Differential Equations, 55 (1984), 59-71.

6. S. M. Chow, J. Mallet-Paret, and J. Yorke, Finding zeroes of maps: homotopy methods that are constructive with probability one, Math Comp. 32 (1978), 887-889.

7. S. N. Chow, J. Mallet-Paret, and J. Yorke, A homotopy method for locating all zeroes of a system of polynomials, in Functional Differential Equations and Approximation of Fixed Points, H.O. Petgen and H. O. Walther, ed., Springer Lecture Notes in Math #730 (1979), pp. 228-237.

8. F. J. Drexler, A homotopy method for the calculation of all zero-dimensional polynomial ideals, in Continuation Methods, H. Wacker, ed., op. cit., pp. 69-93.

9. B. C. Eaves and H. Scarf, The solution of systems of piecewise linear equations, Math Oper. Res. 1 (1976), 1-27.

10 H. Hethcote and J. Yorke, Gonorrhea Transmission Dynamics and Control, Springer Lecture Notes in Biomathematics, #56, 1984.

11. M. Hirsh, A proof of the nonretractability of a cell onto its boundary, Proc. Amer. Math. Soc. 14 (1963), 364-365.

12. R. B. Kellogg, T. Y. Li, and J. Yorke, A constructive proof of the Brouwer fixed-point theorem and computational results, SIAM J. Numer. Anal., vol. 13, No. 4 (1976), 473-483.

13. T. Y. Li, J. Mallet-Paret, and J. Yorke, Regularity results for real analytic homotopics, Numer. Math., 46 (1985), 43-50.

14. T. Y. Li and J. Yorke, Path following approaches for solving nonlinear equations: homotopy continuous Newton projection, in Functional Difrential Equations and Approximation of Fixed Points, H. O. Peitgen and H. O. Walther, ed., op. cit., 228-237.

15. J. Milnor, Topology from the Differentiable Viewpoint, University Press of Virginia, Charlottesville, 1965.

16. P. Rabinowitz, Some global results for non-linear eigenvalue problems, J. Funct. Anal. 7 (1971), 487-513.

17. S. Smale, Convergent processes of price adjustment and global Newton methods, J. Math. Econom. 3 (1976), 1-14.

18. L. Watson, C. Y. Wang, and T. Y. Li, The elliptic porus slider--a homotopy method, J. Appl. Mechanics 45 (1978), 435-436.

UNFOLDING COMPLEXITY AND MODELLING

ASYMPTOTIC SCALING BEHAVIOR

R. Badii*, M. Finardi* and G. Broggi•

* Paul Scherrer Institut, 5232 Villigen, Switzerland
• Physik-Institut der Universität, Winterthurerstr. 190, 8057 Zurich, Switzerland

ABSTRACT

We discuss the problem of quantifying complexity in the framework of a hierarchical modelling of physical systems. The analysis is first performed on the set of all symbolic sequences which label non-empty regions in phase-space. A dynamical process represented by a shift map is associated with each admissible doubly-infinite sequence. The "grammatical" rules governing it are unfolded by using variable-length prefix-free codewords and described by means of allowed transitions on a "logic" tree. The derived model is employed to make predictions about the scaling behaviour of the system's observables at each level of resolution. The complexity of the system, relative to the unfolding scheme, is evaluated through a generalization of the information gain by comparing prediction and observation. Rapidly converging estimates of thermodynamic averages can be obtained from the logic tree in the general incomplete-folding case using a transfer-matrix technique, related to the theory of scaling functions.

1 INTRODUCTION

A large variety of physical processes (propagation of topological defects in fluids, formation of neural networks, life) exhibits features that are clearly distinct from both (space- or time-) periodicity and complete disorder [1]. A high degree of organization can be observed at different resolution scales, which cannot be easily traced back to the microscopic dynamical rules governing the system's evolution. Mathematical models like cellular automata [2], spin glasses [3] and nonlinear dynamical systems [4] are able to produce hierarchies of apparently irregular (but actually strongly structured) spatio-temporal patterns as the result of the repeated application of some basic scheme of either sequential or parallel type. A meaningful characterization of such behaviour cannot be obtained by evaluating entropy-like quantities, but rather by discovering the nature of the generating rules and their mutual relations, steps that correspond to providing an interpretation and a description of the dynamics [5].

The investigation can be carried out efficiently by means of symbolic (discrete) methods, in such a way that the system's topology can be ordered hierarchically on a logic tree. The vertices of the latter are labelled by symbol-sequences S, the length of which increases with the level. This initial model (a bare frame) can be dressed by considering the relationships among the values $Q(S)$ of a physical observable associated to each vertex S of the tree, thus adding metric features to the description. The meaning to be attributed to the transitions (branches) on the tree depends on the chosen interpretation: it may refer either to sequential mechanisms (concatenations of sequences) or to parallel substitutions (Montecarlo updates of spins, cellular-automaton rewriting rules, etc.) or to more general length-increasing transformations on symbolic strings. For continuous systems, it is necessary to perform a preliminary partition of phase-space and to discretize time, in order to obtain a symbolic encoding of the orbits [6]. Eventually, in such a case, the model obtained as sketched above can be complemented by coordinate-dependent mappings (one for each vertex at a given tree-level) in order to describe and predict not only the scaling behaviour of macroscopic observables in the infinite resolution limit (scaling dynamics [7]) but also the actual time-dynamics.

The accuracy of the model is tested, at each level of resolution l, by making predictions about the composition (values of the observable Q) of the tree at the next level. The complexity of the system, relative to the unfolding scheme, is evaluated through a generalization of the information gain by comparing prediction and observation [5]. Systems which present, at increasingly finer levels of resolution, properties that cannot be accurately predicted from those measured at coarser scales are accordingly classified as complex. The predictors used in this process can be cast into the form of transfer matrices acting on symbol-sequences (spin configurations), the coefficients of which represent conditional expectations for the values assumed by $Q(S)$ for each sequence S and are terms of generalized scaling functions [7]. The largest available memory extent is always considered. As a consequence, rapidly converging estimates of thermodynamic averages [8] involving the quantity $Q(S)$ can be obtained from the logic tree, whenever the complexity (referred to the asymptotic scaling of Q) is small.

2 SYMBOLIC DYNAMICS AND HIERARCHICAL MODELLING

When the data are not directly available in the form of a symbolic pattern (such as a spin or a DNA chain) it is necessary to discretize them before starting the analysis. Let the state of the system be represented by a point x in a d-dimensional phase-space X and assume that the time evolution is governed by the discrete-time ($n \in Z$) dynamical law $\mathbf{F} : X \to X$ (e.g., a Poincaré map): that is,

$$x_{n+1} = \mathbf{F}(x_n) \,. \tag{1}$$

The successive values x_n, for all n, constitute the trajectory of the (nonlinear) dynamical system (1). Although the equations of motion are completely deterministic, the time behaviour, depending on the parameter values which specify the function $\mathbf{F}$, may exhibit stochastic properties in close analogy with random processes [4]. Indeed, the points x_n can be interpreted as the set of admissible states for a system in thermodynamical equilibrium as well [8]. The discretization is obtained by partitioning the phase-space X into elements Δ_i, each labelled by a symbol s_i from an alphabet $A_0 = \{0, 1, \ldots, r-1\}$ ($r < \infty$): $X = \cup_{i=0}^{r-1} \Delta_i$. For simplicity, the domains Δ_i are considered as disjoint. Every element, in turn, can be split into several subsets which are identified by "words" w obtained by concatenating the label s_i of the parent set with symbols belonging to the same alphabet A_0. A generic sequence $S = s_1 s_2 \ldots s_n$, of length (number of symbols) $|S| = n$, represents a microstate of the system at the n-th level of resolution, characterized by the value of some local observable $Q(S)$, like the probability $P(S)$ of the sequence itself, or the length $|S|$, or the energy $E(S)$ for a spin system. Successive refinements of the original partition (or "covering") $\Delta = \{\Delta_0, \ldots, \Delta_{r-1}\}$ consist therefore of collections of non-intersecting phase-space domains labelled by increasingly longer sequences. For a generic statistical mechanical ensemble represented by a set of points in X, there is no unique prescription for the choice of the partition and of its refinements. Each domain, in principle, can be subdivided into an arbitrary number of subsets and this need not occur simultaneously for all Δ_is. In the case of a dynamical system, instead, the dynamics itself provides a "natural" refinement of the covering. In fact, a trajectory $\{x_1, x_2, \ldots, x_n\}$ can be associated to the symbolic sequence $s_1 s_2 \ldots s_n$ such that $x_i \in \Delta_{s_i}$, for all times $i = 1, \ldots, n$ [6]: that is, $x_1 \in \cap_{i=1}^{n} \mathbf{F}^{-i+1}(\Delta_{s_i})$, where $\mathbf{F}^i$ denotes the $|i|$-th forward (backward for $i < 0$) iterate of the map $\mathbf{F}$. Hence, all points belonging to an element with label $S = s_1 s_2 \ldots s_n$, at time $i = 1$, will produce the same symbols in the next $n - 1$ steps under the action of the dynamical law. A partition Δ which yields, under $\mathbf{F}$, a one-to-one correspondence $\psi : X \to \Sigma$ between points in phase-space X and infinitely long symbolic sequences in sequence-space Σ is called generating (i.e., "self-refining")[6]. The dynamics in Σ is given by a (left) shift homeomorphism $\hat{\sigma}$: $\hat{\sigma}(s_n) = s_{n+1}$.

Every allowed sequence S corresponds to a succession of enlargements in a non-empty region of phase-space X. The structure of the latter is reflected in the set $\mathcal{L}$ (the "language") of all admissible sequences S. In particular, we concentrate on the properties of the set Ω of "non-wandering" points[4]: a point x is non-wandering for the map $\mathbf{F}$ if, for any neighbourhood U of x, there exists an arbitrarily large $n > 0$ such that $\mathbf{F}^n(U) \cap U \neq \emptyset$. Points in Ω are especially relevant in the study of the long-term behaviour of the system and, because of the correspondence with symbolic sequences, of the high-resolution structure of phase-space. Due to the recurrence properties of Ω, a domain $\Delta_i \subset \Omega$ is "connected" by the dynamics with itself and with all other elements which are intersected by its images $\mathbf{F}^n(\Delta_i)$, for all n. For simplicity, we assume that Δ represents a covering of Ω ($\Omega \subseteq \Delta$) and that each element Δ_i contains non-wandering points (in addition to "transient" ones). The symbolic label S of a generic subset Δ_S contains subsequences common to those of the labels of all domains visited by the orbit leading back to Δ_S. If the motion is ergodic in Ω, an infinitely long sequence S contains all information about the structure in Ω: the study of S is then equivalent to that of the language $\mathcal{L}$ (the collection of all finite-length sequences, corresponding to the ensemble Δ and to all its refinements). The same remark is applicable to the case of a generic statistical-mechanical system, provided that the sequence S corresponds to an equilibrium state, representative of the ensemble of all configurations reachable during the time evolution under the effect of the thermal fluctuations (e.g., during a Montecarlo simulation). The ergodicity, in the former case, is caused by the impossibility of determining the exact position x within a partition element Δ_S and by the nonlinearity of the map (1) which expands Δ_S along the unstable directions, so that the image $\mathbf{F}(\Delta_S)$ can intersect more than one domain, when the control parameters have been chosen in the chaotic regime[4]. Hence, the result s_n of an observation at time n is not uniquely determined by the initial value s_1. It is then clear that the symbolic trajectories are unpredictable and can be regarded as random. In the following, we consider a stationary, one-dimensional, symbolic signal $S_0 = s_1 s_2 \ldots$ of length $|S_0| \gg 1$, such as a spin configuration or the output of a nonlinear dynamical system, indifferently. The methods of investigation to be discussed in the next two sections, in fact, are not restricted to continuous physical systems with a phase-space representation, but are common to other disciplines like information theory (data compression and transmission), computer science (formal languages), theoretical biology (DNA chains)[1]. The stationarity guarantees the existence of translation-invariant probabilities, which renders a statistical analysis feasible (some non-stationary signals may also be studied, after a suitable renormalization). Non-ergodic (e.g., periodic) signals will also be studied with the same approach, as well as phenomena at the border of chaos such as the period-doubling (PD)[7] and the quasiperiodic (QP)[4] transitions to chaos.

The action of the (generally unknown) shift $\hat{\sigma}$ on S_0 (a $\hat{\sigma}$-invariant subset in sequence-space Σ) can be modelled, topologically, by means of a sequence-to-sequence transition (Markov) matrix $\mathbf{M}$ with elements M_{ij} equal to one (zero) if the concatenation $S_i S_j$ is allowed (forbidden)[6]. Each row and each column of $\mathbf{M}$ contains at least a 1 (the size of $\mathbf{M}$ will be finite only if S_0 can be decomposed in terms of a finite set of subsequences S_i: see next section). A state S_i is recurrent if there is a chain of states $S_i S_{j_1} \ldots S_{j_n} S_i$ ($n \geq 0$) leading from S_i back to S_i: the corresponding phase-space element Δ_{S_i} contains a periodic point belonging to an orbit of length $|S_{j_1} \ldots S_{j_n} S_i|$. The topological entropy $K(0)$ is defined as the rate of growth of the number $N(n)$ of admissible sequences of length n for $n \to \infty$ ($N(n) \sim \exp[K(0)n]$) and is an upper bound to the analogous rate for n-periodic points (for other properties of topological Markov chains, see[6]). In general, not all concatenations are allowed (some states S are not accessible by the system) and $N(n)$ is not maximal ($N(n) < r^n$). In nonlinear dynamics this is equivalent to saying that the folding of phase-space is incomplete. Although a signal containing all possible r^n subsequences has a larger entropy (i.e., is "more random") than an "incomplete" one, it is described by a simpler physical model (a matrix M of size r with elements $M_{ij} = 1$, $\forall\, i,j$). Notice, in fact, that a physical approach to modelling requires to describe the set of all signals (subsequences of S_0) generated by the system rather than to reproduce exactly (in most cases sequentially) an individual symbolic trajectory by constructing a specific algorithm (as it is done, e.g., in computer science [9,6]). The hierarchical partitioning of phase-space illustrated above suggests to represent the allowed sequences on a tree, in order of "importance". To this aim, the signal S_0 is decomposed into a succession of "primitive" words [5] w_1, w_2, ..., using a suitable decoding method (see section 4), which are associated to the level-1 vertices of the tree. Concatenations of l primitives are represented at level l ($l = 1, \ldots, \infty$). All branches leaving vertex S point to the allowed continuations Sw_j, Sw_k, ..., of sequence S: if the signal is aperiodic, there are branching vertices (i.e., strings which can be extended in more than one way). Such trees describe the dynamics as a sequence of "events" w_i, occurring with measurable probabilities according to unknown rules (given by the dynamical law (1), through the shift $\hat{\sigma}$): therefore, they are equivalent to generalized Markov models, the order of which depends on the average sequence-length per level [5]

$$\nu \equiv \lim_{l \to \infty} \sum_{\text{level } l} |S| P(S)/l \,. \tag{2}$$

The model can be used to predict the behaviour of an observable $Q(S)$ along each allowed tree-path $S = w_1 w_2 \ldots w_l$ (scaling dynamics [7]), for $l \to \infty$ (see section 6). The primitives should be chosen in such a way that the most compact description of the symbolic dynamics is obtained: their number may be either finite or not, depending on the nature of the signal. The structure of the tree reveals the properties of the language $\mathcal{L}$ and is of guidance in the design of higher-level models, whenever the predictive power of the present one is not sufficient. In the next two sections, various properties of symbolic languages will be illustrated and an explicit decoding procedure will be presented.

3 LANGUAGES AND COMPLEXITY

The most severe obstacle in the quest for efficient unfolding (decoding) algorithms is represented by the unknown nature of the mechanism underlying the formation of the observed signal. Several families of languages have been identified in connection with computational problems [10]. Every formal language has a model counterpart in a discrete automaton which may possess explicit stochastic features. A classification based on the size of the memory required by the models can be made [10,11,12], although the features of the various languages are, in part, overlapping and no meaningful "linear" ordering exists. From a physical point of view, the characterization of symbolic signals should concern asymptotic properties such as the scaling behaviour of "thermodynamical" quantities [8], possibly in the presence of long-range correlations, disregarding details with an irrelevant statistical weight or which are not robust under perturbations of the system.

Mainly, physical processes act either in a sequential or in a parallel way (or in both, if the dynamics is explicitly spatio-temporal). In the former case, given a string $S = s_1 s_1 \ldots s_n$, its next extension is given by the concatenation Sw of S with a word $w = s'_1 s'_2 \ldots$, chosen from an ensemble according to some rule. A parallel rewriting scheme (a "grammar" [10,11]), instead, substitutes each symbol s_i in the initial string S (the "axiom") with a word, also drawn from some list (possibly at random). The particular substitution may depend on the symbol s_i and on a number of its neighbours and need not be performed simultaneously over the whole string S (the so-called "terminal" symbols are left unchanged). The infinite iteration of these procedures yields the language $\mathcal{L}$. Mixed sequential-parallel schemes can be constructed: for example, the image of S may be $Sw\bar{S}$, where w is a generic word and $\bar{S}$ the complement of S with respect to some symmetry operation (e.g., $\bar{A} = T$ and $\bar{C} = G$ for DNA). Cellular automata [2] provide an example of a parallel algorithm.

The analysis of the signal, necessarily performed sequentially, does not allow one to understand the nature of the underlying process univocally and it is especially arduous when long-range correlations are present. In general, parallel rewriting rules are expected to produce signals which are most difficult to describe by means of sequential models. An interpretation scheme (sequential, parallel, deterministic or not) must first be chosen, on the basis of physical considerations. The detailed dynamical rules can then be discovered and related to each other to obtain a description of the system, within the chosen scheme.

The main families of formal languages identified in computer science are the following:

Type 0: unrestricted (or recursively-enumerable) languages;

Type 1: context-sensitive languages;

Type 2: context-free languages;

Type 3: regular languages.

These four classes form the Chomsky hierarchy[10], listed above in decreasing order of generality: type-i languages properly include type-$(i+1)$ languages. The corresponding topological models (i.e., language-recognizing machines) are, respectively, universal computers (Turing machines), linear-bounded, pushdown and finite automata. They are characterized by the size of the required memory (indefinitely large, proportional to input word-length, stack memory with a fixed number of elements, no memory). These automata, however, do not carry information about the occurrence probabilities $P(S)$ of the words S in the input signal. Therefore, they are not suited for a full description of physical systems. Moreover, there are many other computational models which do not fit into this ordering scheme[11]. Nevertheless, a few generally valid considerations about complexity can be made with this classification in mind. First of all, the task of understanding the system's dynamics involves a (possibly approximate) language-recognition phase [13] (identification of code-words, decision about sequentiality or parallelism of the generating rules). Secondly, a topologically correct model for that language (e.g., a tree) should be constructed and "dressed" with metric properties (probabilities of sequences, values of a generic observable $Q(S)$). Predictors for the scaling behaviour of the observables can then be designed and included in the model. Finally, the accuracy achieved in this way constitutes a measure of the complexity C associated with the task of describing the system, within the capabilities of the chosen interpretation scheme[5] (see section 6). Obviously, low-level models (machines) do not recognize higher-level languages (both topologically and metrically) and the resulting complexity is high (possibly infinite). This fact is illustrated in figure 1, where complexity (yet undefined) is schematically plotted as a function of the model size (possibly just the size growth-rate with the approximation level l). The optimal description is obtained with the model of minimal extension which yields a vanishing complexity value (in the infinite-resolution limit $l \to \infty$).

The simplest symbolic signals are represented by the subshifts of finite type: they are closed $\hat{\sigma}$-invariant subsets of Σ (the space of all doubly-infinite sequences on a finite alphabet) which are characterized by a finite number of forbidden words. The transition matrix $\mathbf{M}$ defined in the previous section has then finite size. A wider class of subshifts is given by the sofic systems [14]: the typical example is the language in which the forbidden words can be written as $0(11)^n 0$, $\forall\, n \geq 0$, where w^n indicates the n-th consecutive repetition of w. Although in these cases an infinity of prohibitions occurs, a finite automaton suffices for the definition of the language, as for the previous class. The output of a nonlinear dynamical system, such as the logistic map $x_{n+1} = 1 - a x_n^2$ (with $s_n = [1 + \mathrm{sgn}(x_n)]/2$) in the chaotic regime is a regular language only at special parameter values (for which a Markov partition exists [4]): in the generic case, it is not. At the band-merging points [15] it is sofic [12]. At the period-doubling accumulation point and at the golden-mean quasiperiodic transition to chaos, the dynamics is most efficiently described by the infinite iteration of the parallel transformations (D0L grammars [11]) $(0,1) \to (01,10)$ and $(0,1) \to (1,01)$, respectively [16]. In Ref. [17] it has been shown that such schemes can be seen as an extension of context-free Chomsky grammars. Descriptions of them which are based on regular-language interpretations [18,19,20] require an infinite amount of memory and yield positive complexity (possibly infinite, depending on the definition).

The most common complexity measures considered in computer science are based on space-time requirements of algorithms designed to reproduce exactly a given input string and not to unfold the unknown dynamics governing the set of all admissible signals. Moreover, these quantities reduce, for most physical systems, to a measure of randomness, rather than of complexity [18]. More recent studies have been addressed to finding definitions which are relevant in a physical context [18]. A minimal set of necessary properties for a "strong" measure of complexity, about which there is quite a general agreement [5,12,18,21], emerges from these investigations (although it is fully satisfied only by the definition proposed in [5]):

1. The numerical value of complexity is relative to the observer's ability: hence, it cannot be given by an absolute quantity. However, the definition must have a universal character (possibly within a certain class of physical systems), although the investigation procedure may be rather specialized.

2. Complexity lies between order (which is defined by a finite set of prescriptions) and disorder (to be interpreted as fully uncorrelated randomness, also finitely specified, and not as chaos, since the latter is a highly structured process) and vanishes at the extrema.

3. Complexity is not extensive: the complexity $C(\cup_j A^{(j)})$ of a union of objects $A^{(j)}$ (patterns generated by one or more systems) is not given by the sum $\sum_j C(A^{(j)})$ of the single complexities but rather satisfies $C(\cup_j A^{(j)}) \leq \sup_j C(A^{(j)})$ [22].

4. Similarly, complexity cannot be increased by taking direct products of independent signals: $C(\otimes_j A^{(j)}) = \sup_j C(A^{(j)})$ (the direct product $t_1 t_2 \ldots$ of the two sequences $s_1 s_2 \ldots$ and $s'_1 s'_2 \ldots$, with s_i and $s'_i \in \{0,\ldots,r-1\}$, being defined by $t_i = r \cdot s_i + s'_i \in \{0,\ldots,r^2\}$).

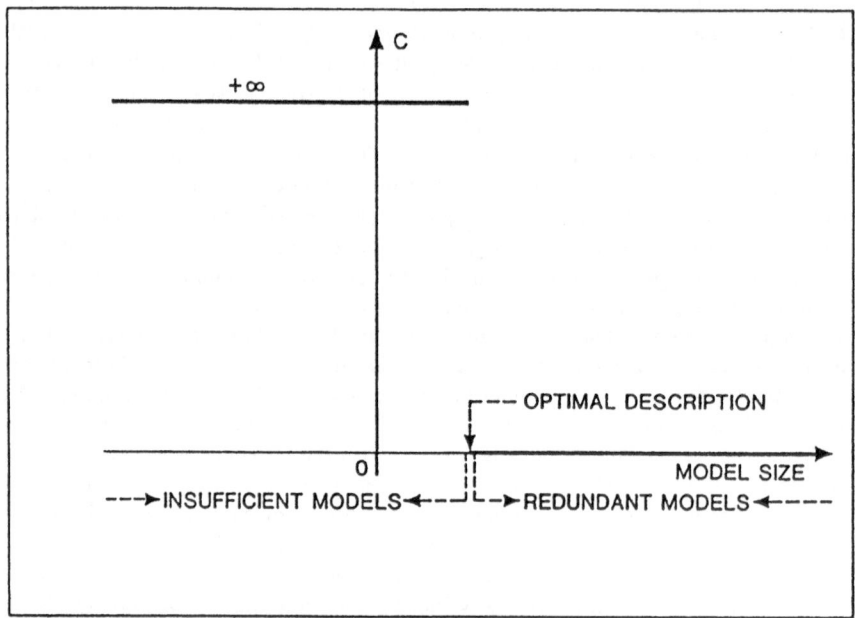

Figure 1. Schematic plot of the complexity C (accuracy of the description) versus the model size or size growth-rate (represented in arbitrary relative units).

These conditions are the "axiomatic" counterpart of the definition of complexity as a measure of the accuracy achieved by the model inferred from the analysis of a given input signal [5]. All other definitions, except for that of Ref. [21], are based on absolute (mostly, entropy-like) quantities, in contrast with point 1. Many others yield positive values for periodic and/or delta-correlated sequences [18,19]. Borderline phenomena like PD and QP have also been attributed infinite (finite) complexity [18,19,20], in terms of the size (growth-rate) of the associated recognizing (regular-language) automata. The measure proposed in Ref. [5] classifies all these signals as simple ($C = 0$), using lowest-order predictors in connection with a logic tree. In the latter two cases, the complexity C is not identically zero but vanishes as $1/N(l)$, where $N(l)$ is the number of orbits allocated at the l-th level of the tree (see section 6).

In order to discriminate further among all existing signals, it has been suggested [23] to introduce a succession of increasingly strong definitions of complexity, the weakest one being the topological entropy $K(0)$, which distinguishes periodic and quasiperiodic motion from chaos or randomness. The second indicator considered in [23] is just the convergence rate of the topological complexity C_0 of Ref. [5] (see section 6). None of these two quantities however satisfies all four "axiomatic" properties listed above. Finally, it may be interesting to evaluate the "computational cost" \$ of the whole unfolding-modelling program in terms of the product between the accuracy (finite-resolution complexity) $C(n)$ achieved after consideration of all sequences with length $|S| \leq n$ and the corresponding machine size $L(n)$. If $\$(n) \equiv C(n)L(n)$ converges to zero, when $n \to \infty$, a finite $L(\infty)$ indicates that the system is unconditionally simple, whereas an infinite $L(\infty)$ is a symptom that the model may be redundant (if not even optimal). Finite or infinite \$ with $C = \lim_{n\to\infty} C(n) = 0$ are even stronger indications of a model-redundancy, although optimality could not be excluded. Viceversa, positive values of C always characterize an insufficient model. The opportunity of adopting a more "expensive" unfolding scheme can then be evaluated on the basis of the values $\$(n)$ and $C(n)$ versus n: a larger size $L(n)$, in the same class of models, is justified if the consequent decrease of $C(n)$ is such that $\$(n)$ does not increase. The choice of a higher-level description (e.g., parallel vs. sequential) may lead to smaller values $L(n)$ compatibly with a reduced complexity $C(n)$ (as for PD or QP).

4 DECODING SYMBOLIC SIGNALS

Consider the stationary symbolic signal $S_0 = s_1 s_2 \ldots$ of length $|S_0| \gg 1$. The intrinsic difficulty of obtaining an accurate and concise description of the underlying dynamics can be ascribed to two causes: the observer's ignorance of the rules which govern the appearance of the words in S_0 and the arbitrariness in the choice of the basic words themselves. In fact, physical systems may be thought to produce signals according to a code[24] which is unknown to the observer and consists of variable-length words (the string 0110110101, e.g., might have been formed by concatenating either the strings $w_1 = 011$ and $w_2 = 01$ or $v_1 = 101$ and $v_2 = 01$). The choice of a code should be, in principle, simultaneous with that of the kind of model (sequential-stochastic, parallel in some class of grammars, etc.) since one may affect the other: e.g., the "wrong" set of words may lead to an infinitely extended model when, instead, a finite machine would suffice and viceversa. The recognition of the "dictionary" (the code) and of the "grammar" (set of rules in a wide sense, not a formal grammar) should be guided by physical considerations. Depending on the aim of the investigation (enumeration of the allowed transitions, evaluation of thermodynamic averages, etc.), this task may be pursued either exactly or approximately [13].

Primitives can be identified as words which satisfy special conditions, to be classified either as "strong" or "weak". Among the former ones are topological properties which do not depend on tuning parameters, such as simple existence or strict periodicity. The second group concerns metric properties (i.e., continuous-valued quantities) and, more in general, consists of acceptance criteria based on suitable threshold parameters. For example, the primitives of length n might be words with probability larger than a given value p_n, for all n. As anticipated in section 2, recurrent orbits play a relevant role in the study of asymptotic behaviour. In particular, periodic orbits are known to be dense in the hyperbolic non-wandering sets of Axiom-A diffeomorphisms [4] and can be used to characterize strange attractors geometrically and dynamically [25]. In formal-language theory, recurrence properties of words are studied in the so-called "pumping lemmas" [10]. For example, any word S with $|S| \geq n$ belonging to a regular language $\mathcal{L}$ can be written as $S = uvw$, in such a way that $|uv| \leq n$, $|v| \geq 1$, and uv^iw is in $\mathcal{L}$, for all $i \geq 0$. A similar result holds for context-free grammars. Following these considerations, we introduced in Ref. [5] an efficient scheme for the unfolding of symbolic patterns generated by nonlinear dynamical systems. For simplicity, we illustrate it for the example of the piecewise-linear roof map

$$x_{k+1} = \begin{cases} a + 2(1-a)x_k & \text{if } x_k < 1/2 \quad (s_k = 0) , \\ \\ 2(1-x_k) & \text{if } x_k \geq 1/2 \quad (s_k = 1) . \end{cases} \tag{3}$$

At $a = \bar{a} = (3 - \sqrt{3})/4$ a Markov partition exists (the critical point $x = 1/2$ belongs to an unstable period-5 orbit). The unit interval can be divided into three subsets, labelled by the sequences 1 (interval $[1/2, 1]$), 01 (left preimage of element 1) and 001 (left preimage of 01). The symbolic signal contains all (random) combinations of these three strings, apart from the forbidden orbit 0011. The folding of phase-space is incomplete, for any value of $a > 0$ (for $a = 0$ the trivial binary grammar is observed).

The analysis begins with the estimation of the probability $P(S)$ of each sequence S, with length $|S| \in [1, n_{max}]$, as the frequency of occurrence of S in the signal $\mathcal{S}_0$ ($\sum_{|S|=n} P(S) = 1$, $\forall \ n$) generated by eq. (3). We then define a (n_{min}-periodically-extendable) primitive as a substring w which can be periodically extended at least up to a length n_{min} ($1 < n_{min} \leq n_{max}$) and which does not contain a prefix with the same property. For example, 001 is a primitive if $(001)^{n_{min}/3}$ is allowed and $(0)^{n_{min}}$ is not: choosing n_{min} equal to n_{max} is equivalent to requiring ordinary periodicity (up to the maximum available length n_{max}). Sequences of increasing length n are considered. If no single symbol ($n = 1$) satisfies this condition, all blocks of length 2 are examined, and so on, up to a cutoff length $n_{cut} \leq n_{min}$. If still no primitive is found, n_{min} is reduced by one and the whole process is iterated, until primitives of some length $n \leq n_{cut}$ are identified. All other allowed strings with the same length n are "transient" orbits. The number of primitives in one-dimensional maps at generic parameter values (PD is an exception) is finite, whereas it may be infinite in higher-dimensional cases. The construction of the tree then proceeds by forming new sequences S' as concatenations of any admissible string S (including transient orbits) with a primitive. If sequence S' in turn exists (i.e. if $P(S') > 0$), it is allocated on the tree, at the position determined by its parental relations. Level 1 contains the primitives, whereas a virtual level 0 is attributed to the transient strings. If some level is not complete because the probability of sequences of length larger than n_{max} cannot be evaluated, a normalization to one of the probabilities at the first level is carried out, and the rest of the tree is completed by using suitable predictors for $P(S)$. The errors introduced by this procedure are negligible if n_{max} is sufficiently large.

The full-periodicity condition is the most useful one for the analysis of chaotic systems: in the example of eq. (3) at $a = \bar{a}$, $w_1 = 1$, $w_2 = 01$ and $w_3 = 001$ (0 being transient). At PD or QP, instead, the appropriate regrouping is automatically obtained when n_{min} has decreased to 4 or 2, respectively, yielding the codewords $w_1 = 01$, $w_2 = 10$, $w_3 = 0010$, $w_4 = 1101$, ... (for PD) and $w_1 = 1$, $w_2 = 01$ (for QP). Notice that our unfolding procedure is applicable independently of the existence of a Markov partition. The acceptance criterion for the primitives is based on a topological property the range of which can be arbitrarily extended ($1 < n_{min} \leq n_{max} \leq \infty$) in order to detect regular structures in the signal. The optimal condition, however, depends on the nature of the problem, so that other choices are possible. They may involve, for example, the probabilities $P(w)$ of the codewords (as in the Huffman or Shannon-Fano codes [24]), or the value $Q(w)$ of some observable. The Huffman code for map (3) at $a = \bar{a}$ would be $(w_1, w_2, w_3) = (1, 00, 01)$.

Finally, notice that even if the generation rule of the symbolic pattern is purely sequential, as in nonlinear dynamical systems, the optimal description of the actual dynamics is a mixture (endowed with stochastic ingredients) of sequential and parallel mechanisms. In our approach, the true dynamical behaviour is approximated by means of Markov trees, with an accuracy which grows with the level l: however, if the symbolic pattern is known to be generated by a grammar belonging to a well defined family, a more faithful description is obtained by constructing, instead, a "derivation" tree [10,11] (nevertheless, the identification of the proper codewords may be, in general, still very difficult).

5 ITERATED CODING AND RENORMALIZATION

The hierarchical-logical analysis described in the previous sections implies a progressive refinement of phase-space, corresponding to a coarse-graining of the signal obtained in terms of the primitives w. In general, however, it is not possible to achieve the optimal (or just a meaningful) regrouping of symbols since the appropriate acceptance conditions are unknown. Even in case the chosen criterion is the correct one (e.g., periodicity for hyperbolic systems) the necessarily finite values of n_{min} do not allow a full identification of the codewords (transient orbits might be accepted as periodic). Hence, the resulting description is not the most compact one and the asymptotic properties of the system cannot be accurately inferred. Therefore, it is necessary to resort to a higher-level modelling procedure, endowed with a parallel unfolding mechanism. This is obtained by assigning to the primitive words $w_1, w_2 \ldots$, identified in the analysis of the original signal S_0, symbols from a new alphabet $A_1 = \{0, 1, \ldots\}$. The whole unfolding procedure can be re-applied to the transformed string S_1 thus obtained. If the primitives detected in S_1 do not coincide with the symbols in A_1 themselves, the recoding procedure may continue to an alphabet A_2, and so on. This iteration yields a progressive coarse-graining of the symbolic signal (a process equivalent to observing phase-space with increased resolution). Obviously, the description of the image-signal S_k obtained after the k-th recoding step consists of the tree and of the code which keeps track of the previous block-renamings (relations between each symbol in the alphabet A_k and its pre-image string in S_0). The recoding operation is the inverse of a step in a parallel generation scheme and is equivalent to a renormalization-group transformation (an enlargement) on the nonlinear map [16]. For self-similar parallel-generated languages [11], such as those of PD and QP, the trees obtained at each step are identical, i.e., an exact infinite renormalization is readily achieved. Conversely, for signals of purely sequential nature, there is no such possibility. Usually, however, it is possible and profitable to carry out some recoding steps, the number of which depends on the system: for the roof map (3), the substitution $(w_1, w_2, w_3) \rightarrow (0, 1, 2)$ yields a tree in which the only forbidden concatenation is 20. Notice that in this case no further improvement is possible since the new symbols 0, 1 and 2 are obviously also primitive (the full-periodicity condition was required for the codewords in S_0). A higher redundancy would be obtained using the four-element Markov partition (with the associated 9 prohibitions). The increased compactness of the description is achieved by exploiting the intrinsic asymmetries of the dynamics. That recoding yields more asymptotic estimates of the scaling observable can be evidenced by considering the probabilities $Q(S) = P(S)$ (for which $\sum_{\text{level} l} P(S) = 1$, since to each complete level l corresponds, by construction, a covering of the whole phase-space). The recoding process lets all strings which previously occurred as overlaps between primitives (for the roof map, sequences 10, 100, etc.) simply disappear, so that overlap-free probabilities are computed and unnecessary strings are automatically neglected. An example of how the renaming reduces code redundancy is furnished by the analysis of PD, for which, although the number of primitives is infinite (no periodic subsequence exists), the code consisting of the first two ($w_1 \equiv 01$ and $w_2 \equiv 10$) is complete and the signal can be rewritten in terms of them only. After renaming w_1 and w_2 as 0 and 1, and measuring the probabilities in the recoded signal, one has $P(0) + P(1) = 1$, whereas $P(w_1) + P(w_2) < 1$, in the original signal. Increased asymptoticity and reduced redundancy are essential properties to extract the asymptotic behaviour of observables (like, e.g., nearest-neighbour distances $\delta(S)$ [7]) with non-trivial scaling properties (the probabilities $P(S)$ at PD are, instead, just uniform).

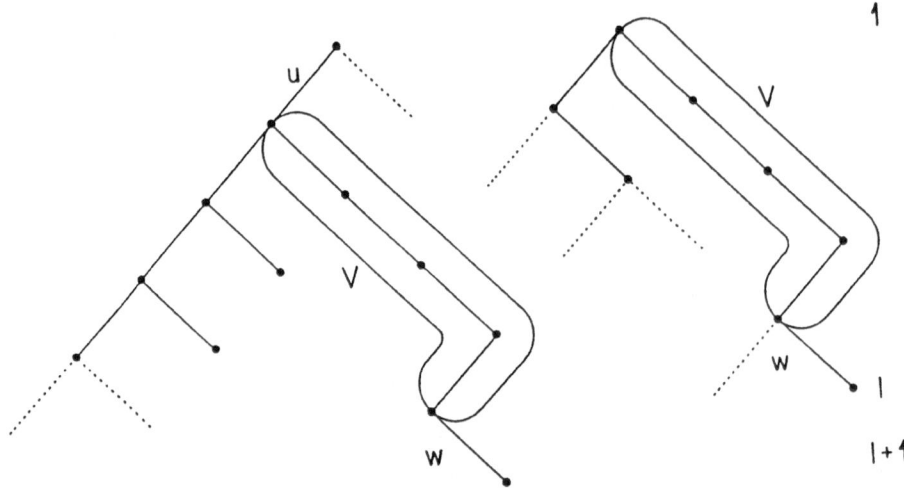

Figure 2. Sequences uVw and Vw involved in the order-l predictor.

6 COMPLEXITY AS A MEASURE OF MODEL ACCURACY

The detailed (microscopic, path-wise) description of the scaling behaviour of an observable $Q(S)$ provided by the logic tree can also be condensed into a macroscopic indicator which measures the predicting power of the derived model. Let us consider, again for $Q(S) = P(S)$, a generic string S at the l-th level of resolution, and all its descendants at level $l + 1$, which are of the form $S' = Sw$, where w is a primitive. Concatenations Sw containing forbidden subwords are discarded a priori. Further strings which are found to have probability $P(S') = 0$ are topological prediction-errors (called "surprises"). The existence and periodicity of the primitives w_k, in fact, do not guarantee that concatenations of them like $w_i w_j$ have the same properties. This is the essence of (topological) complexity. More generally, the actual value $P(S')$ (directly estimated from the signal) is compared with the expected probability $P_0(S')$, assigned to each S' according to some prediction rule. In the lowest-order approximation, P_0 is evaluated simply by assuming factorization of the probabilities: $P_0(Sw) = P(S)P(w)$. An improved predictor with a memory extent of $l - 1$ primitives can be obtained by rewriting $S' = Sw$ as uVw (where u is another primitive and V a string of $l - 1$ primitives), and calculating $P_0(S')$ as $P_0(S') = P(uV)P(w|V) = P(uV)P(Vw)/P(V)$, where the conditional probability of the suffix w, given V, is considered. The metric complexity C_1 of the system, relative to the set of predictions P_0, is thus defined as the information gain

$$C_1 = \lim_{l \to \infty} \sum_{j=1}^{N(l)} P(S) \ln \frac{P(S)}{P_0(S)} \, , \qquad (4)$$

where the sum runs over the $N(l)$ allowed sequences at level l. The metric complexity C_1 is a positive quantity, with the exception of the case of perfectly matching predictions ($P = P_0$, $\forall S$), in which it is identically zero. Analogous definitions can be given for any generic observable $Q(S)$, approximately multiplicative along the path S (i.e., $|Q(S)Q(T)/Q(ST)| = 1 + \varepsilon(S,T)$, with $|\varepsilon(S,T)| < 1$). The finite-size estimates $C_1(l,n)$ (l and $n = |S| < \infty$) provide a global reliability indicator for the associated predictive model, and are therefore also meaningful. If the above described $(l - 1)$-order predictor is used, the terms P/P_0 appearing in eq. (4) are ratios between the values $\sigma^{(l+1)}(uVw) = P(uVw)/P(uV)$ and $\sigma^{(l)}(Vw) = P(Vw)/P(V)$ of a generalized scaling function σ[7] for the probabilities, evaluated at two consecutive levels of resolution (see next section). Therefore, C_1 has the meaning of a global measure of the convergence of the scaling function, and systems with regular scaling properties are simple, if analysed using a sufficiently refined prediction rule [5]. The factorization assumption, for example, yields positive values of C_1 for most non-hyperbolic [4] attractors, since it disregards memory effects, whereas piecewise-linear systems are metrically simple also within this approximation (the natural invariant measure [4] is non-singular). In this sense, C_1 is also a measure of nonlinearity. We have calculated the metric complexity C_1 of the well-known logistic, Lozi and Hénon maps at various parameter values, finding $C_1 < 10^{-2}$: hence, these systems exhibit scaling properties which are sufficiently regular to be estimated with good accuracy. As mentioned above, the usage of more sophisticated estimators (i.e., essentially, the inclusion of memory effects) reduces the value of C_1 and improves therefore the convergence of any "thermodynamic" average [26,27] calculated over the tree structure (see next section).

In order to give a more complete characterization of complexity, eq. (4) can be generalized to a function C_q of a parameter q [5]:

$$C_q \equiv \lim_{l \to \infty} \frac{1}{q-1} \ln \frac{(\sum P^q)(\sum P_0^q)^{q-1}}{\left(\sum P P_0^{q-1}\right)^q} , \tag{5}$$

where the sums again run over the l-th tree level. For $q = 1$, eq. (4) is recovered. For $q = 0$, we obtain the topological complexity

$$C_0 = \lim_{l \to \infty} \lim_{n \to \infty} \ln \frac{N_0(l,n)}{N(l,n)} , \tag{6}$$

where $N_0(l,n)$ is the number of orbits predicted at level l, given the knowledge of all orbits of length $|S| < n$, and $N(l,n)$ the number of those with length $|S| \leq n$ allocated at the same level. Notice that eq. (5) reduces to eq. (6), for any q, if all existing and predicted sequences are separately equiprobable (i.e., $P(S) = 1/N$ and $P_0(S) = 1/N_0$, $\forall$ S). The topological complexity is identically zero if all predicted orbits exist.

One- and two-dimensional maps are topologically simple ($C_0 = 0$). In fact, the relative number $[N_0(l,n) - N(l,n)]/N(l,n)$ of prediction errors vanishes in the limit $(l,n) \to \infty$, since the number $N(n)$ of orbits of length n scales, for large n, as $\exp[K(0)n]$, whereas the quantity $N_0(n) - N(n)$ (for an ordinary tree with constant-length sequences at each level) grows at most like $\exp(\gamma n)$, with $\gamma \leq K(0)/2$ [23] (this result is very likely to hold for any finite dimension). At the period-2 band-merging point of the logistic map (strictly sofic with $0(11)^n 0$ forbidden, $\forall$ n) one error is made every two levels (with primitives $w_1 = 1$ and $w_2 = 01$).

With our definition (5) numerical values are assigned to the complexity C_q only within the frame of a predictive method. At variance with all previous measures (with the exception of that of ref. [21]) C_q is defined as a relative quantity, (trivially) identical to zero in the two limits of completely ordered (periodic, quasiperiodic) and random uncorrelated signals. It also satisfies the "axiomatic" requirements of section 3. Notice that the dependence on q may be very strong: indeed, if there is a path $\bar{S}$ for which P/P_0 is much larger than all other ratios at each level (because the predictions are systematically underestimating P for $\bar{S}$), this will give the dominant contribution to C_q for large q, whereas it will be ininfluent for $q \to 0$. This behaviour may occur in intermittent [4] maps. Therefore, the qualitative plot in Fig. 1, would then display a different discontinuity point for each q.

7   SCALING FUNCTIONS AND THERMODYNAMIC AVERAGES

The unfolding procedure illustrated in sections 2, 3 and 4 provides a hierarchical model of the dynamics which uses the information stored in the tree and in the associated block-renaming translation table (code). This information is supplied to algorithms which estimate the asymptotic scaling behaviour of the system's observables $Q_i(S)$ for $|S| \to \infty$. Such predictors are of paramount importance to achieve rapid convergence of thermodynamic averages[8,26,27] of the type

$$\sum_{|S|=n} P^q(S) \sim e^{-n(q-1)K(q)} , \tag{7}$$

where the sum is over all sequences of length n and $K(q)$ is the generalized metric entropy[4]. The sum in eq. (7) is formally equivalent to a canonical partition function in which $- \ln P(S)$ plays the role of the energy for the spin chain S, $q-1$ corresponds to the inverse temperature and $K(q)$ to the free energy[26]. Analogous sums define fractal dimensions in terms of mass-size scaling relations for the elements of the covering[4]. The evaluation of the free energy in the thermodynamic limit $n \to \infty$ is particularly difficult when phase-transitions[26,28,29] are present. These are caused either by anomalous exponential scaling of the observable $Q(S) \sim \exp[-|S|\kappa(S)]$, with non-generic $\kappa(S)$ for some sequences S, or by mixed exponential-algebraic scaling, in which relations of the type $Q(S) \sim |S|^{-\alpha(S)}$ occur: the former mechanism is ubiquitous in non-hyperbolic dissipative dynamical systems, whereas the latter one is typical of intermittent phenomena and conservative chaos. Accordingly, various kinds of transitions may appear[30]. Rapid convergence can be obtained in hyperbolic systems (for which phase-transitions are also possible[29]) by exploiting the relations between Lyapunov exponents (possibly of periodic orbits[27]), dimensions and entropies[31]. This approach has been so far limited to the complete-folding case, using ordinary fixed word-length trees[27]. Here we present a general scheme which takes into account the full topological complexity of the language, by introducing a grand-canonical formalism.

Since the logic trees introduced in the previous sections to describe the structure of generic languages have variable-length sequences at each level, there is no unique length n which can be singled out in the asymptotic scaling of the thermodynamic averages, as in eq. (7). Therefore, we consider the following level-$(l+2)$ grand-partition sum (again for the case $Q(S) = P(S)$)

$$\Omega_{l+2}(z;q) \equiv \sum_{w_0 w_1 \ldots w_{l+1}} P^q(w_0 w_1 \ldots w_{l+1}) z^{|w_0 w_1 \ldots w_{l+1}|} , \tag{8}$$

where sequences S composed of $l+2$ primitives w appear. The term $z^{|S|}$ has been introduced to provide a detailed compensation for the (generally exponential) decrease of $P(S)$ with $|S|$: in fact, the special value $z(q)$ of z which renders the sum stationary in the limit $l \to \infty$ determines the free energy as $K(q) = \ln z(q)/(q-1)$. Indeed, if $|S| = n$, for all S, we recover relation (7). As it is customary, rather than determining $z(q)$ by equating $\Omega_{l+2}(z;q)$ to 1 (which would introduce prefactor-errors), one compares the partition functions at two consecutive levels:

$$\sum_{w_0 w_1 \ldots w_{l+1}} P^q(w_0 w_1 \ldots w_{l+1}) z^{|w_0 w_1 \ldots w_{l+1}|} = \lambda(z;q) \sum_{w_0 w_1' \ldots w_l'} P^q(w_0 w_1' \ldots w_l') z^{|w_0 w_1' \ldots w_l'|} . \tag{9}$$

In the limit, $z(q)$ is given by the relation $\lambda[z(q);q] = 1$, which states the stability of the sum. For $z > z(q)$, $\Omega_l(z;q)$ diverges (for $z < z(q)$ it vanishes). Relation (10) can be rewritten as an eigenvalue equation in the following way[32]: the argument of the l.h.s. sum in eq. (9) is multiplied and divided by $P^q(w_0 w_1' \ldots w_l') z^{|w_0 w_1' \ldots w_l'|}$, and then a second sum over the indices $w_1' \ldots w_l'$ is taken, after inserting the Kronecker deltas $\delta_{w_1' w_1} \cdot \ldots \cdot \delta_{w_l' w_l}$. Finally, one obtains the eigenvalue equation

$$\sum_{w_0 w_1' \ldots w_l'} \sum_{w_1 w_2 \ldots w_{l+1}} \sigma^q(w_0 w_1 \ldots w_{l+1}) z^{|w_{l+1}|} \delta_{w_1' w_1} \cdot \ldots \cdot \delta_{w_l' w_l} P^q(w_0 w_1' \ldots w_l') z^{|w_0 w_1' \ldots w_l'|} =$$

$$\tag{10}$$

$$= \lambda(z;q) \sum_{w_0 w_1' \ldots w_l'} P^q(w_0 w_1' \ldots w_l') z^{|w_0 w_1' \ldots w_l'|}$$

for a generalized transfer matrix $\mathbf{T}(z)$[8,32] defined by

$$T_{w_0 w_1' \ldots w_l'; w_1 \ldots w_{l+1}} \equiv \sigma^q(w_0 w_1 \ldots w_{l+1}) z^{|w_{l+1}|} \delta_{w_1' w_1} \cdot \ldots \cdot \delta_{w_l' w_l} \tag{11}$$

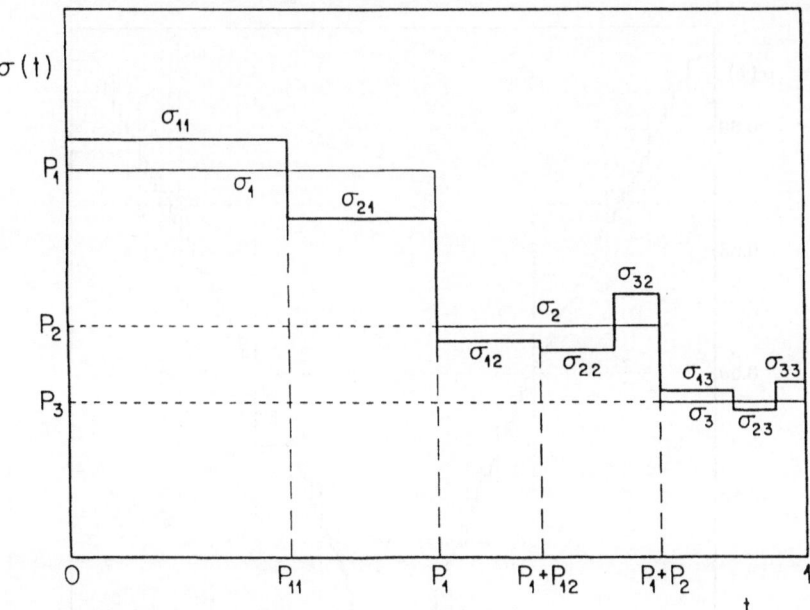

Figure 3. Schematic plot of the probability scaling function for the roof map (3) at $a = \bar{a}$, evaluated at levels 1 and 2.

where $\sigma(w_0 w_1 \ldots w_{l+1}) \equiv P(w_0 w_1 \ldots w_{l+1})/P(w_0 w_1 \ldots w_l)$ is a term of the level-l probability scaling function [7]. The matrix $\mathbf{T}(z)$ describes the conditional probability of the transition between the strings $\ldots w_0 w_1' \ldots w_l' \ldots$ and $\ldots w_1 w_2 \ldots w_{l+1} \ldots$ upon (left) shifting by one primitive. The two sequences are "connected" only if the last l words of the left one coincide with the first l of the right one (the image), as taken into account by the deltas in eq. (11). Moreover, the transition is actually possible if $\sigma > 0$. Using this formalism, the condition for the evaluation of the free energy $K(q)$ can be expressed by saying that the largest eigenvalue $\lambda_1(z; q)$ of $\mathbf{T}$ must be on the unit circle for $z = z(q)$. As an illustration of the method, we consider the limit $q \to 0$ in the case of the roof map (3) for $a = \bar{a}$. The matrix $\mathbf{T}$ for level 1 is

$$
\mathbf{T} = \begin{pmatrix} z & z^2 & z^3 \\ z & z^2 & z^3 \\ 0 & z^2 & z^3 \end{pmatrix}
\tag{12}
$$

since the transition $w_3 w_1$ is forbidden. The value $z(q)$ is then a solution of $z^4 - z^3 - z^2 - z + 1 = 0$. Consideration of higher levels is useless, since there are no other prohibitions appearing and the value $z(q)$ does not change.

In Fig. 3 we plot schematically the probability scaling function $\sigma[t(S)]$ versus the ordering parameter $t(S)$ defined as follows. The sequences w_k are ordered, at level 1, as on the tree, with $t(w_k) = t(w_{k-1}) + P(w_k)$. The generic k-th interval is split, at level 2, into subintervals labelled by all sequences $w_j w_k$ ending with w_k and ordered from left to right according to the order of the w_js at the first level on the tree: $t(w_j w_k) = t(w_{j-1} w_k) + P(w_j w_k)$ (when all js have been scanned, k is increased by one). At level 3, all subintervals of $w_j w_k$ are labelled as $w_i w_j w_k$, with the w_is ordered as usual, and so on. In the figure, the short notation σ_{ij} for $\sigma(w_i w_j)$ and similarly for P_{ij} has been used $(i, j = 1, \ldots, 3)$. With this convention (the widths of the intervals are given by the probabilities of the corresponding sequences), C_1 can be written as $C_1 = \lim_{l \to \infty} \langle \ln(\sigma^{(l+1)}/\sigma^{(l)}) \rangle$, where $\sigma^{(l)}$ is the level-l approximation to σ and $\langle f(t) \rangle = \int_0^1 f(t) dt$. We have also evaluated the scaling function for the well-known Lorenz system at standard parameter values [4]. Notwithstanding the absence of the two period-1 cycles, a complete binary tree description is appropriate, because prohibitions appear only

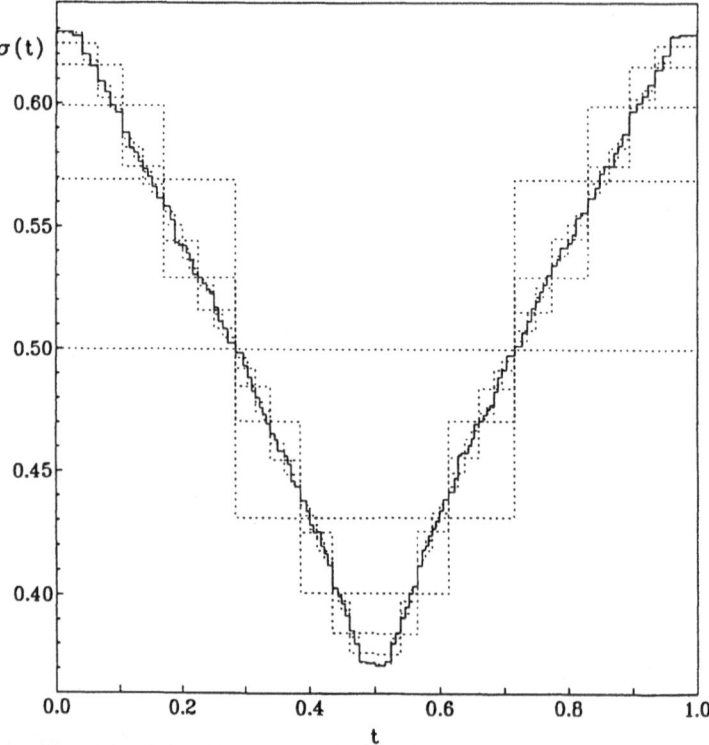

Figure 4. Approximations of the scaling function $\sigma(t)$ for the probabilities versus the ordering parameter t, evaluated on the first 7 levels of resolution for the Lorenz system at standard parameter values.

for $|S| > 25 > n_{max}$. In Fig. 4 we plot $\sigma^{(l)}(t)$ obtained from the first 7 levels ($l = 1, \ldots, 7$): within the statistical fluctuations, a fast convergence is observed for increasing l, indicating that Markov models of order as low as 5 or 6 already reproduce the symbolic dynamics with good accuracy. Correspondingly, the same convergence properties are exhibited by the thermodynamic averages.

The results obtained for the asymptotic behaviour of the conditional probabilities (scaling function) show that the Markov models are able to generate signals with dynamical properties close to those of the original system. This corresponds to performing predictions of the time-dynamics in terms of a succession of deterministic paths (blocks of symbols) which appear at random in time, according to sequence-to-sequence transition probabilities $P(S_j|S_i)$. These are the elements M_{ij} of a metric Markov matrix (at variance with the purely topological one of section 2) and satisfy the relations

$$\sum_{j=1}^{N(l)} M_{ij} = 1 \qquad \text{and} \qquad \sum_{i=1}^{N(l)} M_{ij} = P(S_j) . \tag{13}$$

Accordingly, a bi-infinite sequence $S = \ldots S_{j_{-1}} S_{j_0} S_{j_1} \ldots$ is admissible if $M_{j_n j_{n+1}} > 0$ for all $n \in (\ldots, -1, 0, 1, \ldots)$. The terms M_{ij} (representing a shift of l primitives, at level l) are evaluated from the elements of the l-th power of matrix $\mathbf{T}$ (which advances by one primitive only)[33] as, for example,

$$P(w_1 \ldots, w_l | v_1 \ldots, v_l) \approx P(w_1|v_1 \ldots v_l) P(w_2|v_2 \ldots v_l w_1) \cdots P(w_l|v_l w_1 \ldots w_{l-1}) , \tag{14}$$

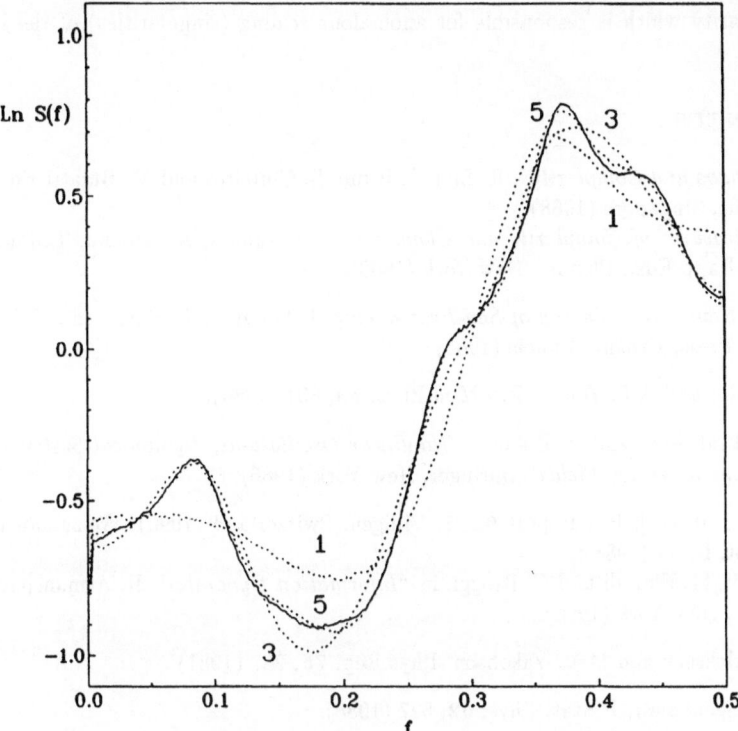

Figure 5. Comparison between the power spectrum $S(f)$ (vs. the frequency f) of the symbolic signal of the logistic map (thick line) and those corresponding to the successive hierarchical reconstructions obtained from levels 1, 3 and 5 of the logic tree (thin lines).

where $P(a_{l+1}|a_1 \ldots a_l) = P(a_1 \ldots a_{l+1})/P(a_1 \ldots a_l)$, as usual. All memory effects available up to level $l + 1$ are included. A comparison between the power spectrum of the symbolic signal produced by the logistic map $x' = 1 - 1.85x^2$ (with primitives 1, 01 and 001) and those obtained from the level-1, -3 and -5 reconstructions is shown in fig. 5, as an illustration of the achieved accuracy. The structure of the power spectra appears to be dependent, in order of decreasing importance, on two ingredients which are invariant under smooth coordinate changes (the topology and the metrics of the logic tree) and on one which is not (the values of the continuous variable x, not considered here)[33].

8   CONCLUSION

We have discussed the concept of complexity in the framework of a hierarchical modelling of physical systems, showing that complexity is "orthogonal" to the usual time-dynamics: the unpredictability of the future time evolution is a much less severe obstacle to the construction of reliable models than the unpredictability of the scaling dynamics, for the understanding of which a parallel unfolding mechanism must be taken into account. With this interpretation, complexity is related to the theory of scaling functions[7] and measures the accuracy achieved by the models in the estimation of suitable thermodynamic limits. The unfolding procedure is always improvable and constitutes the first step in the construction of global models: they will consist, in a first approximation, of piecewise-linear transformations which map each of the phase-space elements at a given hierarchical level to the image-elements. A further refinement requires consideration of continuous maps in phase-space in order to reproduce the nonlinearity which is responsible for anomalous scaling (singularities) of the invariant measure.

REFERENCES

[1] (a) *"Chaos and Complexity"*, R. Livi, S. Ruffo, S. Ciliberto and M. Buiatti Eds., World Scientific, Singapore (1988);
(b) *"Measures of Complexity and Chaos"*, N.B. Abraham, A. Albano, T. Passamante and P. Rapp Eds., Plenum, New York (1990).

[2] J. von Neumann, *"Theory of Self-Reproducing Automata"*, A. Burks ed., University of Illinois Press, Urbana, Illinois (1966).

[3] K. Binder and A.P. Young, Rev.Mod.Phys. **58**, 801 (1986).

[4] J. Guckenheimer and P. Holmes, *"Nonlinear Oscillations, Dynamical Systems and Bifurcations of Vector Fields"*, Springer, New York (1986).

[5] R. Badii, in [1b]; PSI Report **61**, 1, Villigen, Switzerland (1990); Weizmann preprint, Rehovot, Israel (1988);
R. Badii, M. Finardi and G. Broggi, in *"Information Dynamics"*, H. Atmanspacher Ed., Plenum, New York (1990).

[6] V.M. Alekseev and M.V. Yakobson, Phys.Rep. **75**, 287 (1981).

[7] M.J. Feigenbaum, J. Stat. Phys. **52**, 527 (1988).

[8] D. Ruelle, *"Thermodynamic Formalism"*, Vol. 5 of Encyclopedia of Mathematics and its Applications, Addison-Wesley, Reading, MA (1978).

[9] R.J. Solomonoff, Inf. Control **7**, 1 (1964);
A.N. Kolmogorov, Probl.Inform.Transm. **1**, 1 (1965);
G. Chaitin, J. Assoc.Comp.Math. **13**, 547 (1966).

[10] J.E. Hopcroft and J.D. Ullman, *"Introduction to Automata Theory, Languages and Computation"*, Addison-Wesley, Reading, MA (1979).

[11] G. Rozenberg and A. Salomaa, *"The Mathematical Theory of L Systems"*, Academic Press, London (1980).

[12] J.P. Crutchfield, this issue.

[13] R.M. Wharton, Inform. Contr. **26**, 236 (1974).

[14] B. Weiss, Monatshefte für Mathematik **77**, 462 (1973).

[15] P. Collet and J.P. Eckmann, *"Iterated Maps of the Interval as Dynamical Systems"*, Birkhauser, Cambridge, MA, (1980).

[16] I. Procaccia, S. Thomae and C. Tresser, Phys.Rev. **A35**, 1884 (1987).

[17] J.P. Crutchfield and K. Young in *"Complexity, Entropy and Physics of Information"*, W. Zurek Ed., Addison-Wesley, Reading, MA, (1989).

[18] P. Grassberger, Wuppertal preprint B 89-26, (1989).

[19] J.P. Crutchfield and K. Young, Phys.Rev.Lett. **63**, 105 (1989).

[20] D. Auerbach and I. Procaccia, Phys.Rev. **A41**, 6602 (1990).

[21] S. Lloyd and H. Pagels, Ann. of Phys. **188**, 186 (1988).

[22] If system A is described by a number of rules which grows, for increasing resolution, more rapidly than that of system B, the overall descriptive effort is dominated by the properties of A. As a consequence, complexity can be meaningfully defined only in the limit of infinitely extended patterns and characterizes the scaling behaviour of the physical process: it must equal zero, in particular, for systems specified by a finite number of dynamical rules, in agreement with point 1.

[23] G. D'Alessandro and A. Politi, Phys.Rev.Lett. **64**, 1609 (1989).

[24] R. Hamming, *"Coding and Information Theory"*, Prentice-Hall, Englewood Cliffs, NJ (1986).

[25] D. Auerbach, P. Cvitanović, J.P. Eckmann, G. Gunaratne and I. Procaccia, Phys.-Rev.Lett. **58**, 2387 (1987);
P. Cvitanović, Phys.Rev.Lett. **61**, 2729 (1988);
C. Grebogi, E. Ott and J.A. Yorke, Phys.Rev. **A36**, 3522 (1988) and Phys.Rev. **A37**, 1711 (1988).

[26] R. Badii, Riv. Nuovo Cim. **12**, N° 3, 1 (1989).

[27] R. Artuso, E. Aurell and P. Cvitanović, Niels Bohr Institute preprints NBI-89-41 and NBI-89-42.

[28] P. Cvitanovič, in Proceedings of the Workshop in Condensed Matter, Atomic and Molecular Physics, Trieste, Italy (1986);
D. Katzen and I. Procaccia, Phys.Rev.Lett. **58**, 1169 (1987);
P. Grassberger, R. Badii and A. Politi, J.Stat.Phys. **51**, 135 (1988);
G. Broggi and R. Badii, Phys.Rev. **39A**, 434 (1989).

[29] P. Paoli, A. Politi, G. Broggi, M. Ravani and R. Badii, Phys.Rev.Lett. **62**, 2429 (1989);
P. Paoli, A. Politi and R. Badii, Physica **D36**, 263 (1989).

[30] P. Szépfalusy, T. Tél and G. Vattay, Eötvös preprint, Budapest (1990).

[31] R. Badii and A. Politi, Phys. Rev. **35A**, 1288 (1987);
R. Badii and G. Broggi, Phys.Rev. **41A**, 1165 (1990).

[32] M.J. Feigenbaum, M.H. Jensen and I. Procaccia, Phys.Rev.Lett. **57**, 1503 (1986).

[33] M.A. Sepúlveda and R. Badii in [1b]; R. Badii, M. Finardi and G. Broggi, PSI preprint, PSI-LUS-05 (1990).

CONTRIBUTORS

R. BADII
Physik Institut, Univ. Konstanz
D-7750 Konstanz
BRD

L.A. BUNIMOVICH
Inst. of Oceanology
ul. Krasikova 23
SU-117218 Moscow
USSR

B. CHIRIKOV
Inst. of Nuclear Physics
USSR Acad. of Sciences, Siberian division
630090 Novosibirsk
USSR

Y. COUDER
Groupe de Physique des Solides
Ecole Normale Superieure
24 rue Lhomond
75231 Paris Cedex 05
France

R. ECKE
Center for Nonlinear Studies, MS B258
Los Alamos National Laboratory
Los Alamos, NM 87545
USA

J.-P. ECKMANN
Dep. de Phys. Theorique
Universite de Geneve
CH-1211 Geneve
Switzerland

M.J. FEIGENBAUM
Dept. of Physics
Rockefeller Univ., box 75
New York NY 10021
USA

K. KANEKO

Institute of Physics
College of Arts and Science
Univ. of Tokyo
Komaba, Meguro
Tokyo 153, Japan

R.S. MACKAY

Nonlinear Systems Laboratory
Mathematics Institute
University of Warwick
Coventry CV4 7AL
England

Y. POMEAU

Physique Theorique
Ecole Normale Superieure
24 rue Lhomond
F-75231 Paris CEDEX 05
France

I. PROCACCIA

Chem. Phys. Dept.
Weizmann Inst. of Science
76100 Rehovot, Israel

D. SULLIVAN

IHES
35, Route de Chartres
F-91140 Bures-sur-Yvette
France

J. YORKE

Inst. Phys. Sci. Tech.
Mathematics Dept.
University of Maryland
College Pk., MD 20742
USA

INDEX